Nanotechnology in Construction 3

Zdeněk Bittnar, Peter J.M. Bartos,
Jiří Němeček, Vít Šmilauer, Jan Zeman (Eds.)

Nanotechnology in Construction 3

Proceedings of the NICOM3

Prof. Dr. Zdeněk Bittnar
Department of Mechanics,
Faculty of Civil Engineering,
Czech Technical University in Prague,
Czech Republic

Dr. Vít Šmilauer
Department of Mechanics,
Faculty of Civil Engineering,
Czech Technical University in Prague,
Czech Republic

Prof. Dr. Peter J.M. Bartos
ACM Centre, University of Paisley,
Scotland

Dr. Jan Zeman
Department of Mechanics,
Faculty of Civil Engineering,
Czech Technical University in Prague,
Czech Republic

Dr. Jiří Němeček
Department of Mechanics,
Faculty of Civil Engineering,
Czech Technical University in Prague,
Czech Republic

Additional material to this book can be downloaded from http://extras.springer.com.

ISBN 978-3-642-00979-2 e-ISBN 978-3-642-00980-8

DOI 10.1007/978-3-642-00980-8

Library of Congress Control Number: Applied for

Typeset: Scientific Publishing Services Pvt. Ltd., Chennai, India.

Cover Design: WMXDesign GmbH, Heidelberg

Printed in acid-free paper

9 8 7 6 5 4 3 2 1

springer.com

Preface

Nanoscience has been with us ever since 'features' on nano-scale were first seen under a microscope. Nanotechnology came about much more recently, when first tools were developed for characterisation of the 'nano-features' and for their manipulation. Coming soon after the 'dot-com' IT bubble had burst, nanotechnology became the new holy grail for venture capitalists and focus of media. Fantastic developments affecting all aspects of life were proposed. However, it was clear that returns on investment in this case could not be instantaneous and the media hype was, perhaps, not as great as in the past (e.g. regarding superconductivity).

First applications of nanotechnology in construction research occurred in mid-1990s. There were few centres of such research; the novel nano-instrumentation was very expensive, often only custom-built. However, when first products exploiting nanotechnology entered construction market, need arose for a forum to review the research and evaluate its realistic potential. This led me to propose the Intl. Symposium on Nanotechnology in Construction (NICOM1), held in Paisley, Scotland in mid-2003. It was very successful; it attracted a very wide spectrum of participants. In addition to researchers in construction and engineers, there were architects, seeing applications in 'nano-houses' of future, physicists and other scientists who came to examine application of their know-how in the broad and economically significant construction industry. Industrialists and end-users were there too, to learn and to separate reality from the media driven publicity.

The NICOM2, organised by Dr A Porro and his team at Labein, was held in Bilbao, Spain in 2005. The event already indicated that exploitation of nanotechnology in construction was less than expected, very few new nanotechnology-based products appeared on the construction market.

A decade after the peak of the nanotechnology media hype, six years after the NICOM1, the 3rd Symposium on Nanotechnology in Construction (NICOM3) will discuss developments again and analyse reasons for the uneven advances across different sectors of construction. Predictions of progress will be now more reliable due to greater knowledge and amount of evidence in hand. However, the global financial crisis presents a new factor, impact of which no-one can accurately foresee. Papers for NICOM3 indicate that the initially very wide interest has narrowed (cement-based materials tend to dominate) and confirm that the main advance was in knowledge and understanding, followed by instrumentation. Aspects such as health & safety and metrology have now acquired much higher significance but commercial exploitation remains slow.

Global interest in NICOM3 confirms that the NICOM Symposia are an established series, each providing a valuable discussion forum for nanotechnology in construction. However, this has been achieved only through the initiative and untiring efforts of Prof Z Bittnar, Dean of the Faculty of Civil Engineering and his team (Dr J Nemecek et al.) at the Czech Technical University (CVUT) in Prague, who organised the NICOM 3 and edited the Proceedings.

Peter J.M. Bartos,
Co-Chairman of the NICOM3 Scientific Committee

Organization

Organizing Committee

Jiří Němeček	Chairman
Zdeněk Bittnar	
Vít Šmilauer	
Jan Zeman	
Kateřina Forstová	
Alexandra Kurfürstová	

Scientific Advisory Committee

Co-chairmen

Zdeněk Bittnar	CTU, Prague
Peter JM Bartos	UWS, Paisley

Members

Paul Acker	Lafarge Cement, France
Klaas van Breugel	TU Delft, The Netherlands
Ignasi Casanova	UPC Barcelona, Spain
Wolfgang Dienemann	Heidelberg Cement, Germany
Christian Hellmich	TU Wien, Austria
Hamlin M. Jennings	Northwestern Univ., IL, USA
Richard Livingston	Federal Highway Administration, McLean, VA, USA
Bernhard Middendorf	TU Dortmund, Germany
Manfred Partl	EMPA, Switzerland
Antonio Porro	Labein-Technalia, Bilbao, Spain
Marco di Prisco	Politecnico di Milano, Italy
Daniel Quenard	CSTB Grenoble, France
Laila Raki	NRC of Canada, Ottawa, Canada
Gian Marco Revel	Universita Politecnica delle Marche, Ancona, Italy
Karen Scrivener	EPFL, Switzerland

Ake Skarendahl	BIC, Sweden
Konstantin Sobolev	Univ. of Wisconsin, WI, USA
Pavel Trtik	EMPA, Switzerland
Franz J Ulm	MIT, Boston, MA, USA
Johan Vyncke	BBRI, Belgium
Wenzhong Zhu	UWS, UK

Supporters

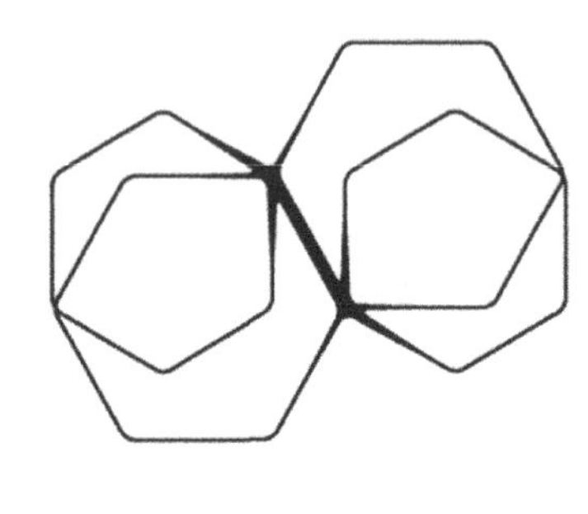

Contents

Regular Papers

Potential Environmental and Human Health Impacts of Nanomaterials Used in the Construction Industry

J. Lee, S. Mahendra, and P.J.J. Alvarez

Abstract. Nanomaterials and nanocomposites with unique physical and chemical properties are increasingly being used by the construction industry to enable novel applications. Yet, we are confronted with the timely concern about their potential (unintended) impacts to the environment and human health. Here, we consider likely environmental release and exposure scenarios for nanomaterials that are often incorporated into building materials and/or used in various applications by the construction industry, such as carbon nanotubes, TiO_2, and quantum dots. To provide a risk perspective, adverse biological and toxicological effects associated with these nanomaterials are also reviewed along with their mode of action. Aligned with ongoing multidisciplinary action on risk assessment of nanomaterials in the environment, this article concludes by discerning critical knowledge gaps and research needs to inform the responsible manufacturing, use and disposal of nanoparticles in construction materials.

1 Introduction

The nanotechnology revolution has enhanced a variety of products, services, and industries, including the construction sector. A comprehensive assessment of their effects on human and environmental health is essential for establishing regulations and guidelines that allow the numerous benefits of nanomaterials while providing adequate protection to ecosystems. Due to the dimensions controlled in the transitional zone between atom and molecule, the nanosized (1 to 100 nm) material gains novel properties compared to the corresponding bulk material. The unique properties achieved at the nanoscale enable the material to show highly-promoted performances in catalysis, conductivity, magnetism, mechanical strength, and/or optical sensitivity, enabling a wide applications including electronic devices, biomedical agents, catalysts, and sensors [8,13,78].

J. Lee, S. Mahendra, and P.J.J. Alvarez
Department of Civil & Environmental Engineering, Rice University, Houston, TX, USA

Keeping pace with nanotechnology applications in diverse industries, engineered nanomaterials are being increasingly used by the architectural and construction industries [19,58,88]. The incorporation of nanomaterials in construction is expected to improve vital qualities of building materials (e.g., strength, durability, and lightness) [19,47,75], offer new collateral functions (e.g., energy-saving, self-heating, and anti-fogging) [28,39,88], and provide main components for maintenance instruments such as structural health sensors [70,87]. In terms of the foregoing advantages of nanomaterials, nanotechnology in construction was selected as one of 10 targeted applications of nanotechnology able to resolve the developing world's biggest problems [2]. Nevertheless, many examples in modern history illustrate the unintended environmental impacts of initially promising technologies, including the deliberate release of "beneficial" chemicals, such as DDT, which was use to control malaria and other water-borne diseases but was later found to be carcinogenic to humans and toxic to several bird species [6,80]. Thus, it is important to take a proactive approach to risk assessment and mitigate the potential impacts of nanoparticles in construction materials to ecosystem and human health.

2 Applications of Nanomaterials in Construction

Table 1 summarizes some ongoing applications of nanomaterials in the construction industry, including high performance structural materials, multifunctional coatings and paintings, sensing/actuating devices. Representative applications are described briefly.

Concrete, having the largest annual production among other materials, undergoes drastic enhancement in mechanical properties by the addition of carbon nanotubes (CNTs) or nanosized SiO_2 (or Fe_2O_3) to the concrete mixtures consisting of binding phase and aggregates [14,19,47,75]. Addition of 1% CNTs (by weight) efficiently prevents crack propagation in concrete composites by functioning as nucleating agents [14,19], while silica and iron oxide nanoparticles (3 to 10% by weight) serve as filling agents to reinforce concrete [47,48,75].

Steel, commonly used in building and bridge constructions, faces challenges related to strength, formability, and corrosion resistance, which may be successfully addressed by introduction of metal nanoparticles (NPs) [19]. Particularly, nanosized copper particles reduce the surface roughness of steel to impart higher weldability and anti-corrosion activity [19].

Window glass can accomplish various additional functions by incorporation of TiO_2 and SiO_2 nanoparticles. TiO_2 coated on window photochemically generates reactive oxygen species (ROS) with sunlight or indoor light, effectively removing dirt and bacterial films attached on window [28,64]. Light-excited superhydrophilic properties of TiO_2 make window glass anti-fogging and easily washable by decreasing contact angle between water droplet and the glass surface [28,39]. On the other hand, nanosized silica layers sandwiched between two glass panels can make windows highly fireproofing [58].

Table 1 Selected Nanomaterial Applications in the Construction Industry

Construction Materials	Nanomaterials	Expectations	References
Concrete	Carbon Nanotubes SiO_2 Fe_2O_3	Reinforcement Crack Hindrance	[47,75]
Steel	Copper Nanoparticles	Weld Ability Corrosion Resistance	[19]
Window	TiO_2 SiO_2	Self-Cleaning Anti-Fogging UV and Heat Blockings Fire-Protective	[39,58,64]
Coatings/ Paintings	TiO_2 Silver Nanoparticles	Anti-Fouling Biocidal Activity	[28,41]
Solar Cells	Dye/TiO_2 C_{60} and Carbon Nanotubes CdSe Quantum Dots	Solar Energy Utilization	[5,20,88]
Cement	Carbon Nanotubes Polypropylene Nanofiber	Strength Fire Resistance	[58]
Sensor	Carbon Nanotubes	Real-Time Monitoring of Structures	[87]

In addition to the building materials, nanomaterials are utilized for other construction-related products. TiO_2 **coating** on pavements, walls, and roofs plays a role as an anti-fouling agent to keep roads and buildings dirt-free with sunlight irradiation [28,88]. Silver nanoparticles (nAg) embedded in **paint** add biocidal properties by exploiting the antimicrobial activity of nAg [41]. Silicon-based photovoltaic or dye-sensitized TiO_2 **solar cells** can be made flexible enough to be coated on surfaces such as roofs and windows (referred to as energy-coating), to enable production of electric energy under sunlight illumination [88]. Furthermore, fuel cells and solar cells, accomplishing partial non-utility generation inside of house, were recently reported to include CNTs, C_{60} fullerenes and CdSe quantum dots for enhanced conversion efficiency [5,20]. Alternatively, application of CNTs can improve adhesion of conventional cement, and the resultant material gains enhanced toughness and durability, as CNTs reinforce the mechanical strength of concrete [58].

For real-time, in-place acquisition of data relevant to material/structural damage (e.g., cracking, strain, and stress) and environmental conditions (e.g., humidity, temperature, and smoke), nano-electromechanical and micro-electromechanical systems (NEMS and MEMS), composed of nano- and microsized **sensors** and actuators, have recently drawn much attention [70]. For example, smart aggregates, formed by placing waterproof piezoceramic patches with lead wires into small concrete blocks, are used for early-staged concrete strength monitoring, impact detection, and structural health checking [76]. Additionally, CNT/polycarbonate composites exhibit functionality as strain-sensing devices by generating momentary changes in the electric resistance in response to strain inputs [87].

3 Environmental Release and Exposure Scenarios

As production and use of nanomaterials increase, so does the possibility of their release in the environment, which increases the potential for adverse effects on human and environmental health. Exposure assessment is a critical step towards characterizing risks and preventing and mitigating unintended impacts. Exposure prevention is a priority because, regardless of nanomaterial toxicity, the lack of exposure eliminates health risk. This is easier to accomplish through improved understanding of the fate, transport, and transformation of nanomaterials in the environment, which is needed to estimate the concentrations and forms to which ecological and human receptors will be exposed to. Furthermore, determining whether manufactured nanomaterials retain their nanoscale size, structure, and reactivity or are aggregated or associated with other media (e.g., sorption, acquisition or loss of coatings) is a critical step to assess nanomaterial bioavailability and impact to living organisms.

Engineered nanomaterials can enter the environment during their manufacture, transport, use, and disposal through intentional as well as unintentional releases (Figure 1) and behave as emerging pollutants [37,83]. Despite the growing awareness of potential releases of nanomaterials, efforts to identify and characterize dominant exposure routes have been quite preliminary. The lack of case studies and relevant data also make it difficult to quantify likely release scenarios. Nevertheless, several studies have evaluated the potential hazard posed by selected nanomaterials, by evaluating a limited number of toxicity end-points towards specific targeted biota [21,49,61,89]. Some studies have also addressed environmental implications by considering nanomaterial fate, transport, transformation, bioavailability and bioaccumulation [10,18,32,45]. Although these studies suggest that the engineered nanomaterials have the potential to impact the environment and human health [31,60], they fall short of providing a sufficient basis to establish regulatory guidelines for the safe production, use and disposal of construction nanomaterials. Accordingly, understanding release source dynamics, reactive transport and fate of construction nanomaterials represent critical knowledge gaps for risk assessment. Nonetheless, based on our understanding of construction waste management [35,40,65] and recent findings about the behavior of some nanomaterials (not necessarily associated with construction) [12,37,83], some realistic exposure scenarios can be suggested.

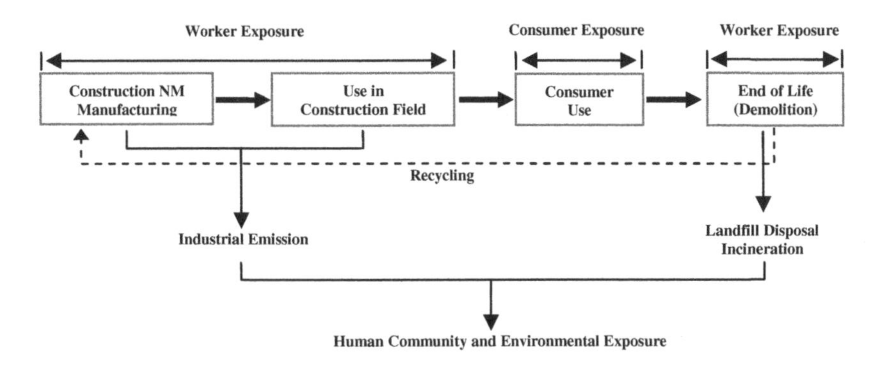

Fig. 1 Possible exposure routes during the whole lifecycle of construction nanomaterials

Manufacturing. Releases of nanomaterials to the environment can occur during the manufacture of building materials, in processes involving coating, compounding, and incorporation of nanomaterials. Occupational exposure to workers can occur through inhalation, which could cause respiratory health problems. Thus, it is advisable to use inhalation protection equipment such as air filters that protect workers against asbestos or ultrafine particles. As contamination originates from point-sources that are easily identifiable, exposure analysis, waste monitoring, and protective equipment installation (e.g., ventilator, air filter) at the workplace can be easily achieved. The challenges associated with this exposure route are that 1) nano-product suppliers are reluctant to disclose the manufacturing processes due to proprietary information and 2) most of them are small start-up companies that can hardly afford to be operated on the basis of the precautionary and very conservative assumption that all nanomaterials are toxic.

Demolition. It is highly probable that demolition, whether partial or complete, results in the environmental release of construction nanomaterials. The standard demolition procedures [40] recommend that trained specialists should dispose of hazardous materials (e.g., asbestos cement, lead-based paint, and some persistent residues) before undertaking extensive demolition. Relatively small-sized construction nano-products such as window, coatings/paintings, and sensor devices can be removed at this stage. Exposure to nanomaterials can be uncontrollable at later stages of demolition because of the use of explosives or heavy mechanical disruption (e.g., wrecking balls, bulldozer). In addition, the random crushing gets the residual debris mixed to make it difficult to separate nanomaterial-associated wastes afterwards. The wastes generated from the demolition are sorted and transported to landfills, which could be prevalent sources of the environmental release of nanomaterials.

Construction. The wastes containing nanomaterials are mainly generated during **repair, renovation, and construction activities**. In addition to potential worker exposure and unintentional release at the construction sites, landfill disposal and dumping of construction wastes could be common ways of discharging nanomaterials to the environment.

Long-term Releases. During the lifetime of buildings, damage, wear, and abrasion of infrastructures, whether artificial or natural, can cause nanomaterial releases to the environment. Accidents (e.g., fire) and disasters (e.g., heavy rainfall, flood, and storm) inflict damages on civil structure containing nano-products. For example, fire or incineration could release nanomaterials to the atmosphere, and rainfall can promote dissolution or leaching and drainage of nanomaterials into natural waterways and soils. Characterizing such releases on a long-term basis is very challenging because of current analytical limitations. Challenges include high detection limits that preclude quantifying nanomaterial releases at trace levels and low rates, and the lack of sufficient analytical specificity to discern the concentration and form of nanomaterials in complex environmental matrices. Thus often makes it difficult to delineate the region of influence of a nanomaterial release.

4 Toxicity of Nanomaterials

Nanomaterials embedded in building materials or used in other construction applications and products can cause cellular toxicity via multiple mechanisms (Figure 1). The important mechanisms of cytotoxic nanomaterials include disruption of cell wall integrity (e.g., SWNTs), nucleic acid damage (e.g., MWNTs), generation of reactive oxygen species (ROS) that exert oxidative stress (e.g., TiO_2), release of toxic heavy metals or other components (e.g., QDs), and direct oxidation upon contact with cell constituents (e.g., nC_{60}). Toxicity studies and effects of various nanomaterials used in construction are summarized in Table 2. These range from no damage to sub-lethal effects to mortality. Carbon nanotubes and TiO_2 nanoparticles are the nanomaterials that have been most studied for their potential toxic effects, and are discussed below.

TiO_2 is a photoactive nanomaterial that causes inflammation, cytotoxicity, and DNA damage in mammalian cells either alone or in the presence of UVA radiation due to ROS production [22,34,62,63,66,73,86,89]. TiO_2 morphology significantly affects its mobility inside a cell or through cell membranes, as well as the interactions with phagocytic cells that can trigger the signaling process for ROS generation [50]. The antimicrobial activity of nanoscale-TiO_2 towards *Escherichia coli, Micrococcus luteus, Bacillus subtilis, and Aspergillus niger* has been utilized in accelerated solar disinfection and in surface coatings [67,68,84].

Carbon nanotubes can exert pulmonary toxicity in mammals [16,30,82]. CNTs exert antibacterial activity via direct physical interaction or oxidative stress causing cell wall damage [33,59].

While **buckminsterfullerene (C_{60})** does not dissolve in water [24], its agglomeration though transitional solvents or long term stirring imparts water stability, and consequently enhances potential exposure and toxicity [71,77,79]. Water-stable C_{60} suspensions, referred to as nC_{60} [18], exhibit broad spectrum antibacterial activity [53,54,56]. The mechanism of nC_{60} cytotoxicity in eukaryotic systems was initially attributed to oxidative stress resulting from the ROS production [29,61,72]. However, recent studies have shown that nC_{60} does not produce detectable levels of ROS [26,44], and that the antibacterial activity is mediated via direct oxidation of the cell [17,55]. However, nC_{60} toxicity can be significantly mitigated by dissolved natural organic matters that coat the particle and reduce their availability [46].

Quantum dots are fluorescent nanoparticles that contain heavy metals such as cadmium, lead, and zinc in their core/shell structures, and are functionalized with organic coatings to enhance their stability [85]. Release of core metals is the primary mechanism of toxicity of QDs towards bacteria [38,57] as well as towards mammalian cells [7,15,23,36,52,74]. While surface coatings reduce core degradation and heavy metal releases, some surface coatings themselves have been shown to be toxic to mammalian cells [25,43,69]. In addition to toxicity caused due to dissolved components, QD particles are internalized or membrane-associated in eukaryotic cells, where they could cause oxidative stress, nucleic acid damage, and cytotoxicity [9,49,51].

Copper or copper oxide nanoparticles exert strong oxidative stress and DNA damage in human, mice, algae, and bacterial cells [4,11,34,45].

Table 2 Toxicity of Nanomaterials towards Various Organisms

Nanomaterial	Organism	Toxic Effects	References
Carbon nanotubes	Bacteria	Antibacterial to *E. coli*, cell membrane damage.	[16,33,42]
	Mice	Inhibit respiratory functions, mitochondrial DNA damage	-
SiO_2	Bacteria	Mild toxicity due to ROS production	[1]
	Rats	Cytotoxicity, apoptosis, up-regulation of tumor necrosis factor –alpha genes	[3]
Quantum dots	Bacteria	Bactericidal to *E. coli* and *Bacillus subtilis*	[38,57]
	Human cells	Toxicity from metal release, particle uptake, oxidative damage to DNA	[9,25,69,74]
	Mice	Accumulation of metals in kidneys	[49,81]
	Rat	Cytotoxic due to oxidative damage to multiple organelles	[15,51]
nCu or nCuO	Mice	Acute toxicity to liver, kidney, and spleen	[4,11]
TiO_2	Bacteria, algae, microcrustaceans, fish	Acute lethality, growth inhibition, suppression of photosynthetic activity, oxidative damage due to ROS.	[4,50,53,67,84]

Ultra-fine **SiO_2 nanoparticles** have been classified as human carcinogens [27]. Exposure to nano-sized SiO_2 causes alveolar cell toxicity and induces tumor necrosis genes in rats [3]. Silica nanoparticles at high concentrations in water (~ 5,000 mg/L) have also been reported to damage bacteria [1].

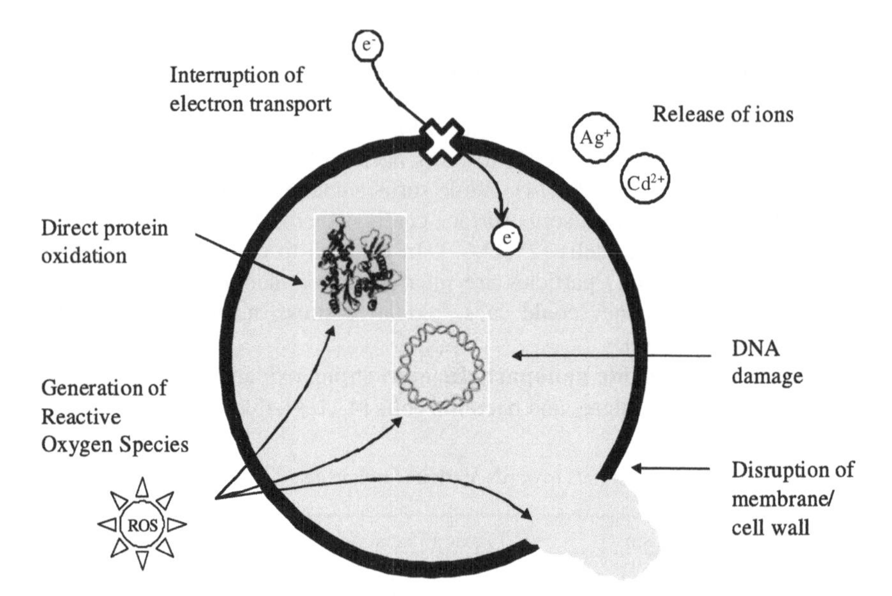

Fig. 2 Possible microbial toxicity mechanisms of nanomaterials. Different nanomaterials may cause toxicity via one or more of these mechanisms

5 Critical Knowledge Gaps and Research Needs

Nanomaterials are expected to become a common feature in some building materials due to their novel and remarkable properties. However, concern about their unintended impacts to human and environmental health is motivating research not only on risk assessment, but also on their safe manufacturing and eco-responsible use and disposal.

Research on the toxicity mechanisms of nanomaterials may unveil information that enables the design of environmentally benign nanocomposites. Nano-scale (ultrafine) particles can cause respiratory damages as well as skin inflammation, but their mode of action is not fully understood. In particular it is poorly understood how particle size distribution, chemical composition, shape, surface chemistry and impurities influence uptake, reactivity, bioavailability and toxicity. Thus, developing a mechanistic understanding of structure-reactivity relationships and their connection to immunology and toxicity is a priority research area. Such research should consider not only acute toxicity and mortality, which has been historically the focus of nanotoxicology, but also address sublethal chronic exposure and impact on the behavior of organisms. The potential for bioaccumulation and trophic transfer, leading to biomagnification, is another important but unchartered area of research.

Most toxicity studies have investigated the dose-response characteristics of a few representative nanomaterials on single species under laboratory conditions. The effects of nanomaterial mixtures, organismal differences, and environmental factors such as pH, salinity, and natural organic matter (which may coat or absorb nanomaterials) are yet to be comprehensively evaluated. This is particularly important because nanomaterials in the environment are likely to undergo significant transformation (e.g., coagulation, aggregation, sorption, loss or acquisition of coatings, biotransformation, etc.) which could exacerbate or mitigate their potential impacts.

Current analytical capabilities are insufficient to quantify and discern the form of nanomaterials in complex matrices at environmentally relevant low concentrations. Thus, analytical techniques and advances in nanoparticle metrology are needed to track nanomaterials and learn about their transport, transformation, behavior and fate in different environmental compartments (e.g., atmospheric, terrestrial and aquatic environment). Improved metrology should enable monitoring of short-term workers exposure during manufacturing, construction and demolition processes, as well as long-term monitoring of nanomaterial releases from construction materials (e.g., nanomaterial dissolution and leaching as the construction materials experience aging, abrasion, corrosion and weathering elements). Quantifying such sources is important to understand their region of influence and develop effective strategies intercept predominant exposure pathways. Improved analytical techniques are also needed to calibrate and validate mathematical fate-and-transport models to predict exposure scenarios and enhance risk management.

Safe disposal of nanomaterial-containing construction wastes will also need to consider the potential for leaching and subsequent transport through landfill clay liners and underlying soil. This information is needed to discern the need for additional barriers to ensure nanomaterial containment and minimize the potential for groundwater pollution. Finally, a life-cycle perspective is likely to motivate research on pollution prevention and identify opportunities to remanufacture, reuse and recycle these nanomaterials. Overall, further research will likely enhance the development of appropriate guidelines and regulations to mitigate potential environmental impacts and enhance the sustainability of the construction industry.

References

1. Adams, L.K., Lyon, D.Y., Alvarez, P.J.J.: Comparative eco-toxicity of nanoscale TiO_2, SiO_2, and ZnO water suspensions. Wat. Res. 40(19), 3527–3532 (2006)
2. ARI-News. Nanotechnology in Construction - One of the top ten answers to world's biggest problems, May 3 (2005)
3. Attik, G., Brown, R., Jackson, P., Creutzenberg, O., Aboukhamis, I., Rihn, B.H.: Internalization, Cytotoxicity, Apoptosis, and Tumor Necrosis Factor-alpha Expression in Rat Alveolar Macrophages Exposed to Various Dusts Occurring in the Ceramics Industry. Inhal. Toxicol. 20(12), 1101–1112 (2008)

4. Blaise, C., Gagne, F., Ferard, J.F., Eullaffroy, P.: Ecotoxicity of selected nanomaterials to aquatic organisms. Environ. Toxicol. 23(5), 591–598 (2008)
5. Brown, P., Kamat, P.V.: Quantum dot solar cells. Electrophoretic deposition of CdSe-C_{60} composite films and capture of photogenerated electrons with nC_{60} cluster shell. J. Amer. Chem. Soc. 130(28), 8890–8891 (2008)
6. Carson, R.: Silent Spring. Hamilton, London (1963)
7. Cha, E.K., Myung, H.: Cytotoxic effects of nanoparticles assessed in vitro and in vivo. J. Microbiol. Biotechnol. 17(9), 1573–1578 (2007)
8. Chan, W.C.W., Maxwell, D.J., Gao, X.H., Bailey, R.E., Han, M.Y., Nie, S.M.: Luminescent quantum dots for multiplexed biological detection and imaging. Curr. Opin. Biotechnol. 13(1), 40–46 (2002)
9. Chang, E., Thekkek, N., Yu, W.W., Colvin, V.L., Drezek, R.: Evaluation of quantum dot cytotoxicity based on intracellular uptake. Small 2(12), 1412–1417 (2006)
10. Chen, K.L., Elimelech, M.: Aggregation and deposition kinetics of fullerene (C60) nanoparticles. Langmuir 22(26), 10994–11001 (2006)
11. Chen, Z., Meng, H.A., Xing, G.M., Chen, C.Y., Zhao, Y.L., Jia, G.A., Wang, T.C., Yuan, H., Ye, C., Zhao, F., Chai, Z.F., Zhu, C.F., Fang, X.H., Ma, B.C., Wan, L.J.: Acute toxicological effects of copper nanoparticles in vivo. Toxicol. Lett. 163(2), 109–120 (2006)
12. Colvin, V.L.: The potential environmental impact of engineered nanomaterials. Nature Biotechnol. 21(10), 1166–1170 (2003)
13. Daniel, M.C., Astruc, D.: Gold nanoparticles: Assembly, supramolecular chemistry, quantum-size-related properties, and applications toward biology, catalysis, and nanotechnology. Chem. Rev. 104(1), 293–346 (2004)
14. de Ibarra, Y.S., Gaitero, J.J., Erkizia, E., Campillo, I.: Atomic force microscopy and nanoindentation of cement pastes with nanotube dispersions. Phys. Stat. Sol. A - Appl. Mater. Sci. 203(6), 1076–1081 (2006)
15. Derfus, A.M., Chan, W.C.W., Bhatia, S.N.: Probing the cytotoxicity of semiconductor quantum dots. Nano Lett. 4(1), 11–18 (2004)
16. Ding, L.H., Stilwell, J., Zhang, T.T., Elboudwarej, O., Jiang, H.J., Selegue, J.P., Cooke, P.A., Gray, J.W., Chen, F.Q.F.: Molecular characterization of the cytotoxic mechanism of multiwall carbon nanotubes and nano-onions on human skin fibroblast. Nano Lett. 5(12), 2448–2464 (2005)
17. Fang, J., Lyon, D.Y., Wiesner, M.R., Dong, J., Alvarez, P.J.J.: Effect of a fullerene water suspension on bacterial phospholipids and membrane phase behavior. Environ. Sci. Technol. 41(7), 2636–2642 (2007)
18. Fortner, J.D., Lyon, D.Y., Sayes, C.M., Boyd, A.M., Falkner, J.C., Hotze, E.M., Alemany, L.B., Tao, Y.J., Guo, W., Ausman, K.D., Colvin, V.L., Hughes, J.B.: C_{60} in water: Nanocrystal formation and microbial response. Environ. Sci. Technol. 39(11), 4307–4316 (2005)
19. Ge, Z., Gao, Z.: Applications of nanotechnology and nanomaterials in construction. In: First Inter. Confer. Construc. Develop. Countries, pp. 235–240 (2008)
20. Girishkumar, G., Rettker, M., Underhile, R., Binz, D., Vinodgopal, K., McGinn, P., Kamat, P.: Single-wall carbon nanotube-based proton exchange membrane assembly for hydrogen fuel cells. Langmuir 21(18), 8487–8494 (2005)
21. Griffitt, R.J., Weil, R., Hyndman, K.A., Denslow, N.D., Powers, K., Taylor, D., Barber, D.S.: Exposure to copper nanoparticles causes gill injury and acute lethality in zebrafish (Danio rerio). Environ. Sci. Technol. 41, 8178–8186 (2007)

22. Handy, R.D., Henry, T.B., Scown, T.M., Johnston, B.D., Tyler, C.R.: Manufactured nanoparticles: their uptake and effects on fish-a mechanistic analysis. Ecotoxicol. 17(5), 396–409 (2008)
23. Hardman, R.: A toxicologic review of quantum dots: Toxicity depends on physicochemical and environmental factors. Environ. Health Perspect 114(2), 165–172 (2006)
24. Heymann, D.: Solubility of C_{60} and C_{70} in seven normal alcohols and their deduced solubility in water. Fuller. Sci. Technol. 4, 509–515 (1994)
25. Hoshino, A., Fujioka, K., Oku, T., Suga, M., Sasaki, Y.F., Ohta, T., Yasuhara, M., Suzuki, K., Yamamoto, K.: Physicochemical properties and cellular toxicity of nanocrystal quantum dots depend on their surface modification. Nano Lett. 4(11), 2163–2169 (2004)
26. Hotze, E.M., Labille, J., Alvarez, P.J.J., Wiesner, M.R.: Mechanisms of photochemistry and reactive oxygen production by fullerene suspensions in water. Environ. Sci. Technol. 42, 4175–4180 (2008)
27. IARC, Silica, some silicates, coal dust and para-aramid fibrils. IARC Monographs on the Evaluation of Carcinogenic Risks to Humans 68, 41 (1997)
28. Irie, H., Sunada, K., Hashimoto, K.: Recent developments in TiO_2 photocatalysis: Novel applications to interior ecology materials and energy saving systems. Electrochem. 72(12), 807–812 (2004)
29. Isakovic, A., Markovic, Z., Todorovic-Markovic, B., Nikolic, N., Vranjes-Djuric, S., Mirkovic, M., Dramicanin, M., Harhaji, L., Raicevic, N., Nikolic, Z., Trajkovic, V.: Distinct cytotoxic mechanisms of pristine versus hydroxylated fullerene. Toxicol. Sci. 91(1), 173–183 (2006)
30. Jia, G., Wang, H.F., Yan, L., Wang, X., Pei, R.J., Yan, T., Zhao, Y.L., Guo, X.B.: Cytotoxicity of carbon nanomaterials: Single-wall nanotube, multi-wall nanotube, and fullerene. Environ. Sci. Technol. 39(5), 1378–1383 (2005)
31. Kanarek, S., Powell, C.: Nanotechnology health risk assessment. Epidemiol. 17(6), S443–S443 (2006)
32. Kandlikar, M., Ramachandran, G., Maynard, A., Murdock, B., Toscano, W.A.: Health risk assessment for nanoparticles: A case for using expert judgment. J. Nanopart. Res. 9(1), 137–156 (2007)
33. Kang, S., Pinault, M., Pfefferle, L.D., Elimelech, M.: Single-walled carbon nanotubes exhibit strong antimicrobial activity. Langmuir 23(17), 8670–8673 (2007)
34. Karlsson, H.L., Cronholm, P., Gustafsson, J., Moller, L.: Copper oxide nanoparticles are highly toxic: A comparison between metal oxide nanoparticles and carbon nanotubes. Chem. Res. Toxicol. 21(9), 1726–1732 (2008)
35. Kartam, N., Al-Mutairi, N., Al-Ghusain, I., Al-Humoud, J.: Environmental management of construction and demolition waste in Kuwait. Waste Management 24(10), 1049–1059 (2004)
36. Kirchner, C., Liedl, T., Kudera, S., Pellegrino, T., Javier, A.M., Gaub, H.E., Stolzle, S., Fertig, N., Parak, W.J.: Cytotoxicity of colloidal CdSe and CdSe/ZnS nanoparticles. Nano Lett. 5(2), 331–338 (2005)
37. Klaine, S.J., Alvarez, P.J.J., Batley, G.E., Fernandes, T.F., Handy, R.D., Lyon, D.Y., Mahendra, S., McLaughlin, M.J., Lead, J.R.: Nanomaterials in the environment: Behavior, fate, bioavailability, and effects. Environ. Toxicol. Chem. 27(9), 1825–1851 (2008)
38. Kloepfer, J.A., Mielke, R.E., Nadeau, J.L.: Uptake of CdSe and CdSe/ZnS quantum dots into bacteria via purine-dependent mechanisms. Appl. Environ. Microbiol. 71(5), 2548–2557 (2005)

39. Kontos, A.I., Kontos, A.G., Tsoukleris, D.S., Vlachos, G.D., Falaras, P.: Superhydrophilicity and photocatalytic property of nanocrystalline titania sol-gel films. Thin Sol. Films 515(18), 7370–7375 (2007)
40. Kourmpanis, B., Papadopoulos, A., Moustakas, K., Stylianou, M., Haralambous, K.J., Loizidou, M.: Preliminary study for the management of construction and demolition waste. Waste Management Res. 26(3), 267–275 (2008)
41. Kumar, A., Vemula, P.K., Ajayan, P.M., John, G.: Silver-nanoparticle-embedded antimicrobial paints based on vegetable oil. Nature Mater. 7(3), 236–241 (2008)
42. Lam, C.W., James, J.T., McCluskey, R., Arepalli, S., Hunter, R.L.: A review of carbon nanotube toxicity and assessment of potential occupational and environmental health risks. Critical Rev. Toxicol. 36(3), 189–217 (2006)
43. Lee, H.A., Imran, M., Monteiro-Riviere, N.A., Colvin, V.L., Yu, W.W., Riviere, J.E.: Biodistribution of quantum dot nanoparticles in perfused skin: Evidence of coating dependency and periodicity in arterial extraction. Nano Lett. 7(9), 2865–2870 (2007)
44. Lee, J., Fortner, J.D., Hughes, J.B., Kim, J.-H.: Photochemical production of reactive oxygen species by C_{60} in the aqueous phase during UV irradiation. Environ. Sci. Technol. 41, 2529–2535 (2007)
45. Lee, W.M., An, Y.J., Yoon, H., Kweon, H.S.: Toxicity and bioavailability of copper nanoparticles to the terrestrial plants mung bean (Phaseolus radiatus) and wheat (Triticum aestivum): Plant agar test for water-insoluble nanoparticles. Environ. Toxicol. Chem. 27(9), 1915–1921 (2008)
46. Li, D., Lyon, D.Y., Li, Q., Alvarez, P.J.J.: Effect of natural organic matter on the antibacterial activity of a fullerene water suspension. Environ. Toxicol. Chem. 27(9), 1888–1894 (2008)
47. Li, G.Y.: Properties of high-volume fly ash concrete incorporating nano-SiO_2. Cement Concrete Res. 34(6), 1043–1049 (2004)
48. Li, H., Xiao, H.G., Yuan, J., Ou, J.P.: Microstructure of cement mortar with nanoparticles. Composites Part B-Engineering 35(2), 185–189 (2004)
49. Lin, P., Chen, J.-W., Chang, L.W., Wu, J.-P., Redding, L., Chang, H., Yeh, T.-K., Yang, C.-S., Tsai, M.-H., Wang, H.-J., Kuo, Y.-C., Yang, R.S.H.: Computational and ultrastructural toxicology of a nanoparticle, quantum Dot 705, in mice. Environ. Sci. Technol. 42(16), 6264–6270 (2008)
50. Long, T.C., Saleh, N., Tilton, R.D., Lowry, G.V., Veronesi, B.: Titanium dioxide (P25) produces reactive oxygen species in immortalized brain microglia (BV2): Implications for nanoparticle neurotoxicity. Environ. Sci. Technol. 40(14), 4346–4352 (2006)
51. Lovric, J., Cho, S.J., Winnik, F.M., Maysinger, D.: Unmodified cadmium telluride quantum dots induce reactive oxygen species formation leading to multiple organelle damage and cell death. Chem. Biol. 12(11), 1227–1234 (2005)
52. Lu, Z.S., Li, C.M., Bao, H.F., Qiao, Y., Toh, Y.H., Yang, X.: Mechanism of antimicrobial activity of CdTe quantum dots. Langmuir 24(10), 5445–5452 (2008)
53. Lyon, D.Y., Adams, L.K., Falkner, J.C., Alvarez, P.J.J.: Antibacterial activity of fullerene water suspensions: Effects of preparation method and particle size. Environ. Sci. Technol. 40(14), 4360–4366 (2006)
54. Lyon, D.Y., Alvarez, P.J.J.: How a fullerene water suspension kills bacteria: Exploring three possible mechanisms. Chem. Res. Toxicol. 20(12), 1991 (2007)
55. Lyon, D.Y., Brunet, L., Hinkal, G.W., Wiesner, M.R., Alvarez, P.J.J.: Antibacterial activity of fullerene water suspensions nC_{60} is not due to ROS-mediated damage. Nano Lett. 8(5), 1539–1543 (2008)

56. Lyon, D.Y., Fortner, J.D., Sayes, C.M., Colvin, V.L., Hughes, J.B.: Bacterial cell association and antimicrobial activity of a C_{60} water suspension. Environ. Toxicol. Chem. 24(11), 2757–2762 (2005)
57. Mahendra, S., Zhu, H., Colvin, V.L., Alvarez, P.J.J.: Quantum Dot Weathering Results in Microbial Toxicity. Environ. Sci. Technol. 42(24), 9424–9430 (2008)
58. Mann, S.: Nanotechnology and construction. Nanoforum Report, May 30 (2006)
59. Narayan, R.J., Berry, C.J., Brigmon, R.L.: Structural and biological properties of carbon nanotube composite films. Mater. Sci. Eng. B. 123, 123–129 (2005)
60. O'Brien, N., Cummins, E.: Recent developments in nanotechnology and risk assessment strategies for addressing public and environmental health concerns. Hum. Ecol. Risk Assess. 14(3), 568–592 (2008)
61. Oberdörster, E.: Manufactured nanomaterials (fullerenes, C_{60}) induce oxidative stress in the brain of juvenile largemouth bass. Environ. Health Perspect. 112(10), 1058–1062 (2004)
62. Oberdorster, G., Gelein, R., Ferin, J., Weiss, B.: Association of particulate air-pollution and acute mortality - Involvement of ultrafine particles. Inhal. Toxicol. 7(1), 111–124 (1995)
63. Park, S., Lee, Y.K., Jung, M., Kim, K.H., Chung, N., Ahn, E.K., Lim, Y., Lee, K.H.: Cellular toxicity of various inhalable metal nanoparticles on human alveolar epithelial cells. Inhal. Toxicol. 19, 59–65 (2007)
64. Paz, Y., Luo, Z., Rabenberg, L., Heller, A.: Photooxidative self-cleaning transparent titanium-dioxide films on glass. J. Mater. Res. 10(11), 2842–2848 (1995)
65. Poon, C.S.: Management of construction and demolition waste. Waste Management 27(2), 159–160 (2007)
66. Reeves, J.F., Davies, S.J., Dodd, N.J.F., Jha, A.N.: Hydroxyl radicals are associated with titanium dioxide (TiO_2) nanoparticle-induced cytotoxicity and oxidative DNA damage in fish cells. Mutat. Res. -Fundam. Mol. Mech. Mutagen. 640(1-2), 113–122 (2008)
67. Rincon, A., Pulgarin, C.: Bactericidal action of illuminated TiO_2 on pure Escherichia coli and natural bacterial consortia: post-irradiation events in the dark and assessment of the effective disinfection time. Appl. Catal. B: Environ. 49, 99–112 (2004)
68. Rincon, A., Pulgarin, C.: Effect of pH, inorganic ions, organic matter and H_2O_2 on E. coli K12 photocatalytic inactivation by TiO_2 Implications in solar water disinfection. Appl. Catal. B: Environ. 51, 283–302 (2004)
69. Ryman-Rasmussen, J.P., Riviere, J.E., Monteiro-Riviere, N.A.: Surface coatings determine cytotoxicity and irritation potential of quantum dot nanoparticles in epidermal keratinocytes. J. Investig. Dermatol. 127(1), 143–153 (2007)
70. Saafi, M., Romine, P.: Nano- and microtechnology. Concrete Inter. 27, 28–34 (2005)
71. Sayes, C.M., Fortner, J.D., Guo, W., Lyon, D., Boyd, A.M., Ausman, K.C., Tao, Y.J., Sitharaman, B., Wilson, L.J., Hughes, J.B., West, J.L., Colvin, V.L.: The differential cytotoxicity of water-soluble fullerenes. Nano Lett. 4(10), 1881–1887 (2004)
72. Sayes, C.M., Gobin, A.M., Ausman, K.D., Mendeza, J., West, J.L., Colvin, V.L.: Nano-C_{60} cytotoxicity is due to lipid peroxidation. Biomater. 26(36), 7587–7595 (2005)
73. Sayes, C.M., Wahi, R., Kurian, P.A., Liu, Y.P., West, J.L., Ausman, K.D., Warheit, D.B., Colvin, V.L.: Correlating nanoscale titania structure with toxicity: A cytotoxicity and inflammatory response study with human dermal fibroblasts and human lung epithelial cells. Toxicol. Sci. 92(1), 174–185 (2006)

74. Shiohara, A., Hoshino, A., Hanaki, K., Suzuki, K., Yamamoto, K.: On the cytotoxicity caused by quantum dots. Microbiol. Immunol. 48(9), 669–675 (2004)
75. Sobolev, K., Gutierrez, M.F.: How nanotechnology can change the concrete world. Am. Ceram. Soc. Bull. 84, 16–20 (2005)
76. Song, G.B., Gu, H.C., Mo, Y.L.: Smart aggregates: multi-functional sensors for concrete structures - a tutorial and a review. Smart Mater. Struct. 17(3), 1–17 (2008)
77. Spesia, M.B., Milanesio, M.E., Durantini, E.N.: Synthesis, properties and photodynamic inactivation of Escherichia coli by novel cationic fullerene C_{60} derivatives. Euro. J. Med. Chem. (2007)
78. Tans, S.J., Verschueren, A.R.M., Dekker, C.: Room-temperature transistor based on a single carbon nanotube. Nature 393(6680), 49–52 (1998)
79. Tsao, N., Luh, T., Chou, C., Chang, T., Wu, J., Liu, C., Lei, H.: In vitro action of carboxyfullerene. J. Antimicro. Chemother. 49(4), 641–649 (2002)
80. Turusov, V., Rakitsky, V., Tomatis, L.: Dichlorodiphenyltrichloroethane (DDT): Ubiquity, persistence, and risks. Environmental Health Perspectives 110(2), 125–128 (2002)
81. Voura, E.B., Jaiswal, J.K., Mattoussi, H., Simon, S.M.: Tracking metastatic tumor cell extravasation with quantum dot nanocrystals and fluorescence emission-scanning microscopy. Nature Med. 10(9), 993–998 (2004)
82. Wei, W., Sethuraman, A., Jin, C., Monteiro-Riviere, N.A., Narayan, R.J.: Biological properties of carbon nanotubes. J. Nanosci. Nanotechnol. 7(4-5), 1284–1297 (2007)
83. Wiesner, M.R., Lowry, G.V., Alvarez, P., Dionysiou, D., Biswas, P.: Assessing the risks of manufactured nanomaterials. Environ. Sci. Technol. 40(14), 4336–4345 (2006)
84. Wolfrum, E.J., Huang, J., Blake, D.M., Maness, P.C., Huang, Z., Fiest, J., Jacoby, W.A.: Photocatalytic oxidation of bacteria, bacterial and fungal spores, and model biofilm components to carbon dioxide on titanium dioxide-coated surfaces. Environ. Sci. Technol. 36(15), 3412–3419 (2002)
85. Yu, W.W., Chang, E., Falkner, J.C., Zhang, J.Y., Al-Somali, A.M., Sayes, C.M., Johns, J., Drezek, R., Colvin, V.L.: Forming biocompatible and nonaggregated nanocrystals in water using amphiphilic polymers. J. Amer. Chem. Soc. 129(10), 2871–2879 (2007)
86. Zhang, Q.W., Kusaka, Y., Sato, K., Nakakuki, K., Kohyama, N., Donaldson, K.: Differences in the extent of inflammation caused by intratracheal exposure to three ultrafine metals: Role of free radicals. J. Toxicol. Environ. Health Part A.-Current Issues 53(6), 423–438 (1998)
87. Zhang, W., Suhr, J., Koratkar, N.: Carbon nanotube/polycarbonate composites as multifunctional strain sensors. J. Nanosci. Nanotechnol. 6(4), 960–964 (2006)
88. Zhu, W., Bartos, P.J.M., Porro, A.: Application of nanotechnology in construction - Summary of a state-of-the-art report. Mater. Struct. 37(273), 649–658 (2004)
89. Zhu, X.S., Zhu, L., Duan, Z.H., Qi, R.Q., Li, Y., Lang, Y.P.: Comparative toxicity of several metal oxide nanoparticle aqueous suspensions to Zebrafish (Danio rerio) early developmental stage. J. Environ. Sci. Health Part A.-Toxic/Hazardous Substances & Environ. Eng. 43(3), 278–284 (2008)

Nanotechnology in Construction: A Roadmap for Development

P.J.M. Bartos

Abstract. Roadmaps were originally developed as tools for finding surface routes for getting from one place to another. Recently, the scope of a roadmap has been extended to cover tools used for indication of pathways for reaching predicted future developments, for assessing progress and indicating trends. This was applied to developments in application of Nanotechnology within the broad domain of Construction. The Roadmap for Nanotechnology in Construction (RoNaC) was first developed in 2003 as an aid for forecasting research and investment directions, with a timescale of 25 years. Five years have elapsed and progress has been achieved along a few pathways indicated in the original RoNaC. However, construction industry continues to lag behind in both the awareness of the potential and the expected commercial exploitation of nanotechnology. This paper provides an updated version based on the three original sectorial charts, indicating where tangible progress has been made, where research is active and where advance along the predicted pathways has slowed down or stopped altogether.

1 Introduction

The purpose of a Roadmap is to chart trends and developments, which, in this case, link nanotechnology and construction. It provides a useful tool, a template, for their predictions. The Roadmap for Nanotechnology in Construction (RoNaC) has been aimed at facilitating identifications of desirable aims/destinations for construction RTD over a short-medium timescale (up to 25-years).

The need for development of a Roadmap for Nanotechnology in Construction arose during the 5th FP European project "NANOCONEX" (2002-2003) [1] as one of its deliverables. It reflected pioneering work exploiting early developments in application of nanotechnology to construction materials at the Advanced Concrete and Masonry Centre (1994-) and the attached Scottish Centre for Nanotechnology in Construction Materials (2000-) at the University of West of Scotland (formerly

P.J.M. Bartos
The Queen's University of Belfast & University of West of Scotland
e-mail: p.bartos@ntlworld.com

the University of Paisley) where the first State of the Art report [2,3] was also produced. The RoNaC was developed to aid forecasting RTD directions and to inform and guide not only the 'end-users' in construction industry but also investors and national / international bodies supporting research and development. The RoNaC showed the diverse pathways towards nanotechnology-linked expectations, aims and targets in the very large and economically very significant domain of construction, envisaged in 2004.

The content of the RoNaC was also linked to work of the RILEM International Technical Committee TC 197-NCM on Nanotechnology in Construction Materials (2002-2007). The earliest version of the RoNaC had been presented at the E-CORE & ECCREDI conference on "Building for a European Future – Strategies & Alliances for Construction Innovation", held in Maastricht in October 2004. Subsequently it was presented and discussed at the 2nd International Symposium on Nanotechnology in Construction (NICOM2) in Bilbao, in November 2005 and at the ACI Seminar on Nanotechnology of Concrete: Recent Developments and Future Perspectives in Denver in November 2006 [4].

Construction industry differs from many other sectors of manufacturing industry in that it adopts and exploits the new nano-scale tools, which have been developed in the more fundamental scientific rather than engineering domain. Compared to many other industrial sectors, instances where Nanotechnology has been already successfully exploited and a construction related major product has already reached open markets still remain few in numbers. Awareness of the potential for exploitation of Nanotechnology in construction has been improving over the last decade, but *expectations of a more practical exploitation have not been fulfilled*, much more remains to be done. Nano-related RTD in construction has been established in a few sectors; however, it can be still described overall as an 'emerging' trend, *often concentrating on new knowledge rather than on an application*. Advances are very non-uniform, leading to a particularly pronounced fragmentation and often to a distinct isolation of current centres of nano-related construction research and development. These are very important, construction-industry specific circumstances, accounted for in this RoNaC. Detailed analysis of nano-related RTD in construction, which was published in the NANOCONEX / RILEM TC197-NCM State-of-the-Art report [2,3] and which has been updated in the final report from the TC 197 NCM (publication expected in early 2009) is still applicable. The report indicated two factors which severely impact on RFTD in construction inn general and on exploitation of nanotechnology in particular:

(a) An inherently different nature of construction compared with other sectors of manufacturing industry. Final products of construction, tend to be very complex, non-mass produced and possess a relatively long service lives. This makes them very different from common products of microelectronics, IT or even aerospace/automotive industries. Construction generally acquires and adapts many inventions from other industries or from related sciences, rather than inventing them. Construction therefore tends to be much more an exploiter of ideas and inventions than their creator.

(b) Historically ***very low levels of investment by construction industry into research*** represent a major hindrance in exploitation of nanotechnology. National levels of RTD investment in construction are often the lowest amongst all sectors of the manufacturing industry. This, together with the very high initial capital investment, invariably required in nano-related RTD, combines to generate a major obstacle for development of an adequate, essential research infrastructure. Overcoming such an obstacle is not helped by the very low margins of profitability within construction industry and the mid-long term timescale for any commercial returns to arise from such investments. Recent decline of economic activity worldwide is likely to worsen the situation.

2 Routes and Pathways

The 'quality' and usefulness of existing and new, developing, roads and pathways for progress reflect the availability and 'quality' of relevant supporting infrastructure. The infrastructure has to be upgraded with passage of time if it were to fulfill its role in supporting research, development and practical exploitation of nanotechnology. The rate of advance towards highly desirable medium-long term goals would slow down, perhaps even stop entirely, if the necessary infrastructures were not adequately maintained and periodically improved. Nanotechnology related research infrastructures, which 'pave the way' include:

- *Instrumentation* and associated methodologies for nano-scale investigations. (characterisation of properties at nano-scale, nano-assembly and nano-fabrication, analytical techniques and imaging at molecular /atomic scale).
- Descriptive, preferably also genuinely *predictive, numerical models*, which include a linkage across the whole scale, from nano-to-macro size.
- Standardisation of basic *nano-scale metrology* equipment and provision of means for an assessment of their performance. Development of new, more effective tools for a meaningful nano-scale characterisation of materials.

Instrumentation and metrology at nano-scale are developing at a very fast pace, which inevitably brings with it a rapid rate of obsolescence. The rate of obsolescence is comparable to that seen in the IST area, however 'hardware' costs of nano-scale instrumentation and costs of its maintenance / calibration / upgrading, even if only on a moderate scale, are higher than for IST research.

It is impracticable to produce one, all-encompassing, single map for the whole of construction in which all the existing and potential routes and /pathways of progress linked with nanotechnology would be shown. A single comprehensive map would include a multitude of 'pathways', criss-crossing each other. To show them all in one chart/roadmap, with all the numerous possible intersections, interactions and feedbacks would lead to a tangled mass of connections resulting in an illegible and incomprehensible document.

A solution has been found in the creation of a number of simpler charts in which the nano-related construction research pathways and / routes, heading towards specific 'destinations', have been clustered. An example is the Chart 2: 'Buildings of Future', where the overall destination has been defined by a number of highly desirable goals/ and outcomes, presented in the chart.

Some of the pathways shown may appear in another chart(s). This is to be expected, as Nanotechnology is fundamentally an 'enabling technology', where one advance, especially when it is a major advance, is very likely to underpin progress towards desirable goals in more sectors than the original one. Simplicity and 'legibility' of a chart restricted the scope for presentation. It was not practicable to indicate the current status of the routes/pathways, e.g. showing which roads were 'under construction' or just 'planned'. Activities shown in the specimen charts have colour coding to indicate Existing pathways are shown in red. Many 'connections' are seen today as relatively 'thin', faint and indistinct routes, which is how they are perceived today, but which, in future, may (or may not!) become well-established, densely 'trafficked' wide principal routes / highways enabling a much faster progress towards the future goals indicated within the 'destination'.

3 Benchmarks and Timescale

Information and data from the State-of-the-Art report [2] helped to establish the benchmarks for the development of the RoNaC. The original survey had been to support European projects and gave priority to Europe. However this was extended to include an analysis of existing knowledge, proposed and current nano-related R&D activities applicable to any part of the construction sector on a global scale. Contributions presented at the 1st International Symposium on Nanotechnology in Construction, Paisley, UK, June 2003 [5], active links with the TC-197 NCM on Nanotechnology in Construction Materials, established under the auspices of RILEM in 2002, and events organised and documents produced by the UK Institute of Nanotechnology [6] and other organisations were reviewed and their conclusions considered. The 2nd International Symposium on Nanotechnology in Construction (NICON2) in Bilbao, Spain in November 2005 confirmed the emerging trends and indicated several new projects exploiting nanotechnology [7].

The RoNaC charts were plotted against a timescale, which provided a measure of how distant the required or desirable research and development destinations were from a 'benchmark' situation in 2004, its starting point. The timescale extends to 25+ years ahead, with the first five years already covered. It is important to note that the greater is the extension of the time interval into future, the greater are the potential errors and the lesser is the reliability of forecasts in this rapidly and unevenly developing field.

There are three 'specimen' Charts shown. Sufficient information is already available from individual publications and reviews [1–4,7] and from the final documentation from the RILEM TC 197 NCM [8] and other sources, to enable production of new charts in a similar format. It is possible to re-define the baselines and destinations and focus on different aspects linked to nanotechnology.

It is important to appreciate that the 'routes/pathways' shown in the charts are in many instances only now being formed or traced. Additional ones may develop in due course, and may not even exist as yet.

4 Vehicles, Drivers

The choice and understanding of the research 'vehicles' needed to pursue specific goals and the capability of their 'drivers' to steer them along correct routes to desired outcomes and destinations, always require adequate prior *knowledge* of both Nanotechnology and the relevant sector(s) of construction industry. Inadequate knowledge, or its absence mean that not all of the pathways/routes (if any), including their intersections, as shown on the attached Charts, may be clear and recognisable to potential developers and exploiters – and any advance may be slow, sometimes taking wrong paths leading to 'dead-ends' (and wasted resources). If knowledge represented the 'vehicle', then the 'drivers', which represent the motivation to go ahead, can be also identified.

The greatest impact on the construction industry and the economy within the timescale of the Roadmap is expected to come from an *enhancement in performance of materials – a very strong driver of construction RTD*. This, in turn, is likely to arise from an improved understanding and control of their internal structure on micro-to-nano-scale and from a potential improvement of their production processes. The eventual total impact in construction will be almost always very substantial, due not necessarily to radical technological leaps forward but mainly to massive quantities in which basic ('bulk') construction materials are used.

Most of the advances along the specific research routes and pathways of RTD are predicted to be incremental, leading to a relatively steady progress related to existing, namely the 'bulk' materials and technologies. Such a relatively steady progress will be strongly influenced by developments in the research 'infrastructure' It may be that some of the expected improvements will not take place and there will be a waiting period with only a small or even zero advances, until a breakthrough occurs and the physical capability of the equipment (infrastructure) is instantly and significantly upgraded or there is a significant advance in interpretation/understanding of data/results obtained.

An improved understanding of structure-properties relationships and an ability to control the structure of many materials on the nano-scale is expected to provide the greatest opportunities for major advances in construction materials science. Such breakthroughs will enable the 'materials by design' approach to replace the traditional one of a 'trial and error', enabling properties of a material to be tailored for optimum performance in a specific application.

Additional, often mutually complementary drivers are market pull, venture capital involvement, competitiveness and prospects of higher financial returns. Their significance can vary, depending on the availability of the basic 'vehicle' – presence of an adequate knowledge & information.

5 Business Environment

Construction industry is a very substantial contributor to economic performance. In Europe alone, its annual turnover is estimated [9] at about 1000 billion Euro. At the same time, it is estimated that 97% of employees working within it are employed by enterprises with less than 10 staff. The Small and Medium size Enterprises in construction are generally not involved in RTD activities and acquisition of the required minimum knowledge, and awareness of opportunities brought by nanotechnology, is difficult. At present, such knowledge still appears to be inadequate if not entirely absent. The 'vehicle' for advance along the existing and potential pathways is therefore available only to large companies, as is confirmed by the appearance of the first practical, commercial, nano-related products on the market.

The way ahead - an exploitation of the directions such as those shown in the specimen RoNaC charts is proving not to be easy and straightforward because the knowledge required at different levels of construction-related staff still tends to lag behind the continuing advances in nanotechnology achieved elsewhere.

It is possible to conclude that in the absence of an environment conducive to acquisition of an adequate knowledge through education and training, many of the basic and most significant drivers of progress, such as market pull, venture capital involvement, competitiveness and prospects of higher financial returns, will be either weak or will not be established. Numbers of skilled research related personnel at both the scientific/supervisory level and at the supporting staff/technician level, who possess the required combination of knowledge related to construction and nanotechnology, continue to appear to be too low to provide the manpower required to steer and move efficiently forward the 'vehicles' for expansion of nano-related RTD in construction and development of marketable products.

Countries leading the development and exploitation of Nanotechnology, such as the USA, Japan and European Union [8,9], appreciated the necessity of providing major financial support from state/government sources The US government set up the National Nanotechnology Initiative in yr. 2000, which has committed an initial funding of more that one 700 million US$ of public funds to facilitate the additional commitment of private venture capital for specific developments through its different agencies [8]. As has been noted in the first RoNaC, all but a very small fraction of the public funding for nano-related RTD was channeled either into non-construction industries or into development of general nano-scale research infrastructure, which was then used primarily to support non-construction research, perceived to be more 'high-tech'.

Investment into research infrastructure can benefit construction and it can 'smooth/pave' some of the development paths shown in the Charts. However, this would require a significant change in management of such facilities, to enable a few construction-related 'vehicles' to squeeze into the densely trafficked RTD pathways, already crowded by nano-science related vehicles with no connection to construction. Unless there is such change, benefits to construction from such nanotechnology related infrastructure investments are likely to be disproportionately low compared with the economic significance of construction, which, in EU contributes approximately 10% of GDP. Other significant drivers are emerging, and are likely to gain significance in terms of construction and nanotechnology. These include *climate change and the associated issues*, both on a global scale and in local urban redevelopment.

6 Roadmap and Charts

An overall roadmap covering the whole of construction would be so complicated to make it difficult to understand. The 'scale' would have to be adjusted to make it 'legible', but then the roadmap would become too 'coarse' to make it useful. As a result, such a roadmap would not reveal adequately details necessary for an appreciation and understanding of the links, relationships and dependencies of varied significance, which may exist between the pathways shown.

Instead of a single roadmap, a 'complete' RoNaC would therefore become a 'road-atlas' comprising a collection of 'sectorial' charts such as the examples attached. Each chart would be focused on a coherent cluster of aims and destinations, depending on the purpose of each chart.

Aims and destinations, which are expected to be of significance for construction in the medium-long term include:

- Understanding basic phenomena (interactions, processes) at nanoscale: e.g. cement hydration and formation of nanostructures, origins of adhesion and bond, pore structure and interfaces in concrete, mechanisms of degradation, relevant modelling and simulation etc.
- Bulk 'traditional' construction materials with a modified nano-structure: e.g. concrete, bitumen, plastics modified with nanoparticulate additives, special admixtures and new processing techniques modifying internal nanostructures etc.
- New high performance structural materials: e.g. carbon nanotubes, new fibre reinforcements, nanocomposites, advanced steels and concrete/cement composites, biomimetic materials, etc. Materials with extended durability in extreme service conditions.
- High performance new coatings, paints and thin films: e.g. wear-resistant coating, durable paints, self-cleaning/anti-bacteria and anti-graffiti coatings, smart thin films, etc.
- New multi-functional materials and components: e.g. aerogel based insulating materials, efficient filters/membranes and catalysts,, self-sensing/healing materials, etc.
- New production techniques, tools and controls: e.g. more energy efficient and environmental friendly production of materials and structures, novel processes with more intelligent and integrated control systems, etc.

- Intelligent structures and use of micro/nano sensors: e.g. nano-electromechanical systems, biomimetic sensors, paint-on sensors, and self-activating structures/ components, etc.
- Integrated monitoring and diagnostic systems: e.g. for monitoring structure defects and reinforcement corrosion, environmental changes/conditions, and detecting security risks, etc.
- Energy saving lighting, fuel cells and communication devices: e.g. efficient and cheap fuel cells and photovoltaics, photocatalytic materials etc.

The Charts provided as examples show pathways emerge from a baseline to the left of the Chart. If appropriate, the baseline can be split into separate 'blocks', indicating different sub-areas of research. Each sub-area may consist of several recognised and related 'research streams/directions.

Progress forecast along each of the identified nano-related RTD activity paths is shown against a linear timescale from 0 to 25 (+) years. Activities, which are shown starting within the first five years, are already being pursued or about to commence. Aims/destinations, common to each of the charts, are shown on the right hand side and basic characteristics/parameters of the aims are also listed there.

A simple elongated rectangular box represents each research activity. However, this does not indicate a very precise and sudden start and finish, and a uniform intensity of the activity throughout its expected duration. It is s a simplification, which has been adopted for the sake of clarity, and because it is impossible to predict variations in the intensity/magnitude of research which are inevitable during each of the activities. There may be ups and downs, and even breaks / discontinuities, when funding may drop below the level required to sustain advance. Unexpected and at that time perhaps temporarily insuperable technological problems may be encountered on the path followed, or even an unexpected dead end may be encountered, for example by discovery of a hitherto unknown ecological or health & safety threat or danger. Each 'activity box' is therefore presented in a simplified linear and parallel direction, with only major interactions indicated as 'diagonal' connecting lines. As mentioned before, some of the activities will be using common paths/routes and share the infrastructure. Many overall 'integrating' aspects therefore arise because of the commonality of some of the nano-scale approaches. This is a feature representing the inherent multi-disciplinarity of RTD, which exploits advances in nanotechnology.

A general activity, which is beginning to be more appreciated, and which can be associated with any of the specific ('boxed') activities shown is an *evaluation of health and safety*, both during the process of research and development and regarding practical products / applications. It is now very urgent to provide an adequate national and international legislative cover and avoid potential health & safety problems undermining confidence in nanotechnology in general.

7 Specimen Charts

Original charts are provided; year 2009 equates to year 5 of the timescales shown.

7.1 Chart 1: Traditional Bulk Construction Materials

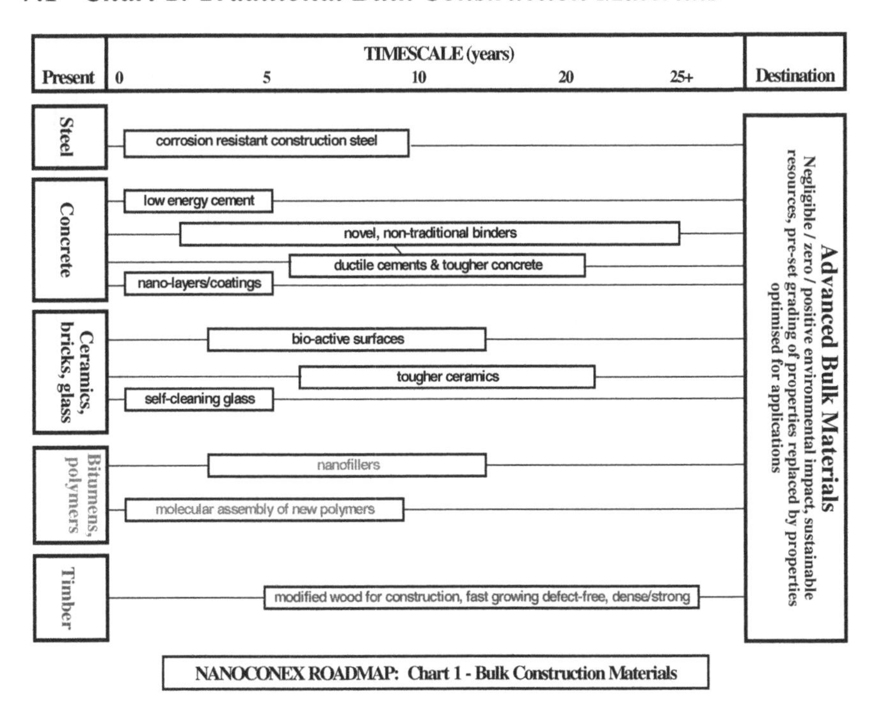

Chart 1 reviews potential for exploitation of nanotechnology where even incremental improvements are likely to lead to big commercial and environmental / societal gains because of the extremely high 'multiplying factors' attached to such materials. Activities shown follow generally the 'top-down' approach, in which an existing material is improved through knowledge and modifications carried out down at the micro-nano level.

Timescale of 5 years have been reached in 2009, with progress in knowledge being achieved in most of the activities but with products appearing only in the areas of steel (very limited) and ceramics (photocatalytic coatings, selfcleaning glass). Concrete coatings have only moved if the photocatalytic surfaces are concerned.

7.2 Chart 2: Buildings of Future

The chart outlines the potential for exploitation nanotechnology leading to a much higher standard and much more environmentally acceptable, and eventually fully

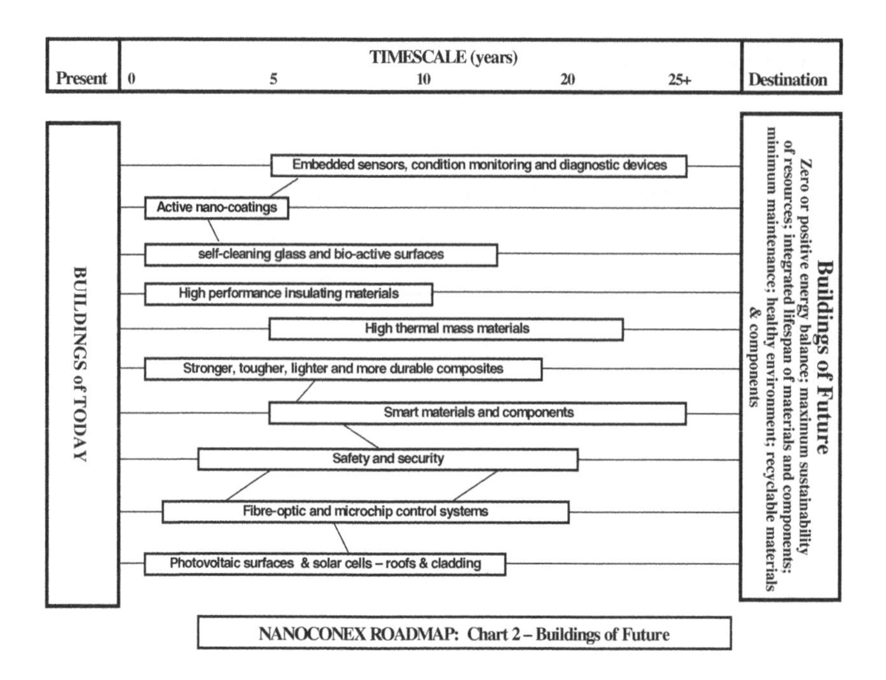

NANOCONEX ROADMAP: Chart 2 – Buildings of Future

sustainable, building construction. Achievements/destinations reached by routes shown in Charts 1 and 3 will be associated with or support activities shown in this Chart.

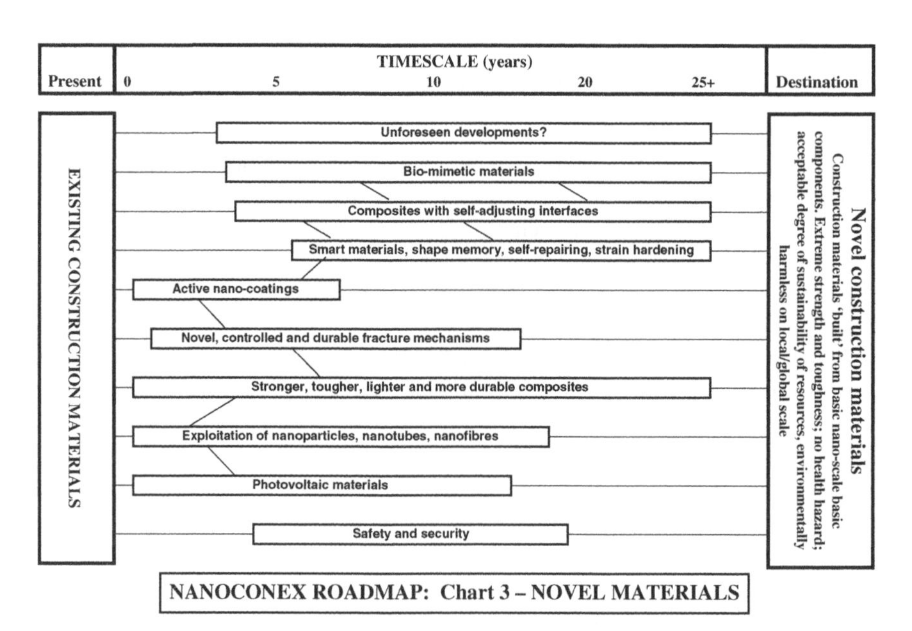

NANOCONEX ROADMAP: Chart 3 – NOVEL MATERIALS

7.3 Chart 3: Novel Construction Materials

The chart follows primarily the 'bottom-up' development process, where materials of substantially altered properties and a much high performance are 'built' or 'assembled' from the basic nano-scale constituents (molecular – atomic level).

It is most important to note the ***'unforeseen developments'***, an activity box that covers developments not even thought of at present.

8 Conclusions

1. Since the first issue of the RoNaC in 2004, the exploitation of the RTD potential in construction through nanotechnology in construction remains in an 'emerging, / early development' stage.

2. The RoNaC presented includes three *specimen charts* covering the most common 'clusters' of construction research activities, where nano-based research and development activities are already carried out and several routes/pathways are already established. Additional charts can be developed for specific activities or aims/destinations.

3. The uneven progress of exploitation of nanotechnology in construction and the specific nature of the construction industry limit the accuracy of the forecasts shown in the RoNaC. Reliability of predictions regarding commencement, duration and end of activity also decreases rapidly as the commencement time becomes more distant from present (year 5 from original time zero).

4. *Integration of the currently fragmented nano-related research in construction* and the use of common 'vehicles' at national or international levels would considerably speed up progress along the pathways outlined in the RoNaC.

5. Most of the overall construction output is through *Small and Medium Enterprises.* The SMEs in construction domain have negligible research and development capabilities. They cannot be therefore considered as effective vehicles for progress in exploitation of nanotechnology in construction, although they need to be associated with the RTD and their awareness of developments should be improved.

6. Activities concerned with monitoring and evaluation *of ecological, environmental and health & safety aspects* of materials and processes both during research at nanoscale and for development of marketable products are expected to become an essential part of all activities considered.

7. The RoNaC charts provide a 'snapshot' of the current situation in only three sub-sectors of construction, after five years from its first publication. It is suggested that similar 'templates' are used to develop further charts focused on different sub-sectors or topics, and the charts are updated every 3-5 years.

Acknowledgments. The author wishes to acknowledge the assistance provided by partners in the FP5 "Nanoconex" European Project and to thank colleagues in the RILEM Committee TC197 NCM "Nanotechnology in Construction Materials" for their collaboration.

References

1. European 5th FP project: Towards setting up of a Network of Excellence on Nanotec nology in Construction - project NANOCONEX. 1.12.2002-30.11.2003; Contract No.G1MA-CT-2002-04016; project No. GMA1-2002-72160; Coordinator: Dr A Porro, Labein, Bilbao, Spain (2003)
2. Zhu, W., Bartos, P.J.M., Gibbs, J.: Application of Nanotechnology in Construction. State of the Art report. ACM Centre/Scottish Centre for Nanotechnology in Construction Materials. University of Paisley, Technical Report, Project "NANOCONEX", 49 p. (March 2004)
3. Zhu, W., Bartos, P.J.M., Porro, A. (eds.): Application of Nanotechnology in Construction. Mater. Struct. 37, 649–659 (2004)
4. Sobolev, K., Shah, S.P.: Nanotechnology of Concrete: Recent Developments and Future Perspectives. ACI SP 254, 164 p. (2008)
5. Bartos, P.J.M., Hughes, J.J., Trtik, P.: Nanotechnology in Construction. The Royal Society for Chemistry, 374 p. (2004) ISBN 0-85404-623-2
6. Institute of Nanotechnology, `http://www.nano.org.uk`
7. Porro, A., de Miguel, Y., Bartos, P.J.M. (eds.): Nanotechnology in Construction. RILEM Publications s.a.r.l (2007)
8. Trtik, P., Bartos, P.J.M. (eds.): Final report of Rilem TC 197-NCM on Nanotechnology in Construction Materials (to be published) (2009)

The Colloid/Nanogranular Nature of Cement Paste and Properties

H. Jennings

Abstract. The importance of microstructure in materials science rests in its ability to establish links between processing and properties. Many properties of concrete are governed by structure at the nano-scale, notably by the variable and difficult to study calcium silicate hydrate. A basis for recent progress has come from viewing the nanostructure as a colloid or granular material (referred to here as C-G), with certain properties assigned to the grains and other properties assigned to reasonably well defined packing arrangements of the grains. This approach taps into both colloid science and granular mechanics. This approach is rich both conceptually and quantitatively. This paper describes recent progress using the C-G approach to understand drying shrinkage and creep, with a view towards further reconciling a vast literature, and improving quantitative relationships between structure and properties.

1 Microstructure

Properties of materials are controlled largely by their microstructures, and microstructures are defined by imperfections, including surfaces and interfaces, pores, and atomic irregularities of various types. Cement based materials are heterogeneous at many scales, which makes their microstructures particularly difficult to describe quantitatively. For more than fifty years the structure at the very small scale has been the subject of considerable research and several models have been proposed, but the general acceptance of any one model has been frustrated by anomalies and lack of conclusive experimental evidence.

A recent review [1] has emphasized that many important properties, including creep and shrinkage, are governed by the structure at the nanoscale. In particular surfaces and very small pores that contain an aqueous solution, can be defined in

---

H. Jennings
Civil and Environmental Engineering, Materials Science and Engineering,
Northwestern University, Evanston IL. USA
e-mail: h-jennings@northwestern.edu
http://www.civil.northwestern.edu/people/jennings.html

terms of how they change with time, temperature, moisture content and load. Calcium silicate hydrates (C-S-H), which are of variable composition and structure, are of central importance. Their structure is nearly amorphous and can experience irreversible changes at any time making it intrinsically difficult to characterize. One broad question is whether C-S-H is better characterized as a continuous material that contains pores or as a granular material composed of reasonably well-defined particles that pack together into different arrangements, and the gel pores are the spaces between particles.

2 A Case for the Colloid/Granular Approach

Microstructure or, in the case of cement based materials, nanostructure serves as the link between processing and properties. From the chemistry perspective the nanostructure has been studied recently as a colloid [see for example 2,3]. Also from the mechanics perspective C-S-H has been studied recently as a granular material [see for example 4]. This paper further explores some of the advantages and challenges that come from posing certain questions about the nanostructure of cement paste from these perspectives.

In many respects the structure of C-S-H is reasonably well understood. For example at the molecular scale it is composed of layers, much like clay, and water can move in and out of these interlayer spaces. Historically, the main points of contention [1] have centered on the value of specific surface area. On the one hand a lower value of specific surface area, as is measured by nitrogen sorption isotherms, implies that pores 1 – 2 nm are not particularly abundant or important [5]. On the other hand an extremely high surface area, as is measured by water sorption isotherms, implies that the smallest pores are abundant, and important.

These two opposing views have remained [1], although a new hybrid model, called CM-II, has been proposed [2]. Recently, sophisticated neutron scattering experiments have provided definitive information about the density, composition, and size of C-S-H particles [6]. Aggregates of nano-bricks fill space with specific packing densities. The nanobricks have a layered structure, but they pack together as colloids or grains of solid, and the grains have one set of properties, and their larger scale packed arrangement another set of properties. For the sake of brevity this structure will be referred to here as the C-G structure.

It has been shown recently [4] that elastic properties are best described by the principles of granular mechanics. The self-consistent scheme gives a linear relationship between porosity and modulus with the percolation threshold of packed spheres being 50% porosity and maximum stiffness, that of the nanobricks, at zero porosity. Data from nanoindentation experiments supports this trend.

While elastic properties of concrete are important, the visco-elastic properties that occur upon drying or under load are central to deformation, cracking and associated durability. The literature here is particularly vague with no definitive and verifiable mechanisms of irreversible deformation [7]. The three most discussed mechanisms involve thermally activated “creep centers,” the motion of water in and out of hindered spaces (that are not well defined), and water moving in and out of the interlayer spaces.

CM-II taps into both colloid chemistry and granular mechanics in an attempt to further define mechanisms of viscous flow. The rearrangement of nanobricks has the advantage of providing a mechanism for viscous flow and poses new questions that may be investigated experimentally. Aging due to time, increased temperature, load and irreversible drying shrinkage are due to rearrangement of packing of the bricks often with the effect of reducing the volume of pores in the 5 – 12 nm range [8].

2.1 Water in the Smallest Pores

Textbook analysis of drying shrinkage describes possible roles of capillary water, disjoining water, and surface water on shrinkage but these phenomena are totally reversible. The roles of capillary underpressure and adsorbed water on surface tension are fairly easy to conceptualize. The former has been analyzed recently in some detail and a new formula for elastic (reversible) drying shrinkage has been proposed [9], Eq. (1), which accurately predicts elastic shrinkage of cement paste during drying to about 50% rh. If C-S-H is composed of nanoscale particles with a well defined pore system, it can be treated as a partially saturated drained granular material.

$$\varepsilon_v = p_c \left(\frac{1}{K_b} - \frac{1}{\overline{K}} \right) \tag{1}$$

where ε_v is the volumetric strain (contraction defined positive), K_b is the drained bulk modulus of porous material, $\overline{K}$ is the effective modulus of the solid plus empty pores (or solid plus partially filled drained pores), and p_c is capillary pore pressure:

$$p_c = -\frac{RT}{M} \ln h; \qquad r_m = 2\gamma / p_c \tag{2}$$

where R is the universal gas constant, T is the absolute temperature, h is the relative humidity, M is the molar volume of liquid, γ is the surface tension of liquid, and r_m is the size of the largest pore that remains full of liquid. However, it has been also argued that disjoining pressure plays a dominant role and capillary pressure is only secondary [10]. A problem is that disjoining pressure has not been defined well enough to quantitatively evaluate. A key to using Eq. (1) is to determine the degree of saturation, the fraction of water that is subject to capillary forces, and this is not be all of the evaporable water in the paste. It must exclude the water experiencing disjoining pressure. Perhaps this water is thermodynamically similar to the high pressure side of an osmotic system.

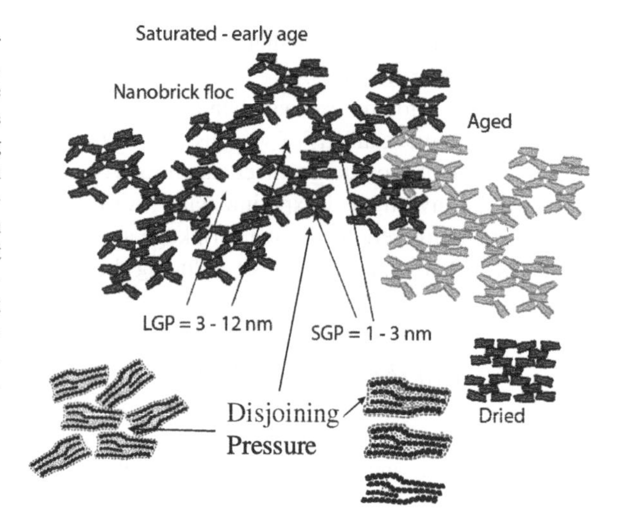

Fig. 1 Schematic of nanobricks and the aging process. In the dried state they are aligned and in the wet state they swell due to disjoining pressure that builds within the smallest pores. Disjoining pressure diminishes on desorption starting at about 20% rh, and increases on adsorption starting at about 80% rh, which is the fundamental reason for the hysteresis in the sorption isotherms in the CM-II model

Water under disjoining pressure must be located in the smallest confined spaces, and is likely to be the water sometimes referred to as "hindered" or "constrained." It is water between surfaces that are separated by a distance such that they are still in the potential well that defines their equilibrium separation. CM-II [2] identifies this as being responsible for the low-pressure hysteresis measured for water sorption. This water is only removed at the lower pressures and only reenters at the highest pressures. Disjoining pressure is a positive pressure that pushes the surfaces apart from their equilibrium separation. When it enters the smallest pores in C-S-H it may have the effect of causing swelling, similar to blowing up a balloon, which also stiffens the solid matrix.

2.2 The Water Sorption Isotherm – A Dilemma and the Role of Disjoining Pressure on Drying Shrinkage

A typical water sorption isotherm is shown in Fig. 2 along with graphs of length change modulus as functions of partial pressure of water. A striking and seldom discussed feature is that at any particular relative humidity the desorption curve exhibits less shrinkage and less water loss than the adsorption curve. This is in spite of the fact that during desorption a capillary is maintained down to about 50% rh whereas during adsorption it is not reestablished until the highest pressures. This is contrary to the obvious fact that the added negative pressure during desorption should lead to greater, not less, shrinkage. Other contradictions [10] include the observation that when the capillary reestablishes itself on adsorption shrinkage should be observed, and when the capillary breaks on desorption expansion should be observed.

Both disjoining pressure and capillary underpressure play a major role in drying shrinkage, and by implication in creep. Additionally, of course, at very low pressures surface energy and associated Bangham forces play the dominant role, but there is general agreement on this. Wittmann and colleagues [see for example 10] have championed the idea that disjoining pressure greatly exceeds capillary underpressure, particularly under the circumstance that the aqueous phase is rich in cations. The view presented here is a hybrid picture. Much of the argument relies on CM-II, which was developed primarily from independent observations of density, moisture content and surface area arguments. Water responsible for disjoining pressure must be located in the smallest pores, the interlayer spaces, the IGP and the SGP (using CM-II terminology). The pressure comes from the force of pushing the walls apart from their equilibrium separation distances.

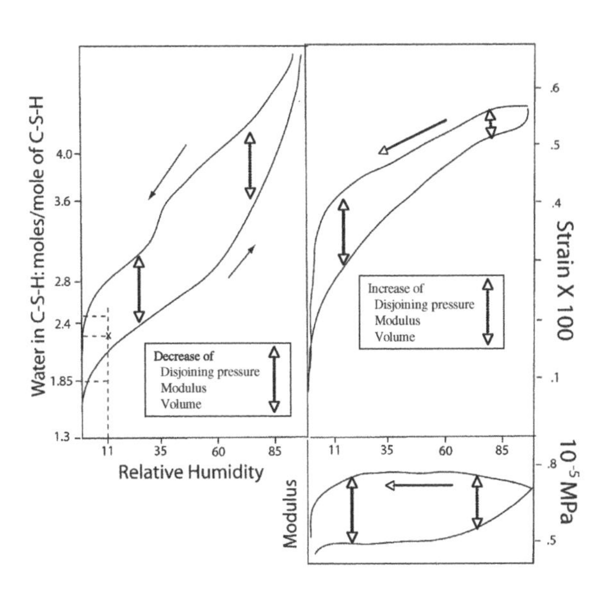

Fig. 2 The dependence of a) water sorption isotherm, b) strain, and c) Young's Modulus on relative humidity. The hysteresis is caused by the "hindered" or "constrained water, which is the water under disjoining pressure. Data from Feldman [11]

A summary of Wittmann's arguments [10] includes an experiment where the separation distance of little quartz balls exhibits a hysteresis with similarities to cement paste except that it only occurs above about 35% rh. . Here there are no complicating tiny pores and the increasing separation with rh must be due to disjoining pressure. Again the presence of a meniscus during drying decreases, instead of increases shrinkage. The difference between desorption and adsorption is the presence or absence of a meniscus, which when present controls the rh of the empty pore making the disjoining pressure independent of rh. During adsorption the disjoining pressure a function of rh, and expansion is governed by this alone until the highest rh.

However cement paste is more complex, with hysteresis at all rh's. If some of the water associated with disjoining pressure is within the nanobricks, this water does not enter and leave the structure reversibly [2]. The nanobricks and their packing arrangement forms the skeleton, and since the nanobricks can be partly

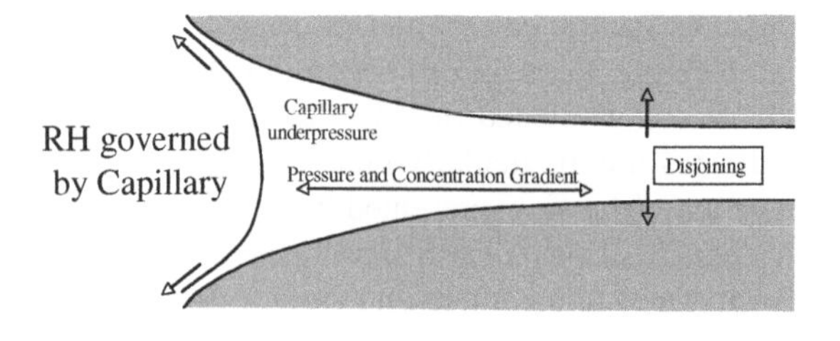

Fig. 3 Schematic of disjoining and capillary forces establishing a pressure gradient. Equilibrium could be maintained by a concentration gradient of cations or by a semipermeable membrane (not shown). The meniscus controls rh according to the Kelvin Laplace equation establishes RH. The disjoining pressure is a function of rh only in the absence of a meniscus. If the capillary is not present, as is the case during adsorption, shrinkage should be reduced, which is opposite to observation

empty during adsorption they must be less stiff than they are during desorption. This explains the modulus measurements as a function of relative humidity as shown in Fig. 2. Unfortunately, by these arguments, the disjoining pressure is not a simple function of relative humidity. It is constant during desorption until the meniscus becomes unstable and it is a function of rh during adsorption until the meniscus forms. Thus, disjoining pressure is active at the scale of the nanobricks, whereas the role of capillary pressure acts over a much larger scale. This argument is fundamentally C-G, in that it assigns one set of properties to the nanoscale as one function of relative humidity and another set of properties to a larger scale with its function of relative humidity. It is important to note that the modulus of a drained porous system depends *only* on the modulus of the skeleton, which in this case is the nanobricks and their packing arrangement.

According to Eq. (1), the shrinkage is a function of degree of saturation, the modulus of the system, the modulus of the solid skeleton, and the capillary pressure of the liquid. According to these arguments the primary reason why the adsorption branch of the length change vs relative humidity isotherms is always below the desorption curve is that the modulus of the particles is less along the adsorption curve, except at the very highest rh's. Furthermore, the lack of observed expansion or contraction as the meniscus breaks or reforms is because the modulus is rapidly changing as a function of rh at these points. Thus bulk modulus of the system in Eq. (1) is not constant, except on the desorption curve down to when the meniscus becomes unstable.

3 Grains and Irreversible Flow

Both drying and creep have irreversible components, but important distinctions give hints about each individual mechanism. These will be discussed in light of the role of moisture in various parts of the structure as discussed above.

3.1 Irreversible Drying Shrinkage

Drying alters the structure of pores [8] and therefore by implication of packing arrangement of the particles. This is almost self-evident from the observed variability of surface area as measured by nitrogen. Higher surface areas are associated with conditions where little time is allowed for particle movement or when surface tension of the aqueous phase is lowered by techniques such as solvent exchange.

During drying most of the irreversible component of shrinkage occurs above 50% rh, and most of the ultimate irreversible component occurs during the first drying. Recent SANS results [12] indicate that the packing becomes more compact, but that very little change occurs in the nanobricks (although their water content and density may change slightly). A granular model where the grains rearrange under the influence of capillary forces best reconciles these observations.

When a colloid is dried several distinct stages have been identified [13]. The first stage is the constant rate period (CRP) when the solid particles pack more tightly at the same rate as water is removed so that water never enters the interparticle region. This is a period of irreversible change in structure. At the "critical point" the water gas interface enters the gel structure and internal pores start to empty. This "first falling rate period" is when stress builds and associated cracking occurs. In the case of C-S-H the largest pores within the gel are about 10 - 15 nm, which dictates a "critical point" at about 85 % rh. Below this rh the liquid gas front can be very irregular and some further rearrangement of particles is possible depending on how well packed the particles are at the critical point.

3.2 Creep

The mechanisms of creep are difficult to investigate and are not well understood. The several models in the literature [7], including a) thermal activation of a deforming micro-volume, redistribution of "hindered" water between particles, and interlayer sliding, all lack direct evidence within the nanostructure. The model described here, however, has some specific implications about certain creep phenomena. For example it has been shown that creep virtually stops if a sample is dried and upon rewetting it only starts again at a relative humidity of 50% and gradually increases as humidity is increased [14].

Granular materials deform under stress with a rate that depends on cohesion and friction between particles. The model described here suggests that during rewetting the water associated with disjoining pressure, within and between nanobricks, begins to build pressure at about 50% rh, which acts to reduce the cohesion and/or friction between particles. This leads to the irreversible component of creep and it gives an new physical interpretation to the seepage of "hindered water" as described by Bazant [15] that controls rate of sliding.

The question becomes why does creep slow with time, something Bazant explained by the exhaustion of creep centers. The stress in a granular material is not distributed evenly; it is carried by columns that transmit most of the load. These stress columns change during flow, and somehow the gel may harden. Also, the nanobricks may become aligned by rotation as the gel deforms and this process may be limiting. This would easily explain why creep can become reactivated by changing the direction of load. The major point here is that exploring the nanogranular nature of the gel leads to specific questions that can lead to a new approach to experimentation.

3.3 Creep, Recovery and Drying

Powers [16] described a small unexplainable effect observed during water adsorption and weight gain experiments, which in one form or another has frustrated simple interpretation of certain experiments in our lab. In Power's experiments the effect is most pronounced when dried samples are resaturated at about 50% rh. First, the sample gains weight (if the paste is ground into fine particles this takes about a day) and then it mysteriously loses a little weight very slowly during the following days. According to CM-II the nanobricks in dried C-S-H are aligned as is observed in an electron microscope [17] and shown schematically in Fig. 1. These aligned nanobricks entrap tiny pores similar in size to IGP size. Upon rewetting these pores fill and disjoining pressure pushes adjacent surfaces apart so that the nanobricks return to the more open pattern of the original saturated structure, as is seen by neutrons [12]. The pores change in size during resaturation, however, the smallest pores fill with water that becomes pressurized and the new equilibrium, which involves the slow diffusion of this water leads to a reduction of the smallest pores, and associated slow weight loss. In this model, therefore, the pore system gradually changes under the influence of disjoining pressure.

Creep recovery, when the applied load is removed, also has a similarly mysterious origin. If, however, the nanobricks align under the influence of load contributing to basic creep, the aligned particles may slowly spring back to the more open structure when load is removed. Again disjoining pressure pushes aligned particles apart.

Change in temperature causes relatively large volume changes in an aqueous phase compared to the solid, which builds pressure gradients in the same way as drying. If a sample is first dried slightly and then subjected to load, creep is reduced, but if a sample is dried during load the combined strain is greater than the linear combination of separate strains due to creep and drying. This is known as the Picket effect. A contributing factor to this effect could be that pressure gradients tend to pull the particles apart, which in effect reduces contact forces, "lubricates" the motion of particles. While the literature has considered the differential shrinkage across larger scales as a force that causes cracking, and has attributed the nonlinear coupling of load and drying to the suppression of cracking (an expansion that reduces the apparent drying shrinkage) under compressive load, this explanation is fundamentally different, and depends on the C-G nature of gel.

4 Thoughts on Design of Concrete and Concluding Remarks

If the mechanisms discussed here control creep and/or shrinkage then strategies for altering these properties must address the C-G nature of cement paste. Thus for example fibers or particles slightly larger (relative to 5 nm grains) than the nano-bricks could potentially disrupt flow. More information about cements that incorporate small particles and reactive mineral admixtures must be developed.

This paper outlines some of the consequences of exploring the structure and properties of cement paste from the nanogranular or colloid perspective, referred to here as the C-G approach. Certain properties are assigned to the grains and other properties to intergranular interactions and pores. It is a start, and these ideas must be translated into quantitative terms and tested against data. This description stimulates experiments from a new perspective and while there remains much to do it is appropriate to share the concepts and invite other investigators to test, criticize, alter, and build on these ideas. Some specific hypotheses that come from this approach, and from CM-II are:

- Pores are not constant and change in volume and size with many variables including drying, load, age etc. The packing arrangement of C-S-H changes upon drying and is also influenced by drying rate. The packing arrangement also changes under external load.
- Disjoining pressure plays a major role in both modulus and viscous flow of C-S-H. Along with capillary underpressure it is responsible for the observed hysteresis in length change rh curves. An explanation for this hysteresis is advanced. It also influences interparticle forces including cohesion and friction.

The value of disjoining pressure is not a state function of relative humidity. It depends on relative humidity history and is responsible for the large hysteresis in sorption isotherms, modulus and other properties.

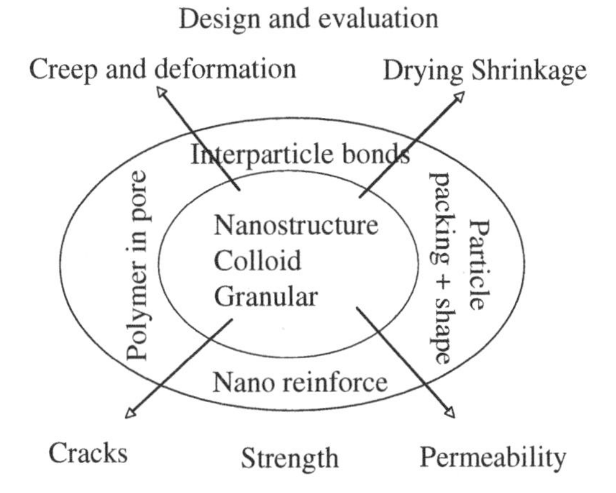

Finally this figure provides a broad overview of strategies that might be explored using the concepts briefly discussed here. For example the C-G approach identifies specific pores into which polymers with specific properties might be introduced, with consequent impact on a variety of properties. Similarly small particles or fibers might alter the viscous properties of gel with impact similar to altering strength, which is the main property usually measured. The major idea of this paper is that engineering the nanostructure will open the door to specific properties, particularly visco-elastic properties into cement based materials.

References

1. Jennings, H.M., Bullard, J.W., et al.: Characterization and Modeling of Pores and Surfaces in Cement Paste: Correlations to Processing and Properties. J. Adv. Concr. Tech. 6, 5–29 (2008)
2. Jennings, H.M.: Refinements to Colloid Model of C-S-H in Cement: CM-II. Cem. Concr. Res. 38, 275–289 (2008)
3. Thomas, J.J., Jennings, H.M.: A colloidal interpretation of chemical aging of the C-S-H gel and its effects on the properties of cement paste. Cem. Concr. Res., 30–38 (2006)
4. Constantinides, G., Ulm, F.J.: The nanogranular nature of C-S-H. Journal of the Mechanics and Physics of Solids 55, 64–90 (2007)
5. Beaudoin, J.J., Alizadeh, R.: A discussion of the paper Refinements to collioidal model of C-S-H in cement: CM-II. Cem. Concr. Res. 38, 1026–1028 (2008)
6. Allen, A.J., Thomas, J.J., et al.: Composition and Density of Nanoscale Calcium-Silicate-Hydrate in Cement. Nature Mat. 6, 311–316 (2007)
7. Mindess, S., Young, J.F.: Concrete. Prentice-Hall, Englewood Cliffs (1981)
8. Jennings, H.M., Thomas, J.J., et al.: A multi-technique investigation of the nanoporosity of cement paste. Cem. Concr. Res. 37, 329–336 (2007)
9. Vlahinic, I., Jennings, H.M., et al.: A Model for Partially Saturated Drying Porous Material. Mech. Mater (2009) doi:10.1016/j.mechmat.2008.10.011
10. Beltzung, F., Wittmann, F.H.: Role of disjoining pressure in cement based materials. Cem. Concr. Res. 35, 2364–2370 (2005)
11. Feldman, R.F.: Sorption and length-change scanning isotherms of methanol and water on hydrated Portland cement. In: 5th International Symposium on the Chemistry of Cement III, pp. 53–66 (1968)
12. Thomas, J.J., Allen, A.J., et al.: Structural changes to the calcium silicate hydrate gel phase of hydrated cement with age, drying and resaturation. J. Am. Ceram. Soc. 91, 3362–3369 (2008)
13. Scherer, G.W.: Theory of Drying. J. Am. Ceram. Soc. 73, 3–14 (1990)
14. Wittmann, F.: Einfluss des Feuchtigkeitsgehaltes auf das Kriechen des Zementsteines. Rheologica Acta 9, 282–287 (1970)
15. Bazant, Z.P.: Thermodynamics of hindered absorption and its implications for hardened cement paste and concrete. Cem. Concr. Res. 2, 1–16 (1972)
16. Powers, T.C., Brownyard, T.L.: Studies of the physical properties of hardened portland cement paste. Journal of the American Concrete Institute 18, 249–336 (1946)
17. Richardson, I.G.: Tobermorite/jennite- and tobermorite/calcium hydroxide-based models for the structure of C-S-H: applicability to hardened pastes of tricalcium silicate, -dicalcium silicate, Portland cement, and blends of Portland cement with blastfurnace slag, metakaolin, or silica fume. Cem. Concr. Res. 34, 1733–1777 (2004)

Nanotechnology and Cementitious Materials

K.L. Scrivener

Abstract. The relevance of nanotechnology and more specifically nanoscience to cementitious materials is discussed. Some examples are given of the influence of nanosciences on our understanding of cementitious materials and its impact on the applications of these materials.

1 Introduction

In recent years notechnology has become *THE* buzz word. In this article I discuss the relevance of nanotechnology to cementitious materials – the most used materials on the planet.

To start with some definition – everyone now knows that nano means very small and more specifically phenomena in the range below 100 nm. We can be perhaps identify 3 main strands of nanotechnology:

1. **Top-down approaches -** seek to create smaller devices by using larger ones to direct their assembly. This applies mainly to technologies descended from conventional solid-state silicon methods for fabricating microprocessors, which are now capable of creating features smaller than 100 nm
2. **Bottom-up approaches -** seek to arrange smaller components into more complex assemblies. Sophisticated examples include the manipulation of base pairs to construct structures out of DNA, but extend to approaches from the field of "classical" chemical synthesis aimed at designing molecules with well-defined shape. [1]
3. **Nanoscience** – the term nanotechnology is also used to refer simply to the study of materials at the nanoscale, usually referring to the use of advanced characterization techniques and atomistic or molecular level modeling.

Therefore it is clear that “nanotechnology” is becoming a catch all phrase to refer to studies which would previous have been considered as branches of chemistry of materials science. In this respect nanotechnology (strands 2 & 3) has an

K.L. Scrivener
Ecole Polytechnique Féderale de Lausanne, EPFL, Switzerland
e-mail : karen.scrivener@epfl.ch

enormous importance for cementitious materials; and while it is unrealistic to think of top- down approaches being applied to materials used in such large quantities it is likely that advances in nanotechnology related to sensing and information processing will also have a huge indirect impact on construction materials.

Before examining some aspects of nanoscience/nanotechnology it is useful to discuss the context in which cementitious materials are used.

2 Context of Cementitious Materials

Cementitious Materials (e.g. concrete) are by far and away the most used materials on the planet. This is not because their properties are intrinsically superior to other materials, but simply because they are cheap, low energy and readily available everywhere. When it is considered that the principal oxides present in cement – CaO, SiO_2, Al_2O_3, Fe_2O_3 – constitute over 90% of the earth's crust it is clear, that solely from a consideration of available resources, they will continue to form the basis of our modern infrastructure.

Despite the fact that the intrinsic properties of concrete such as strength are relatively modest, compared to say steel, cementitious materials have the amazing ability to transform from a fluid suspension to a rigid solid, without any external input at room temperature. We take this for granted, but this process of hydration is very complex, involving tens of chemical species reacting through solution on time scales from seconds to decades and consequently many aspects are still not well understood.

Due to the complexity of reactions in cementitious materials, development to date has been largely based on an empirical approach at the macroscopic scale. The arrival of approaches from nanoscience has the potential to revolutionize this and enable the micro and nanoscale physico-chemical processes which govern macroscopic behavior to be understood and manipulated.

Table 1 EU25 emissions of CO2 in 2007 [2]

Industry	M tonnes
Cement and lime	175
Iron and Steel	121
Glass	17
Ceramics	13
Paper and pulp	27

Such a change in approach is essential to enable us to respond to the challenges of sustainability which confront us today. The press frequently characterizes cement production as one of the highest producers of CO_2. The figures below indicate there is some justification for this. However, this is a direct consequence of the enormous volumes used and substituting cement with other construction materials would almost certainly make the situation worse, without considering that there are simply not sufficient amounts other material available to replace cement in its wide variety of functions.

3 Progress in Nanoscience of Cementitious Materials

The most notable example of progress in the nanoscience of cementitious material is our knowledge about the main hydrates phase in cement paste – calcium silicate hydrate, C-S-H. Due to lack of long range crystalline order, the structure of this phase is difficult to determine by conventional techniques such as X-ray diffraction. Over the last 25 years or so a very clear picture of the atomic level structure has emerged thanks notably to solid state nuclear magnetic resonance (NMR) [3,4,5] and transmission electron microscopy (TEM) [6,7]. Furthermore, it is now possible to model this atomic structure and compute, for example, mechanical properties [8], which show good agreement with experimental results. There are still many open questions on the arrangement of the C-S-H nanocrystals on the meso level and their related growth kinetics – projects within the Nanocem network (www.nanocem.org) discussed later are underway to try and answer these.

Atomic force microscopy (AFM), perhaps the core technique of nanotechnology, is also providing new insights about the reaction of cementitious materials. Figure 1 from the thesis of Helen di Murro [9], shows distinct crystallographic edges on the surface of a reacting grain of tri-calcium silicate, C_3S.

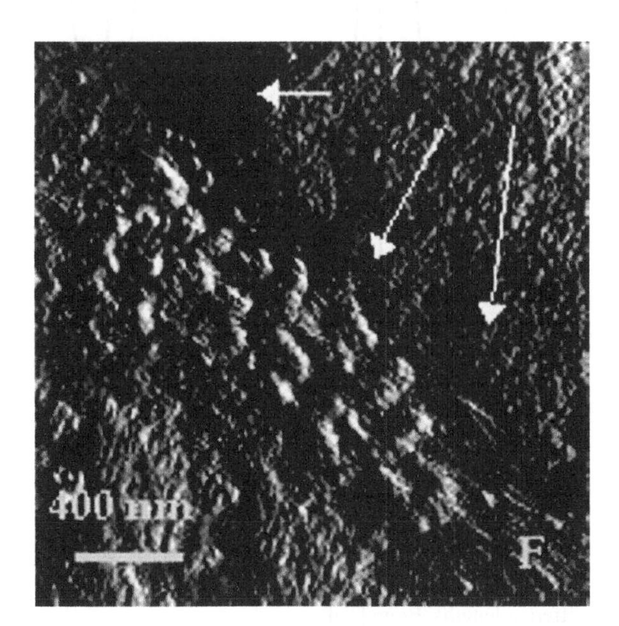

Fig. 1 Surface of reacting C_3S [9]

4 Impact of Nanoscience on Technology of Cementitious Materials

Taking a broad definition of nanotechnology, there are already examples of its impact on the use of cementitious materials. Perhaps the most celebrated is that of the third generation suplasticizers. Early generations of cement dispersants were based on lignosulphonates or sulphonated melamine or naphthalene formaldehyde condensates. These are based on natural products, and there is little control of the basic chemical structure. The most important innovation in recent years has been the introduction of PCE (polycarboxylate ether)- based plasticisers and superplasticizers. The molecular structure of PCE polymers is a comb with a backbone and side branches. By manipulation of the relative lengths of the chain backbone and side branches and the density of the side branches (Figure 2, [10]) the performance can be modified in relation to such concrete properties as workability, retention, cohesion and rate of strength development. The possibility of tailoring additives for specific purposes will likely be one of the most important sources of innovation for the future.

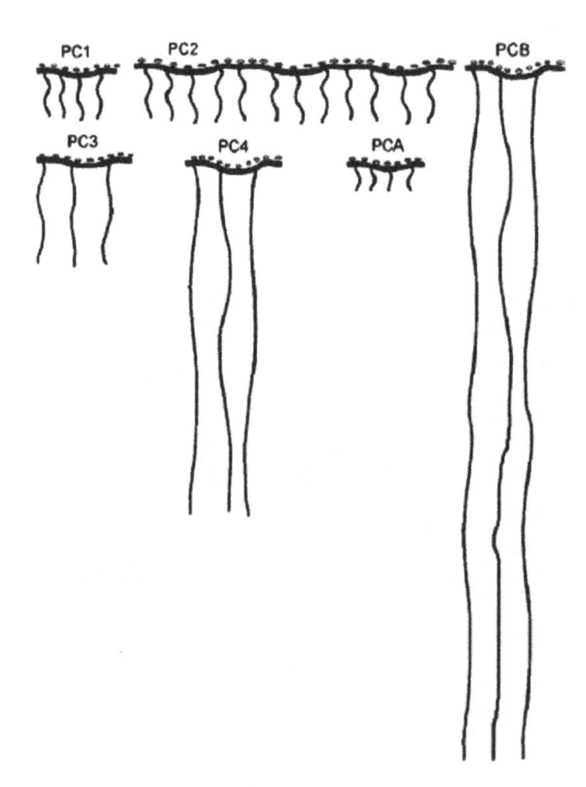

Fig. 2 Schematic illustration of the molecular structure of comb-type copolymers with a negatively charged polycarboxylate backbone with grafted polyethyleneoxide side chains of different lengths. From [10]

These superplasticizers are now widely used in high performance and self consolidating concretes, which are the two main innovations in concrete technology in recent decades [11].

Nanoparticles are an aspect of nanotechnology, which are often discussed in relation to cementitious materials. Indeed most current applications of nanotechnology are limited to the use of "first generation" passive nanomaterials which includes titanium dioxide in sunscreen, cosmetics and some food products [1]. One could imagine the use of nanoparticles as a way of extending the concept of particle packing and manipulation of particle size distribution, which lies behind the technology of ultra high performance concrete (e.g. Ductal®). In fact, silica fume, which may have particles as small as 100 nm, could already be considered a nanomaterial. However, as particles become smaller their relative surface increases, and already with silica fume it is necessary to add significant amounts of superplasticisers to ensure good fluidity. Furthermore, there are now serious questions being posed about the health and safety aspects of very small particles.

The addition of fine anatase, TiO_2 particles [12,13], to provide self cleaning properties is often cited as an example of nanotechnology. Anatase is photocatalytic, and through the absorption of sunlight has a string oxidizing power. This prevents the buildup of dirt and organic growth, preserving the clean appearance of the concrete for longer. This oxidizing power can also breakdown NOx and so contribute to reducing pollution.

Nowadays fibres are an essential part of ultra high performance concrete, which has led some researchers to investigate the addition of carbon nanotubes to concrete. Carbon nanotubes have extremely high intrinsic stiffness and strength [e.g. 14]. However their surfaces have very low friction, so it is very difficult for them to bind together or to matrix materials to realize these extraordinary properties on a macroscopic scale. This is a field of very active development, and the current obstacles of high cost and poor binding are likely to decrease in the future. At present, however, such materials are not practical as an addition to concrete.

5 The Future

It is clear that the impact of nanotechnology on cementitious materials is at present mainly at the research level. These advances in scientific understanding need to be transferred into the field. There is perhaps no more important area for this than in facilitating the introduction of new cements with reduced environmental impact. To date the main approach to reducing CO_2 emissions associated with cement production has been to replace Portland cement clinker with supplementary cementitious materials (SCMs), such as fine limestone, fly ash, slag, silica fume, etc. With our current approach to research on cementitious materials, this technology of substitution is reaching an asymptotic limit. We are lacking knowledge and tools to asses, the reactivity of new possible SCMs, the changes produced in the microstructure and their consequences for durability. Nanoscience can deliver important insights into mechanisms in concrete at the micro and nanolevel in order to provide new performance concepts to allow the use of a wider range of materials and to continue to lower CO_2 emissions per tonne of cement.

Such an effort requires close interaction between cementitious specialists and specialists from other branches and between the industry and academia. The formation of the Nanocem network in 2002, which now encompasses 15 major industrial companies with 24 academic groups in a self-financing structure, has taken the lead in such a pioneering approach. Major progress has already been made on providing a thermodynamic basis for predicting phase assemblages in Portland cement pastes, understanding interactions between superplasticizers and cement, elucidating the fine pore structure and determining the reactivity of SCM in blended systems.

References

1. http://en.wikipedia.org/wiki/Nanotechnology (accessed, February 2009)
2. EU report
3. Cong, X., Kirkpatrick, R.J.: ^{29}Si MAS NMR study of the structure of calcium silicate hydrate. Advn. Cem. Based Mater. 3, 144–156 (1996)
4. Richardson, I.G.: The nature of C-S-H in hardened cements. Cem. Concr. Res. 29, 1131–1147 (1999)
5. Richardson, I.G.: Tobermorite/jennite- and tobermorite/calcium hydroxide-based models for the structure of C-S-H: applicability to hardened pastes of tricalcium silicate, -dicalcium silicate, Portland cement, and blends of Portland cement with blast-furnace slag, metakaolin, or silica fume. Cem. Concr. Res. 34, 1733–1777 (2004)
6. Richardson, I.G.: Electron microscopy of cements. In: Bensted, J., Barnes, P. (eds.) Structure and Performance of Cements. Spon Press, London (2002)
7. Richardson, I.G.: The calcium silicate hydrates. Cem. Concr. Res. 38, 137–158 (2008)
8. Pellenq, R.J.-M., Lequeux, N., van Damme, H.: Engineering the bonding scheme in C–S–H: The iono-covalent framework. Cem. Concr. Res. 38, 159–174 (2008)
9. di Murro, H.: Mécanismes d'élaboration de la microstructure des bétons, These, Universite de Bourgogne (2007)
10. Kjeldsen, A.M., Flatt, R.J., Bergström, L.: Relating the molecular structure of comb-type superplasticizers to the compression rheology of MgO suspensions. Cement Concrete Res. 36, 1231–1239 (2006)
11. Scrivener, K.L., Kirkpatrick, R.J.: Innovation in use and research on cementitious material. Cem. Concr. Res. 38, 128–136 (2008)
12. Cassar, L., Pepe, C., Pimpinelli, N., Amadelli, R., Antolini, L.: Rebuilding the City of Tomorrow. In: 3rd European Conference REBUILD (1999)
13. Cassar, L., Pepe, C., Tognon, G., Guerrini, G.L., Amadelli, R.: Proc 11th ICCC (ICCC) (2003)
14. Srivastava, D., Wei, C., Cho, K.: Nanomechanics of carbon nanotubes and composites. Appl. Mech. Rev. 56, 215–230 (2003)

Probing Nano-structure of C-S-H by Micro-mechanics Based Indentation Techniques

F.-J. Ulm and M. Vandamme

Abstract. This paper summarizes recent developments in the field of nanoindentation analysis of highly heterogeneous composites. The fundamental idea of the proposed approach is that it is possible to assess nanostructure from the implementation of micromechanics-based scaling relations for a large array of nanoindentation tests on heterogeneous materials. We illustrate this approach through the application to Calcium-Silicate-Hydrate (C-S-H), the binding phase of all cement-based materials. For this important class of materials we show that C-S-H exists in at least three structurally distinct but compositionally similar forms: Low Density (LD), High Density (HD) and Ultra-High-Density (UHD). These three forms differ merely in the packing density of five nano-meter sized particles. The proposed approach also gives access to the solid particle properties of C-S-H, which can now be compared with results from atomistic simulations. By way of conclusion, we show how this approach provides a new way of analyzing complex hydrated nanocomposites, in addition to classical microscopy techniques and chemical analysis.

1 Introduction

One of the most promising techniques that emerged from the implementation of nanotechnology in material science and engineering to assess mechanical properties at small scales is nanoindentation. The idea is simple: by pushing a needle onto the surface of a material, the surface deforms in a way that reflects the mechanical properties of the indented material. Yet, in contrast to most metals and ceramics, for which this technique was originally developed, most materials relevant for civil engineering, petroleum engineering or geophysical applications, are

F.-J. Ulm
Massachusetts Institute of Technology, Cambridge
e-mail: `ulm@mit.edu`

M. Vandamme
Ecole des Ponts - UR Navier, Champs-sur-Marne
e-mail: `matthieu.vandamme@enpc.fr`

highly heterogeneous from a scale of a few nanometers to macroscopic scales. The most prominent heterogeneity is the porosity.

Take, for instance, the case of concrete. Groundbreaking contributions date back to the 1950s with the work of Powers and his colleagues [1], who by correlating macroscopic strength [2] and stiffness data [3] with physical data of a large range of materials prepared at different w/c-ratios early on recognized the critical role of the C-S-H porosity (or gel porosity), respectively the C-S-H packing ("one minus porosity") on the macroscopic mechanical behavior; in particular for cement pastes below a water-to-cement ratio of w/c<0.42, for which the entire porosity of the material is situated within the C-S-H (no "capillary water" in Powers' terminology), and for which the hydration degree of the hardened material is smaller than one. Powers considered the C-S-H gel porosity (gel pore volume over total gel volume) to be material invariant and equal to $\phi_0 = 0.28$ independent of mix proportions, hydration degree, C-S-H morphology, etc. The application of advanced microscopy, X-ray mapping and Neutron scattering techniques to cement-based materials later on revealed that the assumption of a constant gel porosity could not be but an oversimplification of the highly heterogeneous nanostructure of cement-based materials, overlooking the particular organizational feature of cement hydration products in highly dense packed "inner" products and loosely packed "outer" products (see, for instance [4]-[14]). The quantitative translation of these morphological observations into a concise microstructure model of the gel microstructure is due to Jennings and co-worker [15]-[19], who recognized that outer and inner products are two structurally distinct but compositionally similar C-S-H phases; that is, amorphous nanoparticles of some 5 nm characteristic size pack into two characteristic forms, a Low Density (LD) C-S-H phase and a High Density (HD) C-S-H phase, that can be associated with outer and inner products. The existence and mechanical importance of these phases have been confirmed by nanoindentation [20]-[22]: LD C-S-H and HD C-S-H were found to be uniquely characterized by a set of material properties that do no depend on mix proportions, type of cement, etc. Instead, they are intrinsic material invariant properties. The link between these mechanical C-S-H phase properties and C-S-H packing density has been established, showing that the C-S-H phases exhibit a unique nanogranular morphology [23,24,18], with packing densities that come remarkable close to limit packing densities of spheres; namely the random close-packed limit (RCP, [25]) or maximally random jammed state (MRJ, [26]) of $\eta \sim 0.64$ for the LD C-S-H phase; and the ordered face-centered cubic (fcc) or hexagonal close-packed (hcp) packing of $\eta = \pi / \sqrt{18} = 0.74$ [27] for the HD C-S-H phase. Far from being constant, the gel porosity of the C-S-H phase is recognized, in this model, to depend on the volume proportions of LD C-S-H (f_{LD}) and HD C-S-H (f_{HD}) [22]:

$$\phi_0 = 1 - (0.64 \times f_{LD} + 0.74 \times f_{HD}) \quad (1)$$

The focus of this conference paper is to review some recent developments in the field of nanoindentation analysis of highly heterogeneous materials. Illustrated through the example of concrete, the approach here proposed can equally be applied to many other highly heterogeneous materials whose material behavior is governed by porosity (bones, shale etc. [28-29]).

2 Micromechanics-Based NanoIndentation Techniques

Consider an indentation test in a porous material, the characteristic size of the porosity being such that scale separability applies. The indenter is at a depth h, and what is sensed in the indentation test is a composite response of the solid phase and the porosity. That is, the hardness H and the indentation modulus M determined from the load – depth (P–h) curve according to classical indentation analysis tools [30] are representative of the particle properties (particle stiffness m_s, hardness h_s, (Drucker-Prager) friction coefficient μ), and of the packing of their particles (packing density η), and some morphological parameters (denoted by η_0) [28]:

$$
\begin{aligned}
H &= \frac{P}{A_c} = h_s \times \Pi_H(\mu,\eta,\eta_0) \\
M &= c\frac{(dP/dh)_{h_{\max}}}{\sqrt{A_c}} = m_s \times \Pi_M(\nu,\eta,\eta_0)
\end{aligned}
\tag{2}
$$

where A_c is the projected contact area, and Π_H and Π_M are dimensionless functions. Linear and nonlinear microporomechanics [31] provides a convenient way to determine these functions.

2.1 Does Particle Shape Matter? [32]

The determination of functions Π_H and Π_M requires the choice of particle morphology. For perfectly disordered materials, the key quantity to be considered is the percolation threshold, that is the critical packing density below which the composite material has no strength, nor stiffness. This percolation threshold depends on the particle shape [33]. Clearly, as seen in TEM images of C-S-H, the elementary particle has an aspect ratio. However, as far as the mechanics response is concerned, it turns out that particle shape does not matter as soon as the packing density of the porous material is greater than 60%. This is obviously the case of cement-based materials (as their industrial success is due to their strength performance), and also of most other natural composite materials like bone,

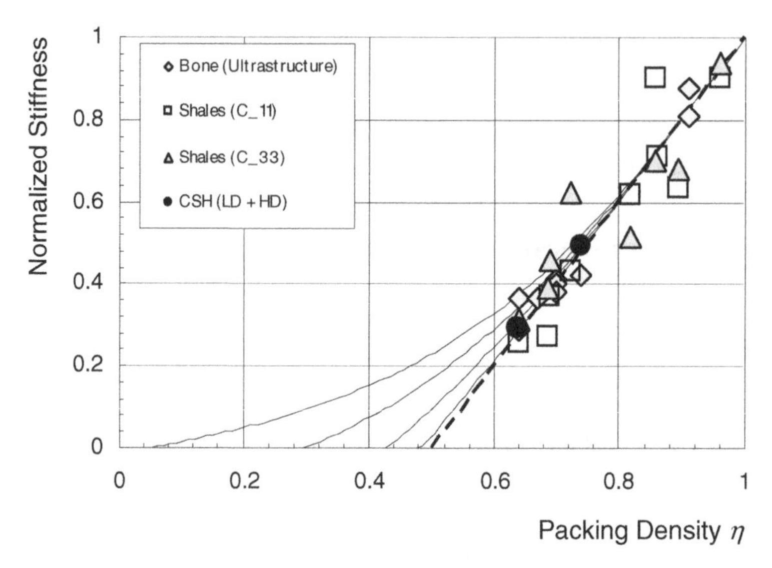

Fig. 1 Normalized stiffness vs. packing density scaling relations. The percolation threshold of 0.5 corresponds to a perfectly disordered material composed of spherical particles, while lower percolation thresholds are representative of disordered materials with particle shape (here ellipsoids). Adapted from [32]

compacted clays (shale), etc. One can thus conclude that the effect of particle shape is negligible as far as the mechanical performance is concerned [32]; see Fig. 1.

2.2 Micromechanics-Based Scaling Functions [34-35]

The negligible effect of the particle shape on the homogenized mechanical properties largely simplifies the micromechanical analysis. It thus suffices, to consider spherical particles and a distinct disordered morphology of the solid phase, similar to a polycrystal, characterized by a solid percolation threshold of $\eta_0 = 0.5$. Based on linear micromechanics of the indentation modulus, a good approximation (which is exact for a solid Poisson's ratio of $\nu = 0.2$) is a linear scaling [23]:

$$\frac{M}{m_s} = \Pi_M = \langle 2\eta - 1 \rangle \tag{3}$$

Using a similar approach for strength homogenization, an appropriate scaling for hardness is [34-35]:

$$\frac{H}{h_s} = \Pi_H = \Pi_1(\eta) + \mu(1-\eta)\Pi_2(\mu,\eta) \tag{4}$$

where:

$$\Pi_1(\eta) = \frac{\sqrt{2(2\eta-1)}-(2\eta-1)}{\sqrt{2}-1}(1+a(1-\eta)+b(1-\eta)^2+c(1-\eta)^3)$$
$$\Pi_2(\mu,\eta) = \frac{2\eta-1}{2}(d+e(1-\eta)+f(1-\eta)\mu+g\mu^3) \tag{5}$$

and where $a = -5.3678$, $b = 12.1933$, $c = -10.3071$, $d = 6.7374$, $e = -39.5893$, $f = 34.3216$ and $g = -21.2053$ are all constants associated with a Berkovich indenter geometry and a polycrystal morphology with percolation threshold of $\eta_0 = 0.5$.

2.3 *Implementation for Back-Analysis of Packing Density [28]*

Consider then a series of *N* indentation tests on a heterogeneous material. What is measured in these tests are *N* times (*M,H*) values at each indentation point representative of a composite response. Assuming that the solid phase is the same, and all what changes between indentation points is the packing density, there are three particle unknowns (m_s, h_s, μ) and *N* packing densities. Hence, for $N >> 3$, the system is highly over-determined, which makes it possible to assess microstructure and particle properties. A typical example of this implementation of micromechanics-based scaling relations to analyze backing density is shown in Fig. 2.

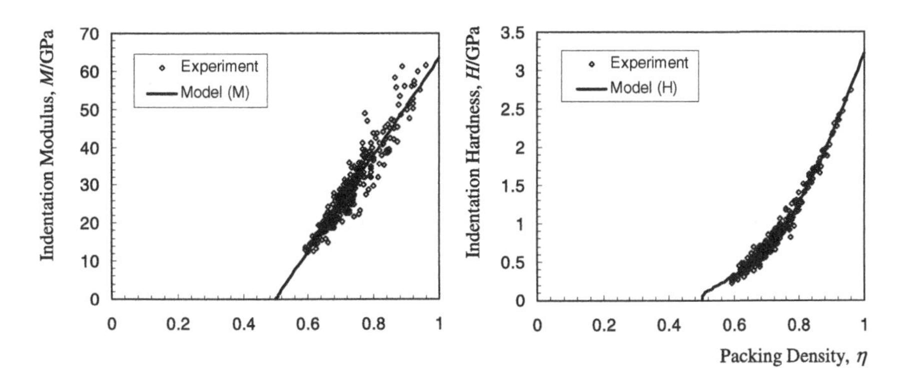

Fig. 2 Micromechanics-based scaling relations of the packing density of indentation modulus and indentation hardness for a w/c=0.4 cement paste. Adapted from [28]

3 Packing Density Distributions of C-S-H

By way of application, we apply this method to analyze packing density distributions for different cement paste materials. In a first step, we deconvolute the

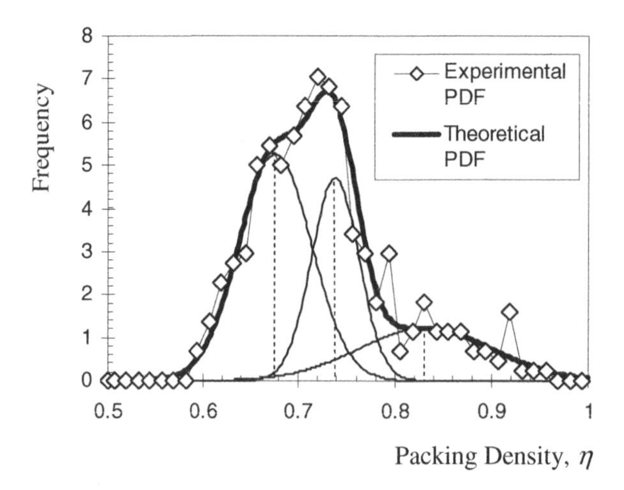

Fig. 3 Packing density distribution for a w/c=0.4 cement paste. Adapted from [28]

packing density together with indentation modulus and indentation hardness to obtain mean values and standard deviations of different phases. This is shown in Fig. 3. This type of analysis is performed for a large range of cement pastes prepared with different w/c ratio. The results, which are shown in Figure 4, call for the following comments:

- The packing density distributions (Fig. 3) for different w/c provides strong evidence of three statistically significant C-S-H phases; namely a Ultra-High-Density (UHD) C-S-H phase, in addition to the already known LD C-S-H and HD C-S-H phases. The nanomechanical properties of the UHD C-S-H phase, *M* and *H*, are found to follow similar packing density scaling relations as LD C-S-H and HD C-S-H. This suggests that the UHD C-S-H phase is structurally distinct but compositionally similar to the other C-S-H phases. That is, it is made of the same elementary building block, the C-S-H solid, and differs from LD and HD C-S-H only in its characteristic packing density. The UHD C-S-H phase has a packing density of $\eta = 0.83 \pm 7\%$, which comes remarkably close to a two-scale limit random packing of $0.64 + (1-0.64) \times 0.64 = 0.87$.
- An increase of the w/c ratio entails an increase in the hydration degree, (Fig. 4a), estimated here as the vol. amount of phases that exhibit an indentation modulus smaller than ~63 GPa. In turn, this increase is due an increase in similar proportions of both the C-S-H solid and the gel porosity; roughly 5% per (w/c)=0.1.
- The increase with the w/c ratio of both the C-S-H solid and the gel porosity in similar proportions has different effects on the microstructure organization of the hydration products into LD C-S-H, HD C-S-H and UHD C-S-H (Fig. 4b): Above w/c=0.2, the volume occupied by UHD C-S-H –among the hydration

products– is almost constant (20%), while the decrease in HD C-S-H is

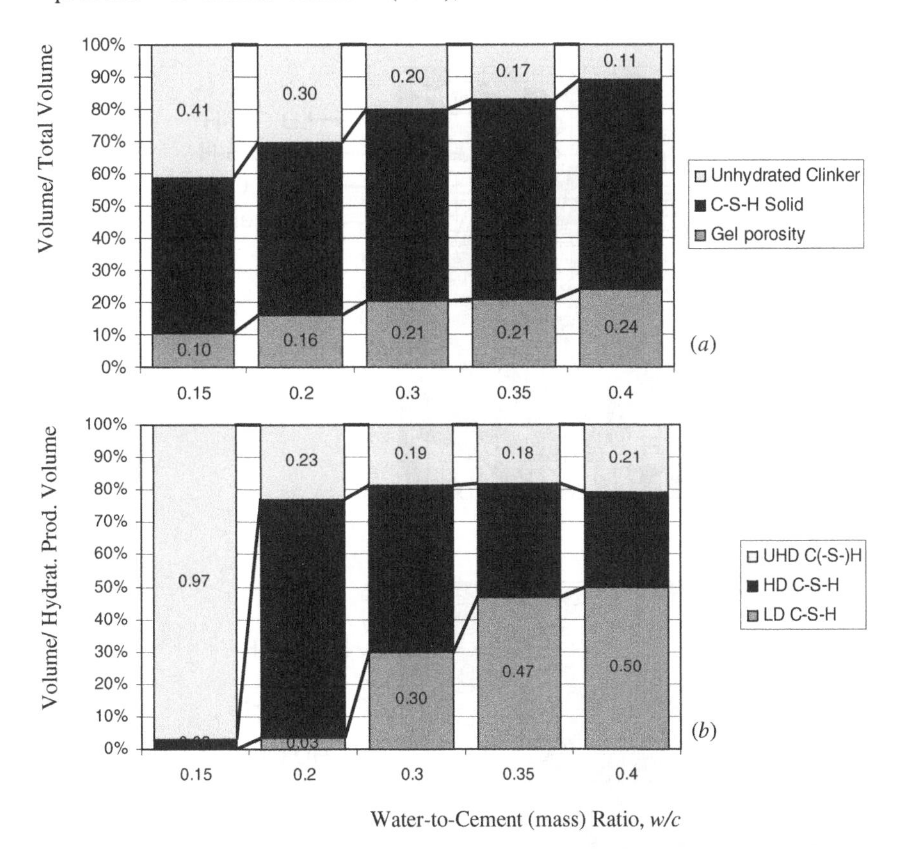

Fig. 4 Volume fraction distributions in the microstructure: (a) volume fractions of the cement paste composite; (b) volume fractions of the hydration phases. Adapted from [38]

compensated by the formation of LD C-S-H: at low w/c, HD C-S-H dominates over LD C-S-H, while it is the inverse at high w/c ratios.

- An extremely low w/c=0.15, characterized by a hydration degree on the order of =0.6, a 10% gel porosity and a 50% C-S-H solid volume fraction, favors (almost) exclusively the formation of the UHD C-S-H phase.

In summary, while the water available for hydration increases the amount of hydration solid, it appears that the concurrent increase in gel porosity favors the formation of looser packed LD C-S-H to the detriment of HD C-S-H; while the UHD C-S-H phase remains almost constant.

Finally, it is interesting to locate these different phases in the microstructure. This can be achieved by spatially mapping the indentation results on a grid. To this end, we performed a series of 900 nanoindentations on a tightly spaced grid, with a spacing of 3μm between nanoindents. After deconvolution of the 900 nanoindentations, and application of the micromechanics-based scaling relations, to each phase (LD C-S-H, HD C-S-H, UHD C-S-H and clinker) is associated an

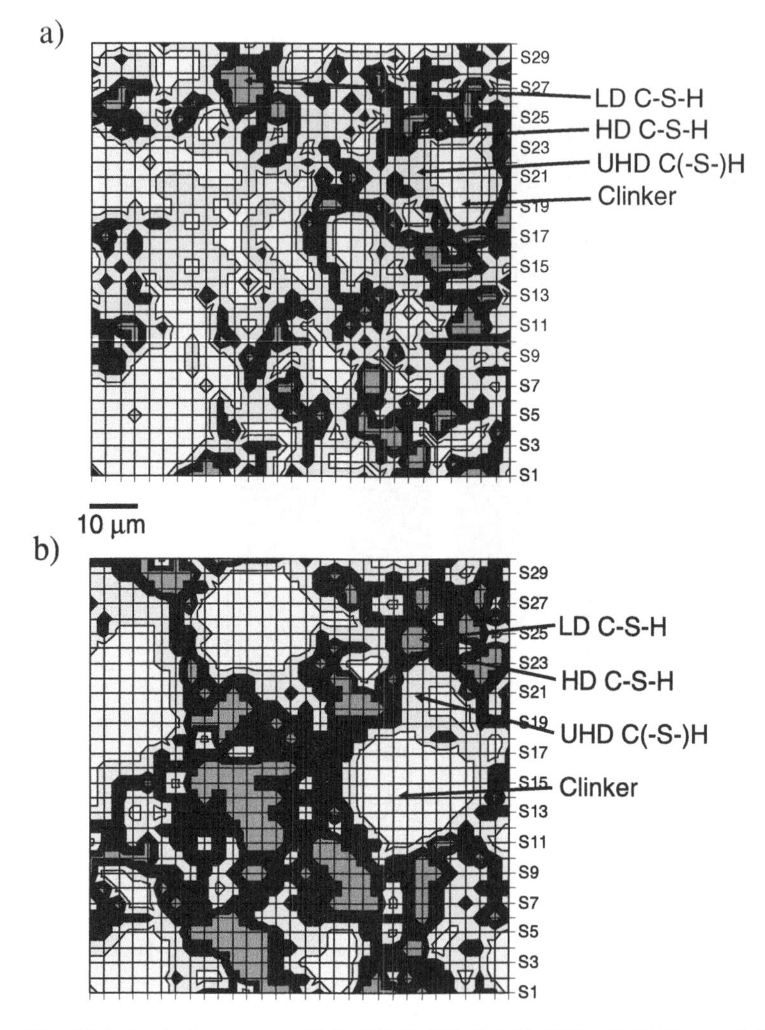

Fig. 5 Mechanical mapping of LD C-S-H, HD C-S-H, UHD C(-S-)H and clinker phases in (a) a w/c=0.2 hardened cement paste; (b) a w/c=0.3 hardened cement paste. Adapted from [38]

interval of packing density. By doing so, each specific point on the grid of nanoindentations can be associated with the phase on which the nanoindentation was performed. Such a technique provides a mechanical mapping of the indented surface with an accuracy limited by the grid spacing. The results of this mapping are displayed in Fig. 5a for a w/c=0.2 material and in Fig. 5b for a w/c=0.3 material.

Similar to HD C-S-H, UHD C-S-H can primarily (i.e. in significant amount) be found close to the clinker grains. A similar observation of a high stiffness phase in the vicinity of clinker grain was also reported in Ref. [36]. Some areas of UHD C-S-H are also visible far from clinker grains. Yet, this occurrence may be attributed to the presence of a clinker grain below the surface; a conjecture that cannot be resolved with the surface mapping technique here employed. In return, a comparison of the amount of the different C-S-H phases present in the two materials clearly shows that HD C-S-H and UHD C-S-H dominate low w/c materials, while higher w/c materials are increasingly dominated by LD C-S-H.

4 Concluding Remarks

Nanoindentation has originally been developed to assess mechanical properties at very fine scales; but its application to highly heterogeneous materials has provided the materials science community with a wealth of new information about morphology of elementary particles; thanks to the application of micromechanics-based indentation analysis tools that link mechanical properties to microstructure and constituent properties.

It is of great interest to compare the particle properties so obtained with particle properties of C-S-H obtained by atomistic simulations. Such atomistic simulations of a realistic C-S-H model provide a particle stiffness of $m_s = 65\,\text{GPa}$ using for the elasticity constants the Reuss-Voigt-Hill average, and a maximum negative isotropic pressure before rupture of the simulation cell perpendicular to the layer plane of $h_s = 65\,\text{GPa}$ [37]. These values are in very good agreement with the asymptotic values obtained from the micromechanics-based scaling relations (Fig. 3). Hence, a combination of nanoindentation, micromechanics-based scaling relations and atomistic simulations, provides a means to bridge between the atomistic structure of C-S-H, its nanomechanical morphology and macroscopic performance.

We are thus one step closer to implementing the Materials Science Paradigm for highly heterogeneous materials like concrete, bones, shale etc., linking composition with microstructure, and to ultimately predicting ("nanoengineering") macroscopic mechanical performance of this class of hydrated nanocomposites based on composition and microstructure.

Acknowledgments. Material and financial support of this research through Lafarge Corporation is gratefully acknowledged. MV also acknowledges the support of Ecole des Ponts, France, enabling his doctoral studies at MIT.

References

1. Powers, T.C., Brownyard, T.L.: Studies of the physical properties of hardened Portland cement paste. Bull. 22, Res. Lab. of Portland Cement Association, Skokie, IL, U.S; J. Am. Concr. Inst. (Proc.) 43, 101–132, 249–336, 469–505, 549–602, 669–712, 845–880, 933–992 (1947) (reprint)

2. Helmuth, R.A., Turk, D.H.: Elastic moduli of hardened portland cement and tricalcium silicate pastes: effect of porosity. In: Symp. Struct. Portland Cem. Paste Concr., pp. 135–144 (1966)
3. Verbeck, G.J., Helmuth, R.A.: Structures and physical properties of cement paste. In: 5th Int. Congress Cement Chemistry, Tokyo, Japan, pp. 1–44 (1969)
4. Taplin, J.H.: A method for following the hydration reaction in portland cement paste. Australian J. Appl. Sc. 10, 329–345 (1959)
5. Dalgleish, B.J., Ibe, K.: Thin foil studies of hydrated cements. Cement Concrete Res. 11, 729–739 (1981)
6. Taylor, H.F.W.: Studies on the chemistry and microstructure of cement pastes. Proc. Brit. Ceram. Soc. 35, 65–82 (1984)
7. Taylor, H.F.W., Newbury, D.E.: An electron microprobe study of a mature cement paste. Cem. Concr. Res. 14, 565–573 (1984)
8. Scrivener, K.L., Patell, H.H., Pratt, P.L., Parrott, L.J.: Analysis of phases in cement paste using backscattered electron images, methanol adsorption and thermogravimetric analysis. In: Mat. Res. Soc. Symp. Proc., vol. 85, pp. 67–76 (1985)
9. Groves, G.W.: TEM studies of cement hydration. In: Mat. Res. Soc. Symposium Proc., vol. 85, pp. 3–12 (1987)
10. Richardson, I.G., Rodger, S.A., Groves, G.W.: The porosity and pore structure of hydrated cement pastes as revealed by electron microscopy techniques. In: Mat. Res. Soc. Symp. Proc., vol. 137, pp. 313–318 (1989)
11. Viehland, D., Li, J.F., Yuan, L.J., Xu, Z.K.: Mesostructure of calcium silicate hydrate (c-s-h) gels in portland-cement paste-a short range ordering nanocrystallinity and local compositional order. J. Am. Ceram. Soc. 79, 1731–1744 (1996)
12. Thomas, J.J., Jennings, H.M., Allen, A.J.: The surface area of cement paste as measured by neutron scattering: evidence for two C-S-H morphologies. Cem. Concr. Res. 28(6), 897–905 (1998)
13. Richardson, I.G.: The nature of C-S-H in hardened cements. Cem. Concr. Res. 29, 1131–1147 (1999)
14. Richardson, I.G.: Tobermorite/jennite- and tobermorite/calcium hydroxide-based models for the structure of C-S-H: applicability to hardened pastes of tricalcium silicate, β-dicalcium silicate, Portland cement, and blends of Portland cement with blast-furnace slag, metakaolin, or silica fume. Cem. Concr. Res. 34, 1733–1777 (2004)
15. Jennings, H.M.: A model for the microstructure of calcium silicate hydrate in cement paste. Cem. Concr. Res. 30, 101–116 (2000)
16. Tennis, P.D., Jennings, H.M.: A model for two types of calcium silicate hydrate in the microstructure of portland cement pastes. Cem. Concr. Res. 30, 855–863 (2000)
17. Jennings, H.M.: Colloid model of C-S-H and implications to the problem of creep and shrinkage. Mater. Struct. 37(265), 59–70 (2004)
18. Jennings, H.M., Thomas, J.J., Gevrenov, J.S., Constantinides, G., Ulm, F.-J.: A multi-technique investigation of the nanoporosity of cement paste. Cem. Concr. Res. 37(3), 329–336 (2007)
19. Jennings, H.M.: Refinements to colloid model of C-S-H in cement: CM-II. Cem. Concr. Res. 38(3), 275–289 (2008)
20. Constantinides, G., Ulm, F.-J., Van Vliet, K.: On the use of nanoindentation for cementitious materials. Mater. Struct. 36, 191–196 (2003)
21. Constantinides, G., Ulm, F.-J.: The effect of two types of C-S-H on the elasticity of cement-based materials: Results from nanoindentation and micromechanical modeling. Cem. Concr. Res. 34(1), 67–80 (2004)

22. Ulm, F.-J., Constantinides, G., Heukamp, F.H.: Is concrete a poromechanics material? - A multiscale investigation of poroelastic properties. Mat. Struct. 37, 43–58 (2004)
23. Constantinides, G., Ulm, F.-J.: The nanogranular nature of C-S-H. J. Mech. Phys. Solids 55, 64–90 (2007)
24. DeJong, M.J., Ulm, F.-J.: The nanogranular behavior of C-S-H at elevated temperatures (up to 700C). Cem. Concr. Res. 37(1), 1–12 (2007)
25. Jaeger, H.M., Nagel, S.R.: Physics of granular state. Science 255(5051), 1523–1531 (1992)
26. Donev, A., Cisse, I., Sachs, D., Variano, E.A., Stillinger, F.H., Connely, R., Torquato, S., Chaikin, P.M.: Improving the density of jammed disordered packings using ellipsoids. Science 303, 990–993 (2004)
27. Sloane, N.J.A.: Kepler's conjecture confirmed. Nature 395, 435–436 (1998)
28. Ulm, F.J., Vandamme, M., Bobko, C., Ortega, J.A., Tai, K., Ortiz, C.: Statistical Indentation Techniques for Hydrated Nanocomposites: Concrete, Bone, and Shale. J. Am. Ceram. Soc. 90, 2677–2692 (2007)
29. Bobko, C., Ulm, F.-J.: The nano-mechanical morphology of shale. Mech. Mater. 40, 318–337 (2008)
30. Oliver, W.C., Pharr, G.M.: An improved technique for determining hardness and elastic modulus using load and displacement sensing indentation experiments. J. Mater. Res. 7, 1564–1583 (1992)
31. Dormieux, L., Kondo, D., Ulm, F.J.: Microporomechanics. J. Wiley & Sons, UK (2006)
32. Ulm, F.J., Jennings, H.M.: Does C-S-H particle shape matter? A discussion of the paper, Modelling elasticity of a hydrating cement paste, by Julien Sanahuja, Luc Dormieux and Gilles Chanvillard. CCR 37, 1427–1439 (2007); Cem. Concr. Res. 38, 1126–1129
33. Sanahuja, J., Dormieux, L., Chanvillard, G.: Modelling elasticity of a hydrating cement paste. Cem. Concr. Res. 37, 1427–1439 (2007)
34. Cariou, S., Ulm, F.J., Dormieux, L.: Hardness-packing density scaling relations for cohesive-frictional porous materials. J. Mech. Phys. Solids 56, 924–952 (2008)
35. Gathier, B., Ulm, F.J.: Multiscale strength homogenization - application to shale nanoindentation, MIT-CEE Res. Rep. R08-01, Dpt. of Civil and Environmental Engineering, Massachusetts Institute of Technology, Cambridge, MA (2008)
36. Mondal, P., Shah, S.P., Marks, L.: A reliable technique to determine the local mechanical properties at the nanoscale for cementitious materials. Cem. Concr. Res. 37, 1440–1444 (2007)
37. Pellenq, R.J.M., et al.: Towards A Consistent Molecular Model of Cement Hydrate: The DNA of Concrete? (forthcoming) (2009)
38. Vandamme, M., Ulm, F.J.: The Nanogranular Origin of Concrete Creep: A Nanoindentation Investigation of Microstructure and Fundamental Properties of Calcium-Silicate-Hydrates, MIT-CEE Res. Rep. R08-02, Dpt. of Civil and Environmental Engineering, Massachusetts Institute of Technology, Cambridge, MA (2008)

Innovative Building Material – Reduction of Air Pollution through TioCem®

G. Bolte

Abstract. In many European cities air quality is a massive problem. Besides the particulate matter, nitrogen oxides (NO_X) and volatile organic compounds (VOC) are mainly responsible for the heavy pollution. Motivation to "do something" to protect the environment and climate is increasing constantly. Pollutants such as nitrogen oxides can be oxidized by means of photolysis. With the help of photocatalytic active particles this effect can be accelerated extensively. Photocatalytic active particles dispersed in the concrete turn it into an air pollutant reducing surface. Pollutants getting in contact with the concrete surface are decomposed or oxidized and therewith rendered harmless. This brand new technique is introduced into building industry with a new label "TX Active®"[1]. A premium brand cement for the production of photo catalytically active concrete products - TX Active® products - is now available in the form of TioCem®. This cement can effectively contribute to air purification by using in numerous concrete components such as pavement, roof tiles, facade plates, concrete road surfaces, mortars etc.

1 Pollution in the Urban Environment

The emission of exhaust gases from vehicles, heating installations and power stations has increased in recent years leading to a rise in air pollution. This is a growing concern for human health as there is a proven link between poor air quality - particularly evident in congested, densely populated urban areas - and an increased risk of respiratory infections and impaired lung function [1]. The main pollutants in the air include fine dust particulates, nitrogen oxides (NO_X), sulphur oxides (SO_2), carbon monoxide (CO), volatile organic compounds (VOC) and Ozone (O_3).

Nitrogen oxides are particularly important as they are the pre-substances for the formation of ozone which is harmful to health in near to ground layers, especially

G. Bolte
HeidelbergCement Technology Center, Leimen
e-mail: gerd.bolte@htc-gmbh.com

[1] TX Active® is a registered trademark of Italcementi S.p.A. Product under license of Italcementi S.p.A.

in summer. In consequence from 2010 on, a mandatory EU directive [2] prescribes a maximum yearly average of 40 µg NO_2/m^3 and a one-hour average of 200 µg NO_2/m^3. Due to the legislative directives, cities are obliged to elaborate air purification plans which in future shall help to observe the critical values. The observance of these limits will not be possible in the vicinity of heavily congested roads, even if all vehicles are conform to the EURO 4 standard. Additional provisions will therefore be necessary.

2 Resolutions for the Reduction of Pollutants

Many compounds and so also air pollutants are decomposed by sunlight, in particular by energy-rich UV radiation. This natural process of the photolysis takes place extremely slowly. With the help of photocatalysts this effect can be accelerated extensively.

Photocatalysts are semi-conductors. Activated by UV-light, an electron is transported from the valence-band to the conduction-band [3]. When oxygen gets in contact with such an activated electron, superoxidion is generated. Gets the valence-band hole in contact with water, the hydroxyl radicals OH• are generated. Both are highly reactive compounds which are able to oxidize most organic compounds and also pollutants such as NO_X.

Scientific proof of the effective reduction of nitrogen oxide exposure has been carried under close to practical conditions by the PICADA (Photocatalytic Innovative Coverings Applications for Depollution Assessment) project supported by the EU [4]. The test apparatus used for this purpose consisted of a simulation of three parallel streets on a scale of 1:5. Cargo containers were used as roadside buildings. The walls of the center street were covered with a photocatalytic active cement mortar. Pollutants were uniformly distributed along the streets by means of a piping system. As a result of the use of the photocatalytic active cement mortar a reduction in NO_X between 40 to 80% was ascertained.

After years of scientific research, photocatalytically active products have meanwhile clearly outgrown the laboratory stage. HeidelbergCement, for example, supplies a highly photocatalytically active cement under the trade name TioCem®.

3 Photocatalytic Activity of Concrete Surface

The photocatalytic activity of surfaces can be demonstrated and quantified in terms of the bleaching of an organic dye [5]. Rhodamine B is applied as a model substance to test objects consisting of a mortar-tile. The color intensity of the test objects is repeatedly measured using a chromameter prior to application of the dye, after drying of the dye, and after a defined period of exposure under a light source (ULTRA Vitalux spotlight). A measure of the activity of the surface is obtained by relating the color difference ΔE* of the test objects exposed under the

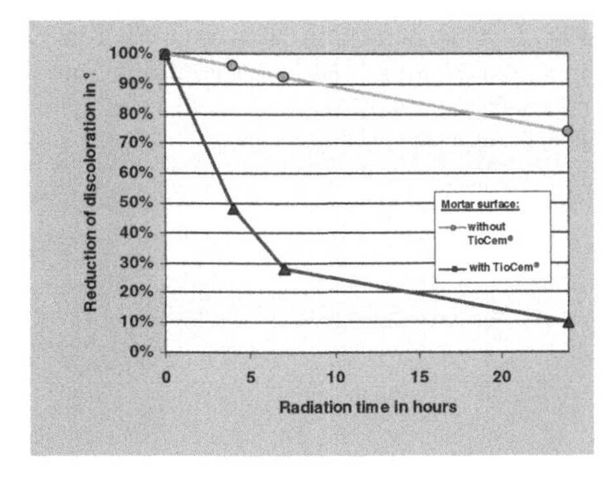

Fig. 1 Discoloration of Rhodamine B on mortar-tiles; UV-A Intensity 600 μW/cm²

spotlight to the ΔE* of the unexposed sample. This test apparatus provides results which are repeatable only on extremely smooth surfaces.

A further measuring method used for detection of photocatalytic activity involves a test apparatus in which a mixture of air and NO, NO_2 or a mixture of the two flows in a chamber over a test object and the NO_X concentration is measured with and without exposure to light. (Fig. 2) These measuring methods are described for example in ISO 22917, Part 1 [6] and UNI11247 [7].

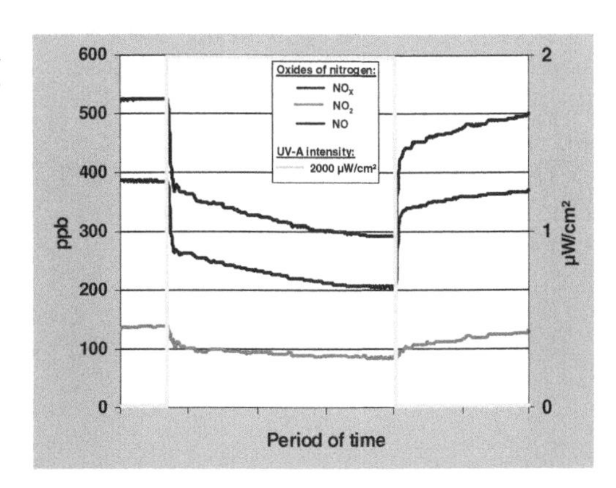

Fig. 2 NO, NO_2, NO_X abatement in accordance with UNI11247 on concrete surface

The degree to which the use of the photocatalytic active cement TioCem® is capable of reducing the oxide of nitrogen concentration in the air is determined at HeidelbergCement's Technology Center in Leimen using a measuring apparatus specially developed for this purpose. This permits variation of a large range of parameters, such as flow velocity, light intensity and the NO and NO_2 concentrations in the feed air, thus allowing simulation of diverse environmental conditions. Figs. 3 and 4 showing for example the extent to which the rates of degradation are influenced by these ambient conditions on a face mix paving block with TioCem® in the top layer.

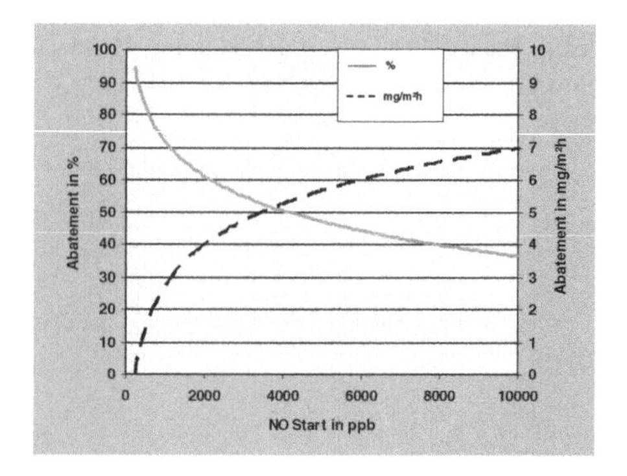

Fig. 3 NO abatement in % vs. mg/m²h as a function of initial NO concentration on concrete paving block; air flow 1 l/min; UV-A Intensity 2000 μw/cm^2

The abatement of NO_x in the flow of air can be declared in % and in mg/m^{2}h. Fig. 3 illustrates that a 70% abatement is achieved from an initial pollutant concentration, for example, of 1000 ppb, which is equivalent to a degradation rate of 2.5 mg/m^{2}h. The amount of NO_x degraded rises to 4.5 mg/m^{2}h at a higher initial pollutant concentration of, for example, 3000 ppb, whereas rate of reduction drops to 55% of the feed flow.

Fig. 4 stated the correlation between photocatalytic activity and UV-A radiation. Degradation of the air pollutant is achieved even at UV-A radiation intensities below those described in the various measuring methods. Breakdown of air pollutants thus starts, so to speak, at sunrise.

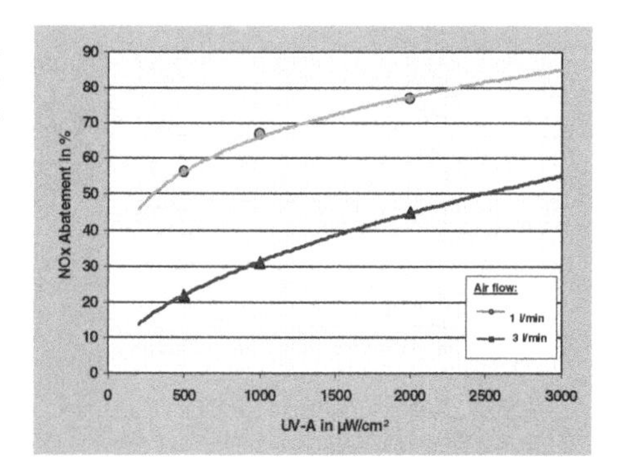

Fig. 4 NO_x Abatement in % as a function of UV-A intensity and air flow on concrete paving block; 550 ppb NO_x in the feed air (400 ppb NO + 150 ppb NO_2)

Many diverse parameters resulting, for example, from fluctuating weather conditions and traffic loadings make evaluation of achieved pollutant reduction extremely difficult in any particular practical application. In the laboratory, on the other hand, it is possible to "merge out" certain parameters, in order to demonstrate that the NO_x content in air which comes into contact with a photo catalytically active concrete surface is significantly reduced.

An attempt to demonstrate this with a decreased number of parameters in an outdoor field test was performed in Stockholm in July, 2008. For this purpose, two identical test chambers consisting of material translucent to visible-light and UV-A radiation were positioned in the inner city. One of the two test chambers contained a surface consisting of a lime-cement render based on TioCem®. Normal ambient air was passed through the test chamber and on to an NO_X analyzer. The daily average on NO_2 in the air fed through the test camber was reduced between 40 to 70%.

The use of paving blocks containing TX Active® implemented by Italcementi in Italy resulted in a reduction in eight-hour average NO_X readings in the Via Borgo Palazzo, Bergamo, between 26 to 56% [9]. A further major project has been implemented in Paris by Ciment Calcia. A concrete street has been constructed using TX Active® in the Rue Jean Bleuzen, in the Vanves district of the city. Air quality was then monitored by an external laboratory (Laboratoire Régional de l'Quest Parisien). Initial results have already been obtained, and the concluding report is expected to appear in 2009.

4 Practical Applications

Climalife, the first "environmentally active roof tile", was unveiled in November, 2007. The surface of this concrete roof tile consists of a TioCem® coating. An external test certificate documents a degradation rate of 1.6 mg NO/m^2h for this roofing tile surface. The test was conducted in accordance with ISO 22917, Part 1, in which a mixture of room air and NO is fed as the analysis substance. NO_2 is degraded to the same degree, as is documented by our own trials. By the end of 2008, more than 200000 m^2 of roof surface area have begun to make an active contribution to reduce air pollution. Roofing of an average detached house (with a roof surface area of around 200 m^2) using the Climalife roofing tile would eliminate, thanks to its photocatalytic surface, the NO_X emissions from three gas central-heating installations. The potentials for improving air quality that are opened up by this innovative product can be roughly estimated when one remembers that around 30 million m^2 of roof area are covered with concrete roof tiles every year in Germany alone.

Fig. 5 George Harrisson Memorial Garden, Chelsea Flower Show 2008, England

Concrete products containing white TioCem® made its debut recently in England. Photocatalytically active white TioCem® was incorporated into concrete elements, in combination with marble, for two show gardens, the George Harrison Memorial Garden (Fig. 5) and the "Sail for Gold" garden, which is dedicated to the English Olympic sailing team at the Chelsea Flower Show.

5 TX Active® – Tried and Tested Quality

A brand new technique based on process photocatalysis was introduced into building industry. The technology is called "TX Active®" and represent advanced feature of cement-bonded building materials. TX Active® has been developed by the licensor Italcementi S.p.A. TX Active® is a quality mark used to validate the photocatalytic activity of building materials and represents a brand new technique and advanced technology. Strict quality standards have been defined, both for the cement and for the end products made from it.

TioCem® - a cement with TX Active® inside - is a cement in conformity to EN 197. Rhodamine B bleaching in accordance with UNI11259 [8] is complemented to the internal production-inspection of the cement. Products made from TioCem® are regularly monitored for their photocatalytic activity at the HeidelbergCement Technology Center. The breakdown rate of the NO_X being determined in the laboratory test procedure under the conditions defined in UNI11247 [7]. The application and durability properties of this cement are, by the way, equivalent to those of a standard cement.

6 Outlook

If a traffic surface for example of the size of a football pitch (˜ 7500 m^2) covered with photocatalytically active paving stones of the quality specified in Fig. 3 exposed on average to 2000 hours of solar radiation annually (> 1000 $\mu W/cm^2$), it will be capable of breaking down the pollutant emissions from more than 190000 car-kilometers, assuming an NO_X emission rate of the EURO 4 pollution category and a ratio of 50% gasoline- and 50% diesel-engines. This calculated figure is greatly exceeded in practice, since photocatalysis occurs not only under exposure to direct sunshine, but also under cloudy skies, and under twilight conditions (Fig. 4). When the many and diverse potential applications are grouped together, it becomes clear that the surface: atmosphere ratio can be increased step-by-step, assuming a not merely sporadic use for refurbishing, repair and construction of new roads, pavements, façades and buildings. In the future, the innovative functional benefit of cement bound building materials and building products will contribute to significant improvements in the quality of life in our towns and cities.

References

1. Hessisches Landesamt für Umwelt und Geologie. Stickstoffdioxid (NO_2) Quellen – Emissionen – Auswirkungen auf Gesundheit und Ökosystem – Bewertungen – Immissionen, http://www.hlug.de
2. Air-Quality Directive 1999/30/EG of the Council of the European Union. April 22 (1999)
3. Fujishima, A., Hashimoto, K., Watanabe, T.: TiO_2 Photocatalysis Fundamentals and Applications. BKC Inc. Tokyo, Japan (1999)
4. http://www.picada-project.com
5. Bolte, G.: Photokatalyse in zementgebundenen Baustoffen. Cement Int. 3, 92–97 (2005)
6. ISO 22917-1 Fine ceramics (advanced ceramics, advanced technical ceramics) – Test method for air-purifi cation performance of semiconducting photocatalytic materials – Part 1: Removal of nitric oxide
7. UNI11247, Diterminazione dell'attività di degradazione di ossidi di azoto in aria de parte di materiali inirganic fotocatalytici
8. UNI11259, Determinazione dell'attività fotocatalitica di leganti idraulici Methodo delle Rodammina
9. Guerrini, G.L., Peccati, E.: Photocatalytic cementitious roads for depollution. In: International RILEM Symposium on Photocatalysis, Environment and Construction Materials., October 8–9, 2007, Florence, Italy (2007)

Nanomechanical Explorations of Cementitious Materials: Recent Results and Future Perspectives

G. Constantinides, J.F. Smith, and F.-J. Ulm

Abstract. Recent progress in experimental and theoretical nanomechanics makes it possible to revisit the response of ubiquitous construction materials, like concrete, reevaluate our existing knowledge and understanding, and device methodologies to optimize their macroscopic performance. Particularly, the advent of instrumented indentation and the advancement of homogenization methods provide the mechanics community an unprecedented opportunity to probe the mechanical behavior of structural materials at the nanoscale (with length-resolution in the nanometer and force-resolution in the nanoNewton) and quantitatively convey these information at the macroscopic scale. Furthermore, the capabilities offered in a spatial and temporal domain by these advanced instruments allow the investigation of a number of additional phenomena: interface mechanics, strain-rate effects, high temperature response, sources of anisotropy, chemo-mechanical effects, etc. We here show the validation of a fluid cell module that allows acquisition of nanomechanical data in liquids.

1 Introduction

Recent advances in modeling [12, 17] allow one to upscale the mechanical response of complex heterogeneous material systems (concrete being an example)

G. Constantinides
Department of Mechanical Engineering and Materials Science and Engineering, Cyprus University of Technology, Lemesos, CY
e-mail: g.constantinides@cut.ac.cy, george_c@mit.edu

J.F. Smith
Micro Materials Ltd, Wrexham, UK
e-mail: nanotest@btinternet.com

G. Constantinides and F.-J. Ulm
Department of Civil and Environmental Engineering, Massachusetts Institute of Technology, Cambridge, MA, US
e-mail: ulm@mit.edu

and obtain effective properties that can be used in structural mechanics applications. Upscaling techniques may vary from analytical, namely continuum micromechanics, to numerical, namely finite element solutions, utilizing in the process physicochemical models that analytically [18] or digitally [1] synthesize microstructures. Such approaches, which have their origin at the level of the individual chemical constituents of the composite material, provide a direct link between physical chemistry and mechanics [9, 11]. Furthermore, they allow one to trace the origin of chemo-mechanical degradation at the length scale where the chemical reactions occur [9]. A common requirement to all modeling approaches is the need for intrinsic mechanical properties of the individual constituents composing the composite material, and their temporal response as chemical softening or stiffening occurs. In the case of concrete, the main constituent phase that governs the macroscopic response (Calcium Silicate Hydrates or in short C-S-H) manifests itself in the nm to μm length scale [14]. This constituent phase cannot be recapitulated effectively *ex-situ*; one has to, therefore, access the mechanical properties of C-S-H *in-situ* at the length scale where it can be naturally found [11]. The advent of instrumented indentation provided an unprecedented opportunity to probe the mechanical response of these phases and incorporate the results in micromechanical models that can deliver the composite response. Herein we show recent results on C-S-H mechanics and recent developments on nanoindentation that might allow further refinements on our understanding and future material optimization.

2 Nanomechanics of C-S-H

The advent of instrumented indentation enabled fundamental studies in the nanomechanical response of metals, ceramics, polymers and composites (see i.e., [2]). Current technology allows for contact-based deformation of nanoscale load and displacement resolution and has been leveraged for both general mechanical characterization of small materials volumes and unprecedented access to the physics and deformations processes of materials. While nanoindentation was originally developed for homogeneous metals and ceramics it was quickly appreciated that nanoscale resolution can be of significant use to the decoding of C-S-H structure, the binding phase of all cementitious materials. However, accurate nanomechanical analysis of natural composites requires advanced analysis that takes into consideration the multi-phase, multi-scale nature of the material and its pressure sensitive mechanical response [8, 10, 11, 13].

A typical nanoindentation test consists of establishing contact between an indenter (typically diamond) and a sample, while continuously measuring the load, P and the penetration depth h. Analysis of the P-h response proceeds by applying a continuum scale model [2] to derive the indentation modulus M and indentation hardness H:

$$M = \frac{\sqrt{\pi}}{2} \frac{S}{\sqrt{A_c}} \tag{1}$$

$$H = \frac{P_{max}}{A_c} \tag{2}$$

where S is the unloading slope at maximum depth h_{max}, P_{max} is the maximum indentation force, and A_c is the projected contact area at h_{max}. Several empirical means to estimate A_c exist, either through post-indentation inspection, geometric idealizations of the probe, or more commonly through analysis of the indentation response for a materials of ostensibly known E and H to determine this area as a function of $A_c = A(h_c)$. These indirect means of geometry estimation work reasonably well but can be subject to errors for nanoscale indentations, h <200nm. A more rigorous approach was recently proposed [6, 22] in which one can determine directly A_c by recourse to atomic force microscopy (AFM) images of the indenter probe. An example of AFM imaging of a Berkovich probe is shown in Fig. 1.

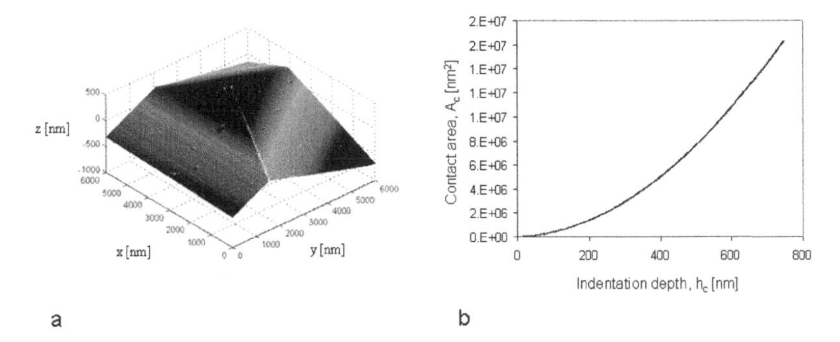

Fig. 1 AFM image of a Berkovich nanoindenters probe (a) and the resulting cross-sectional area as a function of the distance from the apex (b)

Equations 1 and 2 rely on the assumption of a semi-infinite half-space and therefore caution should be taken when testing highly heterogeneous materials. In particular, the number of tests should be significantly increased and the choice of indentation depth should be carefully chosen [7, 8]. The two indentation properties measured during a test (M, H) can then be linked to the elastic M=M(E, ν) and plastic H=(c, φ) properties of the indented materials, through advanced continuum scale models [2, 13]. The extracted mechanical properties of the two types of C-S-H measured on hundreds of specimens where found to be intrinsic to all cement-based materials (see Table 1 [11]).

Table 1 Nanomechanical properties of a low density (C-S-H_{LD}) and a high density (C-S-H_{HD}) Calcium-Silicate-Hydrate found in all cementitious materials

	C-S-H_{LD}	C-S-H_{HD}
Elastic Modulus, E [GPa]	21	31
Angle of Friction, φ [°]	12	12
Cohesion, c [MPa]	130	220

The fundamental mechanical properties of C-S-H together with estimates of the volumetric proportions (f_i) of all constituent phases (i) were incorporated in a multi-scale homogenization scheme that can predict the composite macroscopic elastic (E_{hom},ν_{hom}=F(E_i,ν_i,f_i)) and strength (c_{hom}, $_{hom}$=F(c_i, $_i$,f_i)) behavior [11, 12, 17], where the only input requirements are the elastic (E_i,ν_i) and plastic (c_i, $_i$) properties of the individual constituents and their volumetric proportions (f_i). The morphological arrangement of the phases in space is taken into consideration in the choice of the continuum micromechanical models. It has been found that for cementitious materials a combination of Mori-Tanaka and Self-Consistent schemes appears to deliver robust results [11]. Figure 2 demonstrates the predictive capabilities of the proposed models.

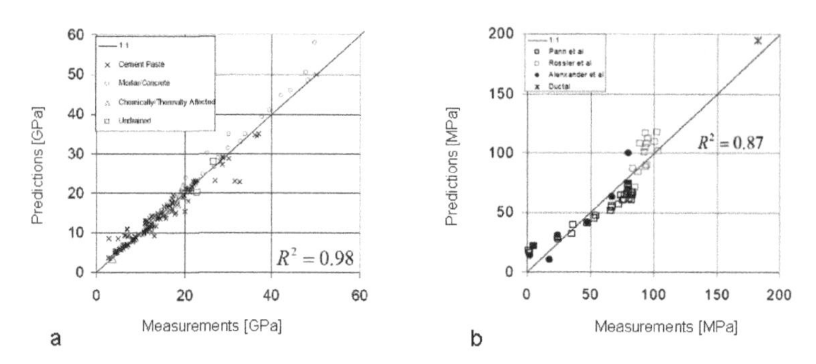

Fig. 2 Micromechanical predictions and experimental data of the elastic (a) and uniaxial strength (b) properties of cementitious materials

3 Advancements on Instrumented Nanoindentation

Instrumented nanoindenters are currently going through a phase of optimization and as a consequence their performance is constantly reevaluated and improved. New developments include the capabilities of high temperature testing (up to 800°C) [5], high strain-rate testing [3, 5], indentation in liquids [4] or controlled humidity environments, etc. On the theoretical side our understanding is improved and new continuum models allow now analysis of indentation data on layered systems, adhesive contact, anisotropic systems, cohesive-frictional materials, visco-elastic-plastic materials, etc [13, 15, 16, 19, 20, 21]. Due to space constrains we here focus on the fluid-cell module extension that allows mechanical measurements of materials in liquid.

Figure 3 presents a straightforward modification of instrumented indentation platform that allows acquisition of nanoscale force-displacement data in liquid media without artifacts of buoyancy or surface tension. The indenter mount for liquid cell applications is comprised of a stiff, corrosion resistant, stainless steel. As shown in Fig.3a the extended indenter is immersed within the liquid cell. In practice, this fluid cell inclusive of mounted samples is maneuvered into position via automated stage displacement, and the indenter automatically contacts the probe. Liquids can be added before or after this operation, and also exchanged intermittently.

We here demonstrate the validity of nanoindentation in fluid via elastoplastic analysis of relatively stiff ($E > 1000$ kPa), water-insensitive materials (Borosilicate Glass and Polypropylene). We then consider the viscoelastic response and representative mechanical properties of compliant, synthetic polymeric hydrogels and biological tissues ($E < 500$ kPa). Examples from indentations on water-saturated synthetic (hydrogels) and natural materials (porcine liver and skin) are presented. The elastic properties of the tested materials are in good agreement with macroscopic data found in the literature, validating the accuracy of the proposed fluid-cell module and demonstrating its ability for nanoscale characterization of hard and soft systems in the kPa to GPa range. These capabilities can be of great importance in the poromechanical and chemomechanical studies of construction materials, including cementitious composites, geomaterials and many more.

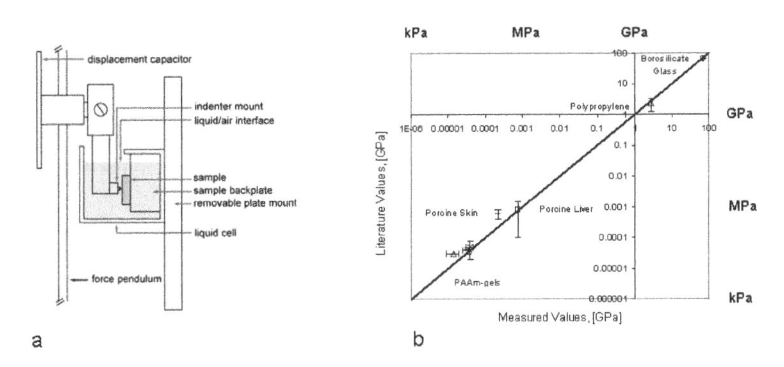

Fig. 3 (a) Schematic diagram of the fluid cell module extension. (b) Measured and literature values of the tested hydrated specimens: borosilicate glass, polypropylene, porcine liver, porcine skin, and PAAm-gels in various mol concentrations

4 Concluding Remarks

The field of nanoscale indentation mechanics is advancing in a rapid pace. As the tools of experimentation and the resulting theoretical frameworks of data analysis are developing, new opportunities for materials understanding and characterization arise. This provides an unprecedented opportunity to probe long-used ubiquitous construction materials that have not been rigorously characterized and modeled in the past and create a science-based platform for material optimization.

References

1. http://ciks.cbt.nist.gov/monograph/
2. Cheng, Y.T., Cheng, C.M.: Scaling, dimensional analysis, and indentation measurements. Mater. Sci. Eng. R. 44(4-5), 91–149 (2004)
3. Constantinides, G., Tweedie, C.A., Savva, N., et al.: Quantitative nanoscale-impulse testing of energy absorption at surfaces Exper Mech. (in press)
4. Constantinides, G., Kalcioglu, I., Smith, J.F., et al.: Probing mechanical properties of fully hydrated gels and biological tissues. J. Biomech. 41, 3285–3289 (2008)
5. Constantinides, G., Tweedie, C.A., Holbrook, D.M., et al.: Quantifying deformation and energy dissipation of polymeric surfaces under localized impact. Mater. Sci. Eng. A. 489(1-2), 403–412 (2008)
6. Constantinides, G., Silva, E.C.C., Blackman, G.S., et al.: Dealing with imperfection: Quantifying potential length scale artefacts from nominally spherical indenter probes. Nanotechnology 18(30), 305503 (2007)
7. Constantinides, G., Ulm, F.-J.: The nanogranular nature of C-S-H. J. Mech. Phys. Solids 55(1), 64–90 (2007)
8. Constantinides, G., Ravichandran, K.S., Ulm, F.-J., et al.: Grid indentation analysis of composite microstructure and mechanics: Principles and validation. Mater. Sci. Eng. A 430(1-2), 189–202 (2006)
9. Constantinides, G., Ulm, F.-J.: The effect of two types of C-S-H on the elasticity of cement-based materials: Results from nanoindentation and micromechanical modeling. Cement Concrete Res. 34(1), 67–80 (2004)
10. Constantinides, G., Ulm, F.-J., van Vliet, K.J.: On the use of nanoindentation for cementitious materials. Mater. Struct. 36, 191–196 (2003)
11. Constantinides, G.: Invariant mechanical properties of calcium silicate hydrates (C-S-H) in cement-based materials: instrumented nanoindentation and microporomechanical modeling. PhD Thesis, MIT (2006)
12. Dormieux, L., Kondo, D., Ulm, F.J.: Microporomechanics. Wiley, Chichester (2006)
13. Ganneau, F.P., Constantinides, G., Ulm, F.-J.: Dual-Indentation technique for the assessment of strength properties of cohesive-frictional material. Int. J. Solids Struct. 43, 1727–1745 (2006)
14. Jennings, H.M., Gevrenov, J.S., Thomas, J.J., et al.: A multi-technique investigation of the nanoporosity of cement paste. Cement Concrete Res. 37(3), 329–336 (2007)
15. Perriot, A., Barthel: Elastic contact to a coated half-space: effective elastic modulus and real penetration. J. Mater. Res. 19(2), 600–608 (2004)
16. Sorelli, L., Constantinides, G., Ulm, F.-J., et al.: The nanomechanical signature of Ultra High Performance Concrete by statistical nanoindentation techniques. Cement Concrete Res. 38, 1447–1456 (2008)
17. Torquato, S.: Random heterogeneous materials. Springer, Heidelberg (2001)
18. Tennis, P.D., Jennings, H.M.: A model for the two types of C-S-H in the microstructure of Portland cement pastes. Cement Concrete Res. 30, 855–863 (2000)

19. Tweedie, C.A., Constantinides, G., Lehman, K.E., et al.: Enhanced stiffness of amorphous polymer surfaces under confinement of localized contact loads. Adv. Mater. 19(18), 2540–2546 (2007)
20. Ulm, F.-J., Constantinides, G., Heukamp, F.H.: Is concrete a poro-mechanics material? A multi-scale investigation of poro-elastic properties. Mater. Struct. 37(265), 43–58 (2004)
21. Ulm, F.J., Delafargue, A., Constantinides, G.: Experimental microporomechanics. In: Dormieux, L., Ulm, F.J. (eds.) Applied microporomechanics of porous materials, CISM Lecture notes no 580. Springer Wien, New York (2005)
22. Van Landingham, M.R., Juliano, T.F., Hagon, M.J.: Measuring tip shape for instrumented indentation using atomic force microscopy. Meas. Sci. Technol. 11, 2173–2185 (2005)

Developments in Metrology in Support of Nanotechnology

J.E. Decker, A. Bogdanov, B.J. Eves, D. Goodchild, L. Johnston, N. Kim, M. McDermott, D. Munoz-Paniagua, J.R. Pekelsky, S. Wingar, and S. Zou

Abstract. Nanotechnology emerges out of fundamental science through capability for accurate, repeatable and reproducible measurements on the nanoscale which allows scientists and engineers to accumulate knowledge. Understanding the measurement science is the first step towards development of new ideas. This paper describes some research initiatives which underpin the development of nanotechnology. Programs underway at the National Research Council of Canada include: development of metrological scanning-probe microscope instrumentation for dimensional calibration, materials characterization, development of artefacts designed specifically for dimensional calibration, investigation of metrology for application to soft materials and investigation of intrinsic length standards for realization of the SI metre at the nanoscale.

1 Introduction

Common terminology, standards and procedures form the basis of fair trade, technical competitiveness and product reliability. Accurate measurements are also important for the advancement of science and acquisition of knowledge as they allow comparability of measurements. Comparability and coherence with the fundamental constants is becoming increasingly relevant to industry as feature sizes diminish and quantum effects are exploited in new devices. This is best achieved through traceability to the SI, which provides a unified basis for measurements.

Metrology and international standards have an important role in taking proof-of-concept ideas to commercialization and trade in our global marketplace. The June '07 Resolution of ISO/TC229 Technical Committee on Nanotechnologies calling for governments to invest in nanotechnology R&D is a testament to the acknowledged large-scale cooperative effort required to establish terminology and

J.E. Decker, A. Bogdanov, B.J. Eves, D. Goodchild, L. Johnston, N. Kim, M. McDermott, D. Munoz-Paniagua, J.R. Pekelsky, S. Wingar, and S. Zou
National Research Council of Canada – Institute for National Measurement Standards,
e-mail: jennifer.decker@nrc-cnrc.gc.ca
http://www.nrc-cnrc.gc.ca/main_e.html

nomenclature, and measurement & characterization techniques – essential to our understanding of health, safety and environmental impacts of nanotechnologies. Consensus standards contribute in a fundamental way to the body of knowledge necessary for realizing benefits, such as predictive toxicology in the area of health and environment. Drawing further on this specific example, terminology, nomenclature and the specification of key physico-chemical properties of nanomaterials are topics of intense interest for the purpose of unique identification and characterization of nanomaterials. There are several lists of measurands deemed as essential currently under consideration by the science and technology community; work is ongoing to determine the most widely-accepted list, which in turn will influence nomenclature of new nano-objects and the development of metrology. Reference materials are a necessary component of reliable measurements since they are used to calibrate instruments and compare testing procedures and measurement results between laboratories. The community of national metrology institutes are launching nanoscale reference materials and measurement protocols with the intention of providing guidance and benchmarks to nanotechnology users and ensuring commutability of measurement results. Evidence that more work is required to understand nanoscale objects and measurement methods is supplied by the case of nanoparticles: the specific value attributed to particle size is determined largely based on the measurement technique applied and end-use of the client.

The quality of the transfer of the definition the metre depends on the design, quality and measurement techniques used to calibrate reference standards. History demonstrates that economic benefit results from high quality manufacturing, which is in turn directly related to highest-level metrological standards that keep pace with the developments of competitive precision mechanics. Nanotechnologies pose new challenges to traditional metrology because our understanding of how nature behaves on nanoscale lengths is in development. This paper outlines the activities and progress of the National Research Council of Canada (NRC) program of research and development, in part drawing upon trusted strategies from macroscale measurements and applying them towards understanding behaviour, and developing new measurement techniques specifically to address nanoscale measurement problems. In the area of length metrology, projects focus on development of length-calibration artefacts, metrological scanning-probe microscope instrumentation for dimensional calibration, materials characterization and investigation of intrinsic length standards for realization of the SI metre at the nanoscale.

2 NRC Metrology for Nanotechnology Program Components

2.1 Nanoimprint Lithography

The Canadian Photonics Fabrication Center (CPFC) was recently established at NRC Institute for Microstructural Sciences (NRC-IMS) to provide an industrial-grade facility for prototype and low volume production of photonic devices for external clients. Many of these devices depend on accurate control of critical dimensions from a few micrometres down to a few nanometres. For example, while the grating features in semiconductor distributed feedback (DFB) lasers may be of the order of 10 nm – 100 nm, their pitch must be controlled to better than 0.1 nm accuracy to achieve best control of wavelength. CPFC has recently established nano-imprint lithography as an attractive means of patterning sub-100 nm features which until now, were usually achieved by direct write e-beam lithography.

Nanoimprinting as opposed to projection lithography or e-beam lithography relies in the physical transfer of a 3-D mold or template into an intermediate polymer which is then used to transfer the pattern into the substrate through dry etching. In some cases the imprinted polymer is the final structure itself. By its very nature this process lends itself to highly reproducible replication of the original structure. One tile of the required pattern is first produced on a quartz template (using e-beam lithography and etching). The template is used to replicate this tile (in UV curable resist) across all the die on a wafer. Once the template is made, the wafer patterning time (ie manufacturing cost of this step) is reduced by more than an order of magnitude. Another key advantage is that the features printed will have the same dimensions on every die, which is replicated from the same template or master. By choosing to use a low molecular weight, UV curable material as the transfer polymer very fine features as small as 10 nm can be replicated at the same time as large micron sized structures. Once this material has been cured it is dimensionally very stable. The key to the success of using nanoimprinting as a method to create nanoscale structures is the fabrication of the 3-D mold or template.

While this Nanoimprint technology provides an attractive route for fabrication of nanostructured devices, it also offers a means of replicating standard structures, which can be of use for metrology. Establishing traceable metrology standards

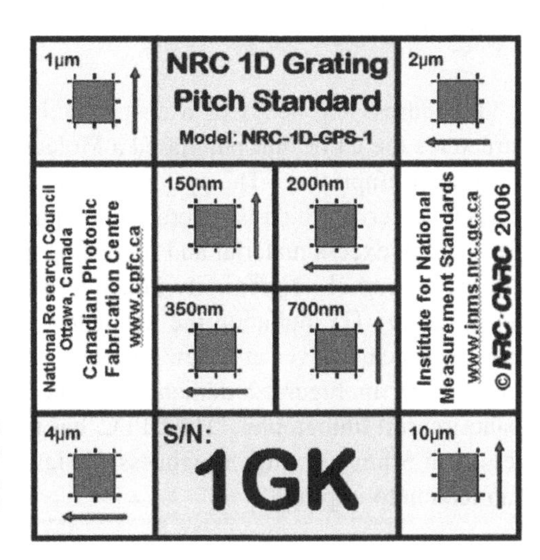

Fig. 1 Design of the NRC grating pitch reference artifact. The 10-mm square chip contains eight grating patches. Nominal grating pitch ranges from 150 nm to 10000 nm

and measurement methods is a prerequisite for manufacture of nanodevices. Figure 1 depicts the high-quality reference artefacts under development which possess features amenable to lowest-uncertainty calibration and validation by inter-laboratory comparison measurements.

The NRC-IMS facility has a JEOL JBX-6000 direct write e-beam system that has a minimum beam diameter of 5 nm. A custom holder was made for the system to hold the template to be exposed. Once the template has been coated with masking materials and has been exposed by the e-beam tool, the pattern is transferred in to the template using dry etch processes to create a 3-D structure that is then used by the imprinting tool.

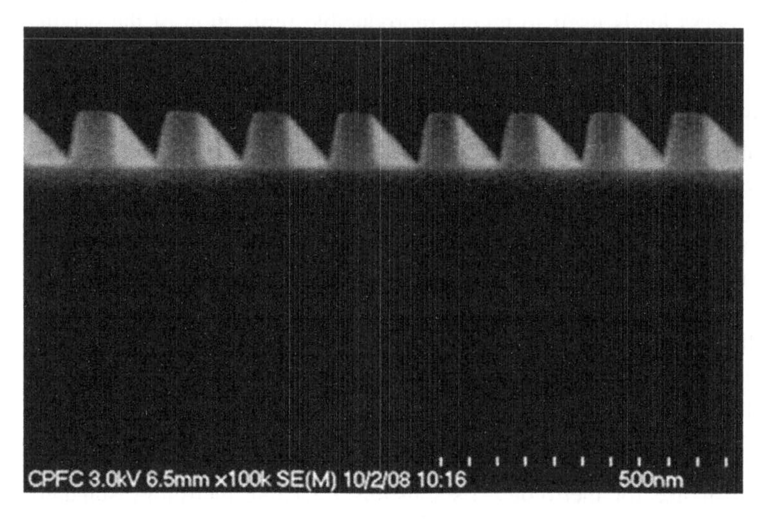

Fig. 2 Scanning electron microscope image of a grating produced by method of nano-imprint lithography

To address the needs of a cost effective method of producing sub 100 nm structures the CPFC has purchased a Molecular Imprints Imprio 100 UV step and flash nano imprinter. The system uses a ink-jet like process to deposit the UV curable material which is tailored to the pattern density of the template, so minimizing any excess material and producing a reproducible imprint as the template is brought into close proximity (touching the polymer only) to the substrate. After curing by UV radiation the template is separated from the substrate and the process repeated over at a new location. An image taken by scanning electron microscopy in Figure 2 demonstrates the quality of the gratings produced by nano-imprint lithography. The CPFC has produced 90 nm 1:1 pitch structures of less than 5 nm line edge roughness (3-sigma) using this process (from template fabrication to imprint).

2.2 Metrological Atomic Force Microscope

A necessary requirement of accurate metrology is precise and unambiguous definition of the measurand. One-dimensional (1D) grating pitch artefacts have the advantage of simple, repeatable geometry amenable for calibration of the lateral scales of microscopes [1-3]. Diffractometers and scanning probe microscopes (SPM) are complementary instruments which are both needed to accurately measure this line spacing. The diffractometer measures the average grating pitch over the spot size of the laser beam (pitch = 350 nm, U = 10 pm) and the SPM tip records the profile of the individual grating lines, providing information on the variation of the line spacing.

The scientific community uses atomic force microscopes (AFM) extensively to generate topographic and material contrast images of nanometre to micrometre scale features. Only a small percentage of AFMs are suitable for extracting dimensions such as distance between features with a small or even known uncertainty [2-7]. The National Research Council of Canada Institute for National Measurement Standards (NRC-INMS) is developing a platform to investigate the performance of dimensional measurements using scanning probe microscopy [8]. A key requirement for the platform is a large measuring volume to increase the diversity of possible applications. The design shown in Figure 3 achieves a measurement

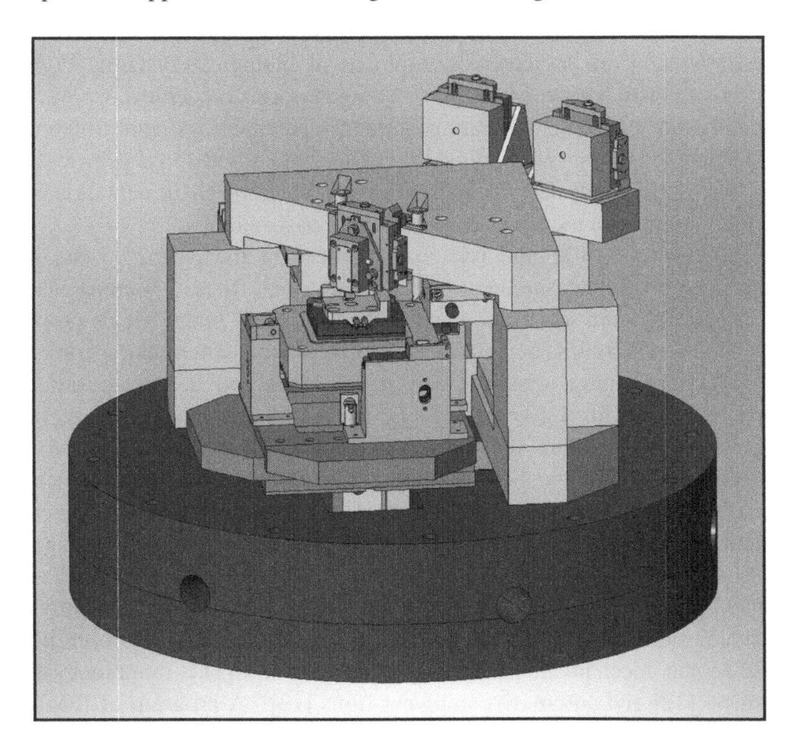

Fig. 3 Schematic drawing of the NRC metrological atomic force microscope

volume of 40 mm x 40 mm x 6 mm and aims to reach a measurement uncertainty of position of 1 nm (k=1). The stated uncertainty target does not include the uncertainty contribution of the probe. The design also maintains an open architecture such that various probe types can be integrated with the precision motion stage ensuring a versatile research instrument.

To attain the large measurement volume the translation of the sample is accomplished via multiple stages which allow for separate *coarse* and *fine* motions. Laser interferometers and autocollimators are used to measure the position and orientation of the sample and enable the correction of Abbe errors and form errors. These errors are expected to dominate the uncertainty budget. The instrument does not attempt to control position via feedback from the interferometers thereby allowing use of readily available commercial translation stages and controllers. Additional length errors which will be characterized and accounted for include: alignment errors, thermal expansion errors, and index of refraction of air.

2.3 Interfacial Force Microscope

The synthesis of novel nanomaterials and nanostuctured films has driven the need for new nanoscale characterization techniques. Nanomechanical properties such as elastic modulus and hardness define the ability of a thin film to resist wear and reduce friction. These properties are also important in nano-electro-mechanical systems (NEMS) and are increasingly explored in biological systems. Low-uncertainty quantitative measures of these parameters can be achieved only via instrumentation with traceable calibration. Influence parameters crucial to the uncertainty of nanoindentation measurements include: measured force sensor calibration, and z-displacement and tip area. Developments in instrumentation and technique aim to minimize the uncertainties in these parameters.

The development of SPM and related techniques in the past 20 years has opened pathways for the measurement of extremely small forces. Nanomechanical testing is growing area where the measurement of small forces yields properties such as adhesion, hardness and elastic modulus. Nanoindentation experiments are used to determine these nanomechanical properties and involve pushing a well-characterized tip with a given force into the test material while monitoring the distance it penetrates into the sample. The integrated cantilever-tip force sensing system of an AFM has been applied in mechanical force measurements; however, quantitative results are compromised by the compliant nature of the cantilever system. In the early 1990s, J. Houston and coworkers introduced the interfacial force microscope (IFM), which employs a microfabricated, mechanically stable, noncompliant and quantitative force sensor [9]. The sensor can provide quantitative adhesion [10] and indention force measurements. IFM has been used to characterize the mechanical properties of metal films [11], monomolecular films, polymers [12] and automotive antiwear films [13]. A program of IFM instrument development is underway at the National Institute for Nanotechnology (NINT).

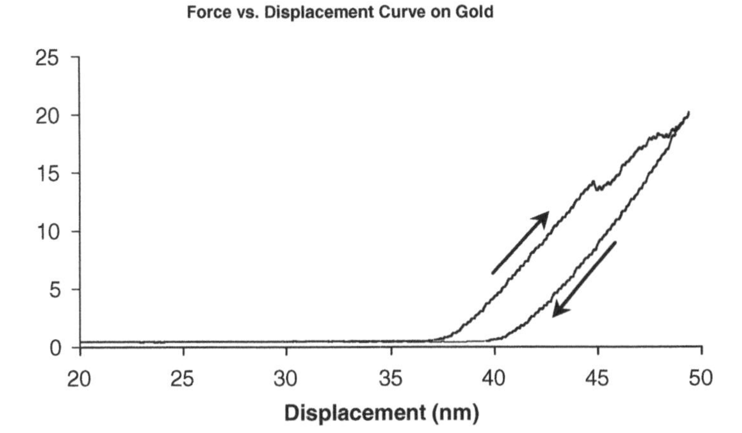

Fig. 4 Force-displacement curve of a Au(111) surface taken with an IFM

The output of an IFM nanoindentation experiment is a force-displacement curve such as that shown in Figure 4. These types of curves are analyzed by comparison to known models to obtain parameters such as hardness and elastic modulus. For quantitative results, traceable calibration and verification of force and displacement data are primary concerns. For example, in order to obtain values for hardness and elastic modulus with accuracy on the order of 5 %, force and displacement data must be accurate to 1 %, and must therefore be calibrated and verified with devices that are accurate at the level of 0.25 % to 0.5 % [14]. In addition, although the IFM provides high force sensitivity relative to commercial indentors, the technique still falls far short of the sensitivity of cantilever based AFM systems that can detect piconewton forces. With a goal of applying IFM to soft materials (e.g., biological), the first steps toward a lower force floor include changing of force detection from capacitance to interferometry allowing for a modified sensor design that will increase force sensitivity.

2.4 Characterization of Soft Materials

The NRC Steacie Institute for Molecular Sciences (NRC-SIMS) has a unique set of imaging tools that range from topographic imaging and force measurements to correlated optical/scanning probe microscopy and advanced linear and non-linear optical microscopies. Scanning probe microscopy expertise is focused on development of multi-modal methods that combine AFM and fluorescence imaging/spectroscopy and their application to studies of molecular interactions at membrane surfaces. Recently the group has demonstrated the capabilities of near-field scanning optical microscopy (NSOM) for the nanoscale visualization of membrane protein receptors on cell surfaces, establishing a powerful new approach for characterizing nanoscale signalling domains [15]. Other studies have

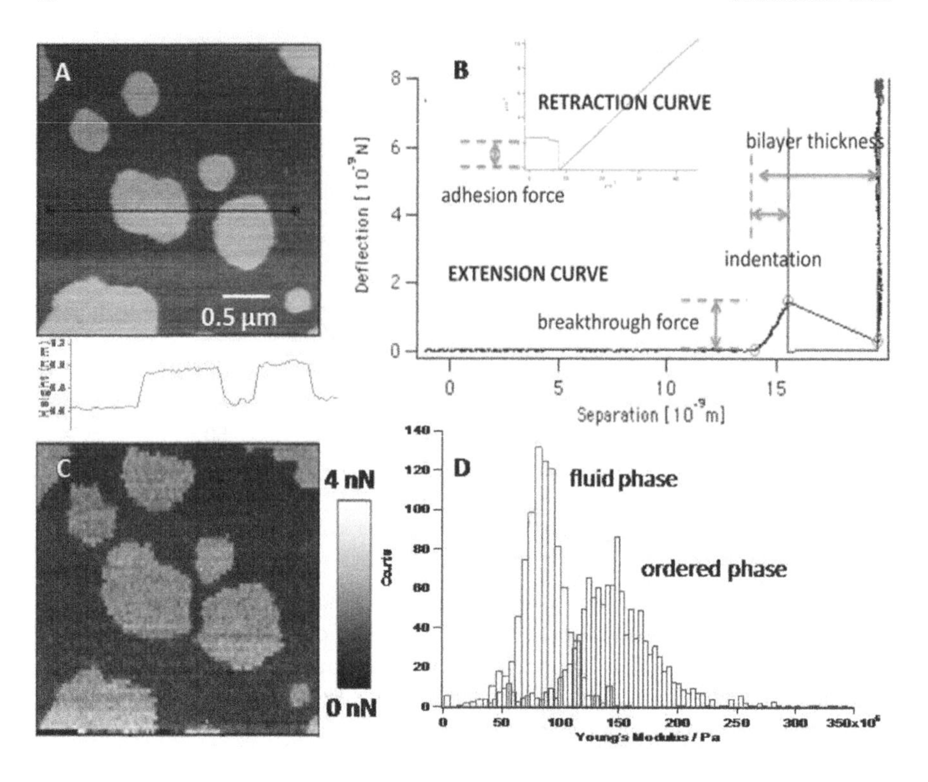

Fig. 5 AFM height image with a height profile (A), force curves (B), adhesion map (C), and histograms of the elastic moduli (D) measured on a lipid bilayer of DOPC/ Sphingomyelin/Cholesterol mixture in a 2:2:1 molar ratio

focused on examining the interaction of amyloid peptides with supported lipid bilayers and probing the enzyme-mediated restructuring of membrane raft domains using a combination of atomic force and fluorescence microscopies [16, 17].

Calibration standards are crucial elements of any measuring technique. Scanning Probe Microscopes (SPMs) are standard instruments at scientific and industrial laboratories and are one of the most suitable instruments for imaging, modification and manipulation of nano-objects and biological samples. Methods of measurement developed for conventional materials in many cases cannot be directly applied to nano-objects or biological systems. For such cases special protocols need to be developed. There are several reference artifacts available on the market that can be applied for calibration of SPMs, including NSOM and AFM. Since international standards for such artifacts are in development, and there is a current shortage of documentary standards, there is a need not only for measuring tools and techniques including calibration artifacts, but also for authoritative guidance on how to interpret measurement results from a practical perspective. The NRC research program includes development of artefact standards applicable for evaluation of mechanical properties such as elasticity and step height, as well as to accurately perform force measurements in the nanonewton range on soft biological samples. Supported lipid membranes that can be produced reproducibly from readily available materials are being examined as standard samples (Fig. 5). The intrinsic nanomechanical properties of the supported bilayers, such as breakthrough forces, elastic moduli, bilayer thicknesses, adhesion and correlation among these properties, were systematically characterized by AFM-based high resolution force mapping method.

3 Conclusions

This paper describes an NRC program of scientific research and development offering a unique opportunity for innovative design and fabrication of standards, development of *metrological* instrumentation and calibration procedures. Goals focus on length and force quantities; establishing a solid foundation for development of nanotechnologies which are expected to have powerful applications not only in physics and engineering, but also in chemistry and biology. Instrumentation and methods developed in this program will be validated via international inter-laboratory comparison experiments. Measurement comparisons offer the only opportunity for performance evaluation of measuring instruments at the highest metrological level, since instruments do not always perform to the specifications anticipated [18, 19]. They also uncover practical aspects of reference standard quality, longevity/durability, etc. Comparison results offer international peer review and validation of research-oriented instruments, and a quality check on production-based industrial instruments.

References

1. Koops, K.R., et al.: Calibration strategies for scanning probe metrology. Meas Sci. Technol. 18, 390–394 (2007)
2. Dai, G., Koenders, L., Pohlenz, F., Dziomba, T., Danzebrink, H.-U.: Accurate and traceable calibration of one-dimensional gratings. Meas Sci. Technol. 13, 1241–1249 (2005)
3. Misumi, I., Gonda, S., Huang, Q., Keem, T., Kurosawa, T., Fujii, A., Hisata, N., Yamagishi, T., Fujimoto, H., Enjoji, K., Aya, S., Sumitani, H.: Sub-hundred nanometre pitch measurements using an AFM with differential laser interferometers for designing usable lateral scales. Meas Sci. Technol. 16, 2080–2090 (2005)
4. Dixson, R., Orji, N.G., Fu, J., Cresswell, M., Allen, R., Guthrie, W.: Traceable atomic force microscope dimensional metrology at NIST. In: Proc. SPIE, vol. 6152, pp. 1–11 (2006)
5. Haycocks, J., Jackson, K.: Traceable calibration of transfer standards for scanning probe microscopy. Prec. Eng. 29, 168–175 (2005)
6. Holmes, M., Hocken, R., Trumper, D.: The long-range scanning stage: a novel platform for scanned-probe microscopy. Prec. Eng. 24, 191–209 (2000)
7. Meli, F., Thalmann, R.: Long-range AFM profiler used for accurate pitch measurements. Meas Sci. Technol. 9, 1087–1092 (1998)
8. Eves, B.J.: Design of a large measurement-volume metrological AFM. Meas Sci. Technol. (2009) (to be published)

9. Joyce, S.A., Houston, J.E.: A new force sensor incorporating force-feedback control for interfacial force microscopy. Rev. Sci. Instrum. 62, 710–715 (1991)
10. Thomas, R.C., et al.: Probing adhesion forces at the molecular scale. J. Am. Chem. Soc. 117, 3830–3834 (1995)
11. Tangyunyong, P., et al.: Nanometer-scale mechanics of gold-films. Phys. Rev. Lett. 71, 3319–3322 (1993)
12. Graham, et al.: Quantitative nanoscale mechanical properties of a phase segregated homopolymer surface. J. Mater. Res. 13, 3565–3570 (1998)
13. Warren, et al.: Nanomechanical properties of ZDDP films by interfacial force microscopy. A Abstr. Pap. Am. Chem. Soc. 213, 211–COLL (1997)
14. Pratt, J.R., Kramer, Newell, D., Smith, D.T.: Review of SI traceable force metrology for instrumented indentation and atomic force microscopy. Meas. Sci. Technol. 16, 2129–2137 (2005)
15. Ianoul, A., et al.: Imaging nanometer domains of beta-adrenergic receptor complexes on the surface of cardiac myocytes. Nat. Chem. Biol. 1, 196–202 (2005)
16. Ira, Johnston, L.J.: Ceramide Promotes Restructuring of Model Raft Membranes. Langmuir 22, 11284–11289 (2006)
17. Johnston, L.J.: Nanoscale imaging of domains in supported lipid membranes. Langmuir 23, 5886–5895 (2007)
18. Meli, F.: Nano 4 Final Report, 34 p. (2000), `http://kcdb.bipm.org/appendixB/`
19. Breil, R., et al.: Intercomparison of scanning probe microscopes. Prec. Eng. 26, 296–305 (2002)

Concrete Nanoscience and Nanotechnology: Definitions and Applications

E.J. Garboczi

Abstract. There are many improvements needed in concrete, especially for use in renewal and expansion of the world's infrastructure. Nanomodification can help solve many of these problems. However, concrete has been slow to catch on to the nanotechnology revolution. There are several reasons for this lag in the nanoscience and nanotechnology of concrete (NNC). First is the lack of a complete basic understanding of chemical and physical mechanisms and structure at the nanometer length scale. Another reason is the lack of a broad understanding of what nanomodification means to concrete, which is a liquid-solid composite. NNC ideas need to profit from, but not be bound by, experience with other materials. As an illustration of these ideas, a specific application will be given of using nanosize molecules in solution to affect the viscosity of the concrete pore solution so that ionic diffusion is slowed. A molecular-based understanding would help move this project towards true nanotechnology. A final section of this paper lists some possibly fruitful focus areas for the nanoscience and nanotechnology of concrete.

1 Introduction - The Need for Research

There are many improvements needed in concrete, especially for use in renewal and expansion of the world's infrastructure, e.g. increased durability, decreased brittleness and increased tensile strength, and routine use of large volumes of non-traditional materials like fly ash. Nanomodification can probably help solve many of these problems. However, concrete has been slow to catch on to the revolution in the nanotechnology that is ongoing in many materials. There are several possible reasons for this lag in the nanoscience and nanotechnology of concrete (NNC).

The first reason is a lack of a basic understanding of chemical and physical mechanisms and structure at the nanometer length scale, without which any attempted modifications at this length scale are only empirically-based and cannot

E.J. Garboczi
National Institute of Standards and Technology
e-mail: edward.garboczi@nist.gov
http://ciks.cbt.nist.gov/monograph

be fully successful. Greater use needs to be made of advances in instrumentation from other fields to help characterize concrete at the nano-scale. In conjunction with this experimental need is the need for improved modeling of concrete at the nanoscale.

To make this point crystal clear, imagine someone who knew absolutely nothing of cement, of chemistry, and of other chemicals being given the task of finding a new chemical admixture that will improve property X of concrete. He doesn't even know how to make concrete, much less apply chemistry to it or measure a property. The cement and concrete industrial and scientific community is not exactly in this situation when it comes to NNC, of course, but perhaps we are not as far from it as we might wish.

Another reason for the lag in NNC is the lack of a broad understanding of what nanomodification means to concrete, which is a liquid-solid composite. Concrete is a material that is quite different from many other materials, and so NNC ideas need to profit from, but not be bound by, experience with other materials. Simply trying to duplicate advances made in other materials without adaptation to concrete is hampering the field of NNC. Further details will be given in subsequent sections of this paper.

As an illustration of these ideas, a specific application will be given of using nano-size molecules in solution to affect the viscosity of the concrete pore solution so that ionic diffusion is slowed, thus increasing service life under diffusion-controlled scenarios (e.g. chloride penetration). The success of this application is an example of what can be done based on the understanding gained by years of fundamental research on transport in concrete pore solution. But a molecular-level understanding, which would lead to more intelligent material selection, is still elusive. A final section of this paper lists some possibly fruitful focus areas for the nanoscience and nanotechnology of concrete.

This is not a review paper, so references will be limited to just a few examples to illustrate points. In order to review the nanoscience of concrete, one would have to synthesize what is known about cement paste at the nanoscale, which would be a large task. To review the nanotechnology of concrete, which is defined here as "using understanding of concrete nanoscience to affect the technology of concrete at the nanolevel and so improve macroscale properties," would unfortunately be a lot easier to review. A fairly recent and useful review exists, as well as a summary of some of the nanoscience of concrete work to date [1,2].

2 The Unique Material - Concrete

To simply speak of "nanomodification" of a material doesn't convey much information. The microstructure, much less the nanostructure, of materials differ widely. The typical nanotechnology application that one hears about in other materials is that of inserting nanosized solid particles into a solid matrix. However, one must broaden this understanding of what nanomodification means when one thinks of concrete.

Concrete is clearly a porous liquid-solid composite, which is quite different from almost all materials used in human technology. The cement paste binder includes water at the nanometer and higher length scales. If any of the aggregates is porous, then they may contain water as well. So for concrete, there are three places where nanomodification can be carried out: in any of the solid phases, in any of the liquid phases, or at any of the water:solid interfaces (see Figure 1). For instance, nano-size additives can be solid particles added to the solid phase, or nano-size molecules added in solution to the liquid phase that can affect either the liquid phase itself or the liquid-solid interface. Since concrete is a material that is quite different from many other materials, NNC ideas should profit from, but not be bound by, experience with other materials.

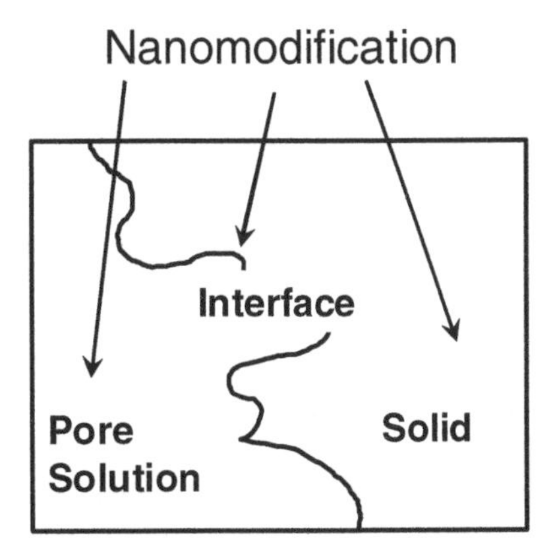

Fig. 1 Schematic of concrete showing liquid-solid composite nature (no length scale is given)

In some sense, polymeric chemical admixtures in concrete, which have been used for decades, should be considered as NNC. For example, polymers that affect the nanoscale interaction forces between cement particles are nanosized molecules that are affecting nanoscale behavior by being added to the liquid phase. Shrinkage reducing admixtures affect the liquid-solid interface by reducing the surface tension in water:air menisci.

But what new types of nanoscale particles can be added to concrete? A fruitful area could be inorganic nanotubes [3,4]. Just adding carbon nanotubes does not seem to have a great effect on concrete properties, except perhaps for electrical conductivity, perhaps because the carbon nature of the nanotubes does not allow them to bond significantly with the calcium-silicate-hydrate (C-S-H) phase. But inorganic nanotubes, made of the correct material to interact strongly with the C-S-H phase, could lead to nanoreinforcement and other such, probably desirable, outcomes. Even carbon nanotubes, if they are appropriately functionalized [5], could also play a role. The point is that any nanoparticles that will be added to concrete must take into account the unique chemistry and physics of concrete.

3 Advances in Material Characterization to Be Used

The list of modern experimental techniques that can give information at the nanoscale include atomic force microscopy (AFM) [6,7], transmission electron microscopy [8,9], nanotomography via dual ion focused beams [10], mechanical nanoindentation [11-13], although this gives information more on the hundreds of nanometer length scale, small-angle neutron scattering [14-16], nuclear resonance experiments [17], and synchrotron x-ray diffraction and other techniques, which are discussed in a recent review [18]. These references all list applications of these techniques to cement and concrete, and there are many more such references. The nanoscience understanding of cement and concrete is developing fairly rapidly. However, these techniques must be employed more routinely and in conjunction with each other, since no one experiment can give all the information necessary and desirable on which to base applications and models.

4 Advances in Models Needed

The experimental techniques of the last section rarely give a complete description of a system, so they must be coupled to computation to develop a self-consistent view. This will require nanoscale and molecular scale models for concrete, which are in their infancy. Not only do such models need to be developed and adapted to cement needs, there also needs to be advances made in the typical kinds of model used (e.g. molecular dynamics [19,20], ab initio [21]) so that they can be usefully employed on the randomness that is encountered in concrete at the nanoscale. The kinds of models that are typically used at this scale have been designed for relatively simple, small systems. To handle the complexities of cement paste, a systematic attempt needs to be made to use greater computer power (e.g. larger numbers of parallel processors) to be able to handle the larger numbers of types of atoms and molecules in these complex materials, the ubiquitity of water, the greater absolute number of atoms and molecules needed to be modeled, molecular reactions (an enormous complication over mere simulation of the movement of atoms and molecules), and the longer times that need to be modeled. The National Institute of Standards and Technology (NIST) has made strides in using massively parallel computers to help model concrete problems like rheology, where the large size range of particles on the micrometer and millimeter scale dictates the computational power needed [22], but not at the nanoscale. But even in the rheology problem, to go from 10 or 20 processors to 500 or 1000 processors required non-trivial changes in the code [22]. Similarly, there are public versions of molecular dynamics codes that are parallel [23], but to go to massively parallel versions will also require new computational structures.

To bring 3-D nanostructure simultaneously into such models, and not just thermodynamics and kinetics, something like HydratiCA [24,25], a 3-D model for simulating diffusion, reaction, and 3-D growth at the micrometer scale, also needs to be developed at the nanoscale. This is an achievable task. Although the Virtual

Cement and Concrete Testing Laboratory (VCCTL) at NIST [26] does not model concrete at the nanoscale but only currently at the micrometer to millimeter length scales, nanoscale models could be incorporated (not easily, but readily) into the computational structure, directly linking nanoscale models with macro properties via multi-scale models.

5 An Example - Doubling the Service Life of Concrete

Past attempts to reduce the chloride diffusivity of concrete have focused on producing denser, less porous concretes. Unfortunately, these concrete formulations have a greater tendency to crack. A different approach was taken at NIST. Rather than change the size and density of the pores in concrete by working on the solid phase, a method was sought to change the *viscosity* of the pore solution in the concrete, using a soluble additive, to reduce the speed at which chloride ions and sulfates diffuse through the pore solution. Continuum transport equations that have been worked on for many years show that the speed of ionic diffusion through the pore solution is inversely proportional to viscosity of the pore solution [27]. These equations, however, do not give any guidance about molecular details.

In this research project, it was quickly found that the molecular size of the additive was critical for it to serve as a diffusion barrier. Larger molecules such as cellulose ether and xanthum gum increased viscosity greatly, but did not cut diffusion rates. Smaller molecules – of only a few nanometers in size – slowed ion diffusion. Several molecules of about the correct size were found to work, though some similar ones did not work, but the lack of molecular models do not allow predictivity so that any new such molecules still have to be found by trial and error. Such additives can be effective by being added directly to the concrete mixture as are current chemical admixtures, but even better performance is achieved when the additives are mixed into the concrete by saturating absorptive lightweight sand [28,29].

6 Some Possibly Fruitful Areas for NNC

Some areas in which concrete nanoscience could spawn some useful applications of nanotechnology are listed and briefly discussed below.

In autogeneous shrinkage, often a cause of early-age cracking, the reaction of cement and cementitious materials with water and each other induces chemical shrinkage because the reaction products take up less space than reactants. Chemical shrinkage results in air-water menisci (self-desiccation), especially in dense mixtures, where the excess water needed cannot get in quickly enough from the outside. These menisci, which are under tension, induce tensile stresses that can cause cracking. It has been shown [30] that shrinkage-reducing admixtures, among other chemicals, affect surface tension via nanometer-scale rearrangement of molecules. Understanding this phenomenon better at the nanometer-scale could lead to better ways of addressing autogeneous shrinkage, which is not a solved problem by any means.

The rheology of concrete is controlled by the rheology of cement paste, and the rheology of cement paste, while strongly dependent on parameters like particle size distribution, particle shape, and volume fraction, is also influenced strongly at the nanometer scale via interparticle forces. If these forces were understood better, the rheology could be controlled better via nanoscale chemical admixtures that were designed for their job, not just found by trial and error.

Another NNC area that has been exploited empirically but would greatly benefit from nanoscience understanding is the area of using chemical admixtures to affect the reactivity of cement particle surfaces. This is an NNC area, since the layer of dissolution/reaction at early ages is of the order of nanometers in size. Models and experiments at this level working together can lead the way to being able to design molecules for various purposes.

The whole area of concrete degradation via alkali-silicate reaction or sulfate attack, for example, involves ionic transport via the high ionic strength and high pH concrete pore solution, with surface complexation, followed by reaction, microstructure degradation that changes transport properties, and so on. Many key steps of these processes are at the nanoscale, and should be fruitful for nanotechnological applications once further nanoscience understanding is achieved.

Finally, probably the most important application of nanoscience and nanotechnology in concrete is affecting the C-S-H "glue" that holds concrete together. This is a nano-porous, nanostructured material that controls almost all the mechanical and transport properties of concrete. For example, the only reason concrete is viscoelastic is because C-S-H is viscoelastic. The mechanical properties of C-S-H set the scale for cement paste and therefore concrete elastic moduli and strength. Marrying inorganic nanotubes and/or particles with profound knowledge of C-S-H nanostructure, obtained from experiments that probe and models that accurately simulate this length scale should bring about significant NNC advancements, e.g. nanoscale tensile reinforcement and toughening. One should note, however, that the whole problem of adding nano-sized particles, reactive or otherwise, and their potential benefit, involves dispersing them properly. Only then does their reactivity really matter.

7 Summary and Conclusions

It is hopefully clear from this paper that one cannot just take a "practical engineering" approach to NNC, pull nanoparticles from other materials and throw them in the mix and hope for a miracle. There are unique aspects of concrete that must be taken into account in order to have successful NNC. Nanoscale research is needed to provide profound knowledge of the nanoscale to guide applications, since we have no intuitive "feel" for this length scale. One needs to let research knowledge be the guide to intelligently design concrete from the nanoscale up.

The NNC of concrete is well underway. There have been two previous NICOM conferences: NICOM 1 at Paisley and NICOM 2 at Bilbao, a US National Science Foundation-funded workshop on the nanomodification of concrete at the University of Florida in 2006, a workshop in March, 2009 in Spain focusing on nanoscale modeling, an upcoming (May, 2010) US Transportation Research Board (TRB) international conference entitled "Nanotechnology in Cement and Concrete" sponsored by the TRB task force on the nanotechnology of concrete, and undoubtedly there have been some other workshops and conference sessions that have been missed in this brief list. Note that the annual Computer Modeling Workshop held each summer at NIST has also touched upon nanoscale experiments and modeling (the 20th in the series is coming up in August, 2009, see http://ciks.cbt.nist.gov/monograph). The Nanocem consortium [31] in Europe has also made some strides in NNC research. Overall, the field of NNC needs to mature, to focus, to develop, and to make progress using experiments and models together in a synergistic way.

References

1. Sobolev, K., Gutiérrez, M.F.: How nanotechnology can change the concrete world. Am. Ceram. Soc. Bull. 84, 14–18, 16–20 (2005)
2. Sobolev, K., Shah, S.P., ACI Committee (eds.): SP-254 Nanotechnology of Concrete: Recent Developments and Future Perspectives, vol. 236. American Concrete Institute, Detroit (2008)
3. Halford, B.: Inorganic Menagerie 83, 30–33 (2005)
4. Tenne, R.: Inorganic nanotubes and fullerene-like nanoparticles. Nat. Nanotechnol. 1, 103–111 (2006)
5. Ciraci, S., Dag, S., Yildirim, T., Gülseren, O., Senger, R.T.: Functionalized carbon nanotubes and device applications. J. Phys-Condens. Mat. 16, R901–R960 (2004)
6. Sáez de Ibarra, Y., Gaitero, J.J., Erkizia, E., Campillo, I.: Atomic force microscopy and nanoindentation of cement pastes with nanotube dispersions. In: Special Issue: Trends in Nanotechnology (TNT 2005), vol. 203, pp. 1076–1081 (2006)
7. Hall, C., Bosbach, D.: Scanning Probe Microscopy: A New View of the Mineral Surface. In: Skalny, J., Mindess, S. (eds.) Materials Science of Concrete VI, pp. 101–128. American Ceramic Society, Westerville (2001)
8. Love, C.A., Richardson, I.G., Brough, A.R.: Composition and structure of C–S–H in white Portland cement – 20% metakaolin pastes hydrated at 25°C. Cement Concrete Res. 37, 109–117 (2007)
9. Zhang, X., Chang, W., Zhang, T., Ong, C.K.: Nanostructure of Calcium Silicate Hydrate gels in cement paste. J. Am. Ceram. Soc. 83, 2600–2604 (2004)
10. Holzer, L., Muench, B., Wegmann, M., Gasser, P., Flatt, R.J.: FIB-nanotomography of particulate systems-Part I: Particle shape and topology of interfaces. J. Am. Ceram. Soc. 89, 2577–2585 (2006)
11. Constantinides, G., Ulm, F.-J.: The effect of two types of C-S-H on the elasticity of cement-based materials: Results from nanoindentation and micromechanical modeling. Cement Concrete Res. 34, 67–80 (2004)
12. Mondal, P., Shah, S.P., Marks, L.: A reliable technique to determine the local mechanical properties at the nanoscale for cementitious materials. Cement Concrete Res. 37, 1440–1444 (2007)

13. Hughes, J.J., Trtik, P.: Micro-mechanical properties of cement paste measured by depth-sensing nanoindentation: A preliminary correlation of physical properties with phase type. Mater. Charact. 53, 223–231 (2004)
14. Jennings, H.M., Thomas, J.J., Gevrenov, J.S., Constantinides, G., Ulm, F.-J.: A multi-technique investigation of the nanoporosity of cement paste. Cement Concrete Res. 37, 329–336 (2007)
15. Nemes, N.M., Neumann, D.A., Livingston, R.A.: States of water in hydrated C_3S (tricalcium silicate) as a function of relative humidity. J. Mater. Res. 21, 2516–2523 (2006)
16. Allen, A.J., Thomas, J.J., Jennings, H.M.: Composition and density of nanoscale calcium–silicate–hydrate in cement. Nat. Mater. 6, 311–316 (2007)
17. Schweitzer, J.S., Livingston, R.A., Rolfs, C., Becker, H.-W., Kubsky, S., Spillane, T., Castellote, M., de Viedm, P.G.: Nanoscale studies of cement chemistry with 15N resonance reaction analysis. Nucl. Instrum. Meth. B 241, 441–445 (2005)
18. Skibsteda, J., Hall, C.: Characterization of cement minerals, cements and their reaction products at the atomic and nano scales. Cement Concrete Res. 38, 205–225 (2008)
19. Dolado, J.S., Griebel, M., Hamaekers, J.: A molecular dynamic study of cementitious Calcium Silicate Hydrate (C–S–H) gels. J. Am. Ceram. Soc. 90, 3938–3942 (2007)
20. Kalinichev, A.G., Wang, J., Kirkpatrick, R.J.: Molecular dynamics modeling of the structure, dynamics and energetics of mineral–water interfaces: Application to cement materials. Cement Concrete Res. 37, 337–347 (2007)
21. Pellenq, R.J.-M., Lequeux, N., van Damme, H.: Engineering the bonding scheme in C–S–H: The iono-covalent framework. Cement Concrete 38, 159–174 (2008)
22. Martys, N.S., Lootens, D., George, W.L., Satterfield, S.G., Hebraud, P.: Spatial-Temporal Correlations at the Onset of Flow in Concentrated Suspensions. In: Co, A., Leal, L.G., Colby, R.H., Giacomin, A.J. (eds.) The XVth International Congress On Rheology. Monterey CA, AIP Conf. Proc., vol. 1027, pp. 207–209 (2008)
23. LAMMPS, http://lammps.sandia.gov/
24. Bullard, J.W.: A determination of hydration mechanisms for tricalcium silicate using a kinetic cellular automaton model. J. Am. Ceram. Soc. 91, 2088–2097 (2008)
25. Bullard, J.W.: A three-dimensional microstructural model of reactions and transport in aqueous mineral systems. Model Simul. Mater. Sc. 15, 711–738 (2007)
26. Bentz, D.P., Garboczi, E.J., Bullard, J.W., Ferraris, C.F., Martys, N.S.: Virtual testing of cement and concrete. In: Lamond, J., Pielert, J. (eds.) Significance of tests and properties of concrete and concrete-making materials. ASTM STP 169D (2006)
27. Snyder, K.A., Marchand, J.: Effect of speciation on the apparent diffusion coefficient in nonreactive porous systems. Cement Concrete Res. 31, 1837–1845 (2001)
28. Bentz, D.P., Snyder, K.A., Cass, L.C., Peltz, M.A.: Doubling the service life of concrete structures. I: Reducing ion mobility using nanoscale viscosity modifiers. Cement Concrete Comp. 30, 674–678 (2008)
29. Bentz, D.P., Peltz, M.A., Snyder, K.A., Davis, J.M.: VERDiCT: viscosity enhancers reducing diffusion in concrete technology. Concrete Int. 31, 31–36 (2009)
30. Bentz, D.P., Jensen, O.M.: Mitigation strategies for autogenous shrinkage cracking. Cement Concrete Comp. 26, 677–685 (2004)
31. The Industrial-Academic Research Network on Cement and Concrete, http://www.nanocem.org

Continuum Microviscoelasticity Model for Cementitious Materials: Upscaling Technique and First Experimental Validation

S. Scheiner and C. Hellmich

Abstract. We propose a micromechanics model for aging basic creep of early-age concrete. Therefore, we formulate viscoelastic boundary value problems on two representative volume elements (RVEs), one related to cement paste (composed of cement, water, hydrates, air), and one related to concrete (composed of cement paste and aggregates). Homogenization of the non-aging elastic and viscoelastic properties of the material's constituents involves the transformation of the aforementioned viscoelastic boundary value problems to the Laplace-Carson (LC) domain. There, formally elastic, classical self-consistent and Mori-Tanaka solutions are employed, leading to pointwisely defined LC-transformed tensorial creep and relaxation functions. Subsequently, the latter are backtransformed, by means of the Gaver-Wynn-Rho algorithm, into the time domain. Temporal derivatives of corresponding homogenized creep tensors, evaluated for the current maturation state of the material and for the current time period since loading of the hydrating composite material, allow for micromechanical prediction of the aging basic creep properties of early-age concrete.

1 Fundamentals of Continuum Micromechanics

In continuum micromechanics [10, 16], a material is understood as a microheterogeneous body filling a macrohomogeneous representative volume element (RVE) with characteristic length l_{II}, $l_{II} >> d_{II}$, d_{II} standing for the characteristic length of inhomogeneities within the RVE, see Figure 1. These inhomogeneities are referred to as material phases, each exhibiting a homogeneous microstructure. The homogenized mechanical behavior of the material on the observation scale of the

---

S. Scheiner and C. Hellmich
Institute for Mechanics of Materials and Structures, Vienna University of Technology, Vienna, Austria
e-mail: Stefan.Scheiner@tuwien.ac.at,
Christian.Hellmich@tuwien.ac.at
www.imws.tuwien.ac.at

RVE, i.e. the relation between homogeneous deformations acting on the boundary of the RVE and resulting macroscopic (average) stresses, can then be estimated from the mechanical behavior of the material phases, their dosages within the RVE, their characteristic shapes, and their interactions. If a single material phase possesses a heterogeneous microstructure itself, its mechanical behavior can be estimated by introduction of RVEs within this phase, with characteristic lengths l_I, $l_I \leq d_{II}$, comprising again inhomogeneities with characteristic length $d_I << l_I$, and so on, see Fig. 1. Such an approach is referred to as multistep homogenization and should, in the end, provide access to "universal" phase properties at sufficiently low observation scales.

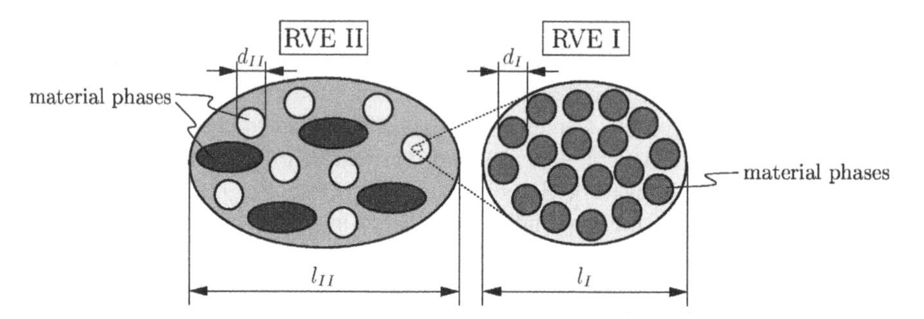

Fig. 1 Separation of scales for multistep homogenization by means of continuum micromechanics, $d_I << l_I \leq d_{II} << l_{II}$ [8]

2 Continuum Microviscoelasticity Model for Concrete

Explanation of the mechanical behavior of concrete by means of continuum micromechanics requires resolution of the material microstructure down to the observation scale of at most 1 μm, where calcium silicate hydrates (CSH), reaction products of cement grains and water, can be discerned [11]. Thereby, either distinction between high-density and low-density CSH [15] as well as portlandite and aluminate is made [5], or, as implemented subsequently, hydration products are introduced as one material phase [9]. Consequently, at the observation scale of cement paste, a typical RVE comprises the phases "hydration products", "unhydrated cement grains", "liquid (capillary) pores", and "air pores (voids)" [5, 9]. Furthermore, at the observation scale of concrete, a typical RVE comprises the phases "cement paste" and "aggregates" [5, 9]. Thus, a two-step homogenization strategy is pursued. In the first homogenization step the mechanical behavior of cement paste is determined under the assumption that, due to the disorder of the hydrate phase, cement paste is reasonably represented as a polycrystal [5, 9], i.e. no dominant material phase can be identified within the RVE of cement paste. Thereby, CSH, unhydrated cement grains, water, and air pores are represented as isotropic spherical inclusions. In the line of Laws and McLaughlin [13], the homogenized mechanical properties of cement paste are obtained by means of the viscoelastic correspondence principle. In detail, the transformation of the viscoelastic constitutive law describing the mechanical behavior of single constituents from the time domain into the Laplace-Carson (LC) domain yields a mathematical structure which is formally identical to the (corresponding) purely elastic case. Thus, according to the correspondence principle, the LC-transformed material properties of cement paste can be homogenized analogously to purely elastic properties [5,9]. In order to finally obtain material properties in the time domain, the LC-transformed homogenized material properties must be back-transformed. This is carried out numerically by means of the Gaver-Wynn-Rho (GWR) algorithm [1]. In the second homogenization step, in turn, isotropic spherical aggregate inclusions are embedded in the polycrystalline cement paste matrix. Again, the LC-transformed material properties of concrete must be numerically backtransformed into the time domain. Furthermore, the microviscoelasticity theory (for details see [14]) considers the experimentally verified hypothesis that CSH is the only viscoelastic constituent of concrete (the remaining constituents are purely elastic) [2], described by the well-known Burgers model [6], whereas this phenomenon can be considered as purely deviatoric [4].

3 Model Validation

The microviscoelasticity model is experimentally validated on the basis of the creep tests of Laplante [12] and of Atrushi [3], both carried out on sealed concrete specimens. Laplante subjected concrete at the ages of 20 hours, 27 hours, 3 days, 7 days, and 28 days to uniaxial compressive loading. Thus, since at these ages

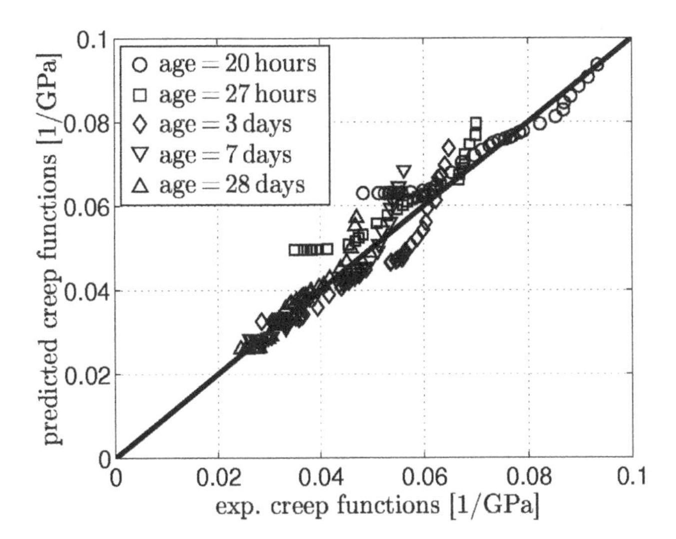

Fig. 2 Model-predicted aging creep functions compared to corresponding experimentally obtained ones [12], r^2_{mean}=91%

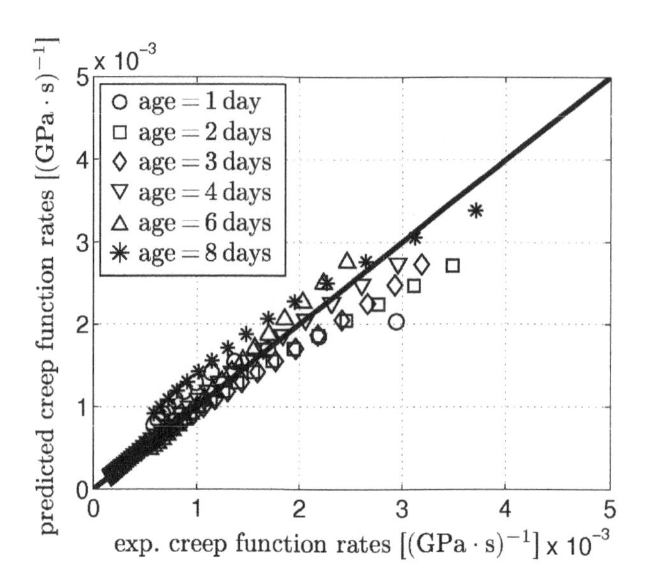

Fig. 3 Model-predicted aging creep function rates compared to corresponding experimentally obtained ones [3], r^2_{mean}=96%

concrete still undergoes hydration, related aging of concrete, leading to a continuously changing microstructure, must be considered. Evaluating the microviscoelasticity model in terms of the aging creep function allows for comparison of model predictions with corresponding experimental values, see Figure 2. The satisfying agreement of model-predicted with experimentally obtained creep functions at different age (r^2_{mean} = 91%) allows for concluding that the presented microviscoelasticity model is capable to predict the aging viscoelastic behavior of cementitious materials. This is further corroborated by comparison of the model-predicted creep function rates with corresponding experimental results of Atrushi [3], who subjected concrete to uniaxial compressive loading at the ages of 1 day, 2 days, 3 days, 4 days, 6 days, and 8 days (r^2_{mean} = 96%), see Figure 3.

Acknowledgments. Financial support by "TUNCONSTRUCT – Technology Innovation in Underground Construction" (IP011817-2), sponsored by the European Commission, is gratefully acknowledged.

References

1. Abate, J., Valkó, P.P.: Multi-precision Laplace transform inversion. Int. J. Numer. Meth. Eng. 60, 979–993 (2004)
2. Acker, P.: Micromechanical analysis of creep and shrinkage mechanisms. In: Ulm, F.-J., Bažant, Z.P., Wittmann, F.H. (eds.) Concreep 6: Creep, Shrinkage and Durability of Concrete and Concrete Structures, Cambridge, USA, pp. 15–25. Elsevier Science Ltd., Amsterdam (2001)

3. Atrushi, D.: Tensile and Compressive Creep of Early-Age Concrete: Testing and Modelling. PhD thesis, Norwegian University of Science and Technology, Trondheim, Norway (2003)
4. Bernard, O., Ulm, F.-J., Germaine, J.T.: Volume and deviator creep of calcium-leached cement-based materials. Cement Concrete Res. 33, 1127–1136 (2003)
5. Bernard, O., Ulm, F.-J., Lemarchand, E.: A multiscale micromechanics-hydration model for the early-age elastic properties of cement-based materials. Cement Concrete Res. 33, 1293–1309 (2003)
6. Burgers, J.M.: First report on viscosity and plasticity. Report, Amsterdam (1935)
7. Dormieux, L.: Applied Micromechanics of Porous Media. CISM Courses and Lectures, vol. 480. Springer, Wien (2005)
8. Hellmich, C.: Microelasticity of Bone. In: [7], ch. 8, pp. 289–331. Springer, Wien (2005)
9. Hellmich, C., Mang, H.A.: Shotcrete elasticity revisited in the framework of continuum micromechanics: from submicron to meter level. J. Mater. Civil Eng-ASCE 17, 246–256 (2005)
10. Hill, R.: Continuum micro-mechanics of elastoplastic polycrystals. J. Mech. Phys. Solids 13(2), 89–101 (1965)
11. Hua, C., Ehrlacher, A., Acker, P.: Analyses and models of the autogeneous shrinkage of hardening cement paste – II. Modelling at scale of hydrating grains. Cement Concrete Res. 27(2), 245–258 (1997)
12. Laplante, P.: Propriétés mécaniques de bétons durcissants: analyse comparée des bétons classique et à trés hautes performances [Mechanical properties of hardening concrete: a comparative analysis of ordinary and high performance concretes]. PhD thesis, Ecole Nationale des Ponts et Chaussées, Paris, France (1993) (in French)
13. Laws, N., McLaughlin, R.: Self-consistent estimates for the viscoelastic creep compliances of composite materials. P. Roy. Soc. A-Math Phy. 359, 251–273 (1978)
14. Scheiner, S., Hellmich, C.: Continuum microviscoelasticity model for aging basic creep of early-age concrete. J. Eng. Mech-ASCE (accepted for publication) (2008)
15. Thomas, J.J., Jennings, H.M., Allen, A.J.: The surface area of cement paste as measured by neutron scattering: evidence for two C-S-H morphologies. Cement Concrete Res. 28(6), 897–905 (1998)
16. Zaoui, A.: Continuum micromechanics: survey. J. Eng. Mech-ASCE 128(8), 808–816 (2002)

Production, Properties and End-Uses of Nanofibres

O. Jirsák and T.A. Dao

Abstract. Nanofibers are produces from organic and inorganic polymers via electrospnning technology. Both polymer solutions and polymer melts can be electrospun. Fiber diameters of 50 to 500 nanometers are typical. An industrial – scale production method of nanofibres production has been developed. Small fibre diameters and great specific surface are the main specific properties of nanofiber assemblies, namely nanofiber layers. Number of specific end-uses of nanofibres have been developed such as filters, sound absorbing materials, wound dressings, scaffolds for tissue engineering etc. Machinery for nanofibres production is produced and offered in the Czech Republic. There is a great potential for utilization of nanofibres in civil engineering.

1 Introduction

Although the idea of nanofibres production is rather old [1], technical interest in this form of material started in 1970's. Nevertheless, scientific and technical activities of a great extent appeared after the year 2000. Number of papers and patents on nanofibres is shown in Fig. 1.

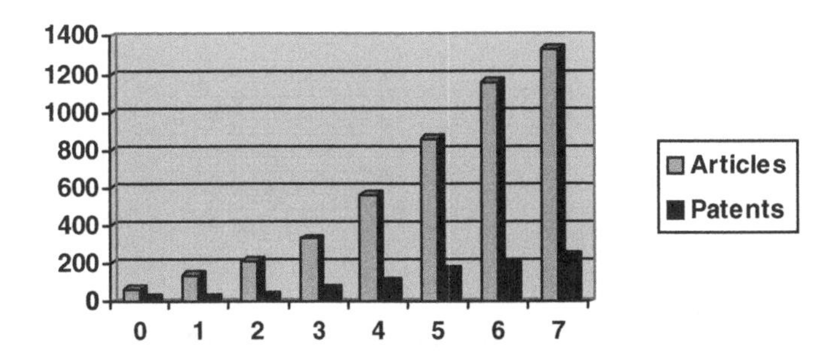

Fig. 1 Number of papers and patents on nanofibers in the years 2000 – 2007

O. Jirsák and T.A. Dao
Technical University of Liberec, Liberec, Czech Republic
e-mail: oldrich.jirsak@tul.cz

Beside papers and patents, some books on nanofibers appeared recently [1-4].

Nanofibers are one of three main types of nanomaterials, beside nanoparticles and nanosurfaces. There are several laboratory methods of laboratory preparation of nanofibres [2], nevertheless the electrospinning method is the most common both in laboratory and industry. A laboratory electrospinning method based on a syringe is shown in Fig. 2.

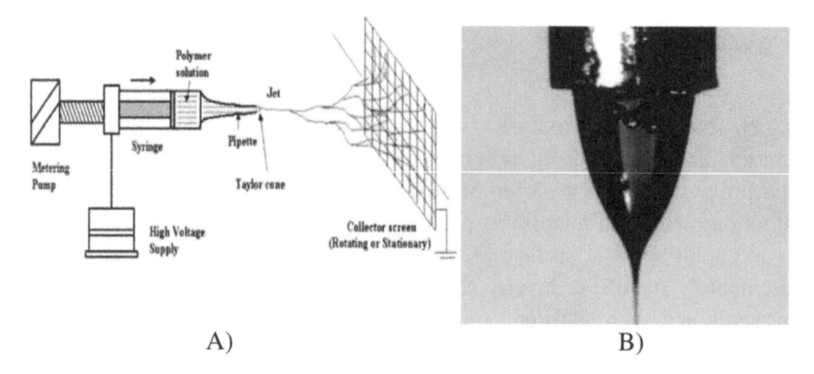

Fig. 2 A laboratory method of nanofibres preparation (A) and a detail of Taylor cone (B)

The device shown in Fig. 2 is used in many laboratories. It is not suitable for industrial purposes for its low production rate, typically 0.1 to 2 gramms of polymer per hour.

An industrial method was developed [5] based on the roller as the spinning electrode (Fig. 3).

Fig. 3 Roller electrospinning principle

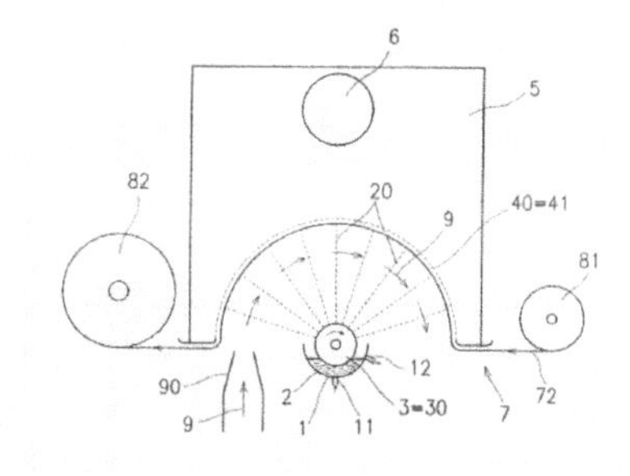

In the roller spinner, the rotating roller (3) is immersed in a polymer solution-which creates a layer on the roller surface. Thousands of Taylor cones are present on the surface of the roller due to high voltage and between the roller and the collector electrode (40). The nanofibres are collected on the textile backing layer which is moving along collector electrode. Thus, a nanofibre layer is produced continuously and the production rate is high, depending on width of the machine, number of spinning rollers and required area weight of nanofibre layer.

2 Properties of Nanofibres and Nanofibre Layers

Typical properties of nanofibres with comparison with conventional textile fibres and special, extremely fine "melt blown" fibres are shown in Table 1.

Table 1 Typical dimensions of conventional fibres, melt-blown fibres and nanofibres

Fibres	Fibre diameter (μ m)	Linear density (dtex)	Specific surface (m^2/g)
Conventional	10-40	1-30	ca. 0.2
Melt-blown	1-5	ca. 0.01	ca. 2
Nanofibres	0.05–0.5	ca 0.0001	ca. 20

Nanofibres are produced from a variety of organic and inorganic polymers such as polyvinylalcohol, polyamides, polyurethanes, polyimides, polystyrene, p-HEMA, chitosan, co-polymers, polymers containing a variety of additives, silica and many others.

Nanofibre formations of various forms can be produced depending on the shape of collector electrode such as planar layers, yarns, nanofibre coated yarns, tubular bodies, 3D scaffolds for tissue engineering and others. Some of them are shown in Fig. 4.

Great specific surface area together with small fibre diameters allow rapid interactions of materials with surrounding media. As an example, release of water

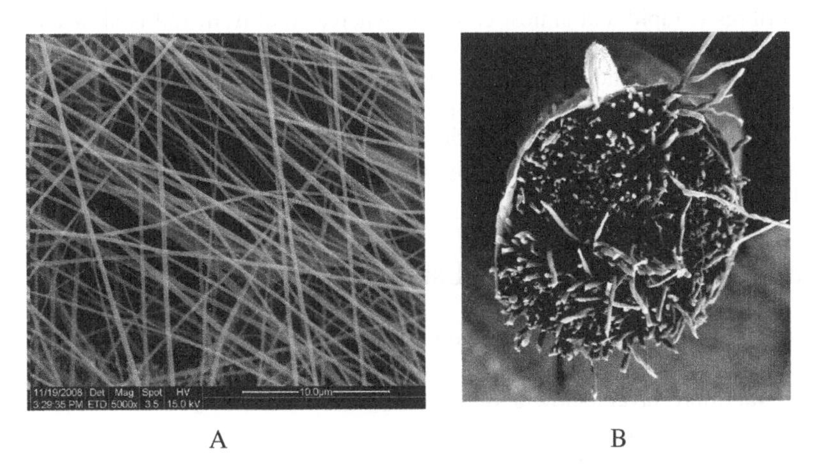

Fig. 4 Planar nanofibre layer (A) and a yarn coated by nanofibres (B)

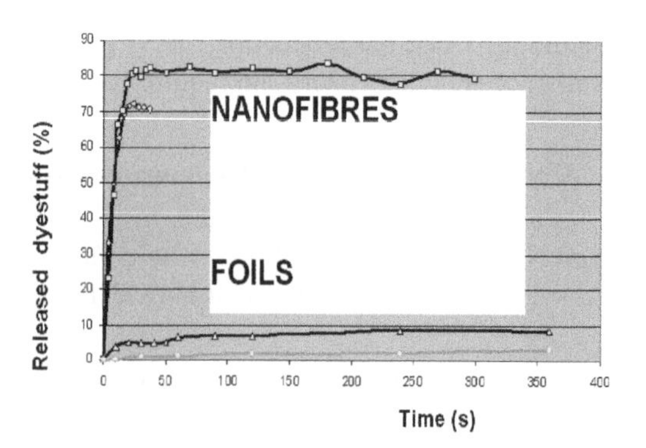

Fig. 5 Release of a dyestuff from nanofibres into water

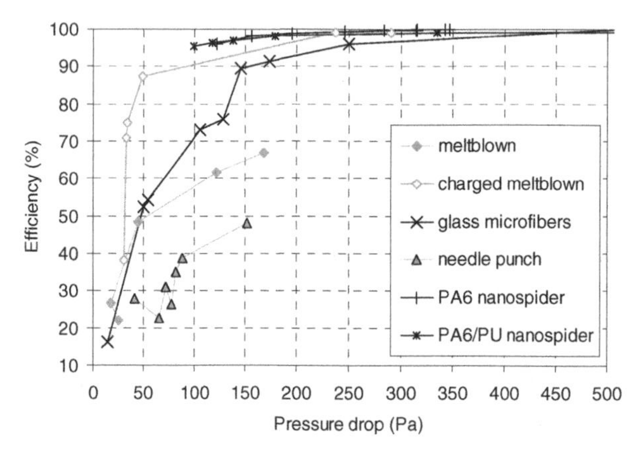

Fig. 6 Filtration efficiency versus pressure drop of various filter materials

soluble dyestuff from nanofibres and from thin foils is compared in Fig. 5. Release from nanofibres is rapid and almost complete whereas that from foil is slow and incomplete. Fig. 5 illustrates differences in interactions with surrounding media between nanomaterials and macroscopic bodies. Variety of specific end-uses of nanomaterials is based on their specific properties such as filters, semipermeable membranes, scaffolds in tissue engineering, wound dressings, chemical and biological protective clothing, energy storing, sensors, composite reinforcements and many others.

3 Examples of Nanofibre End-Uses

Excellent filtration properties of extremely thin (0.05 – 0.1 grammes per square meter) nanofibre layers are shown in Fig. 6 in comparison with other typical filter materials. Nanofibre layers show a very high filtration efficiency whereas maintain low values of pressure drop.

Voluminous materials composed of nanofibres and textile fibres layers (Fig. 7) show specific sound absorbing properties. In comparison with conventional sound absorbing materials such as fibrous layers or polyurethane foams, the composites containing nanofibre layers absorb sound effectively at lower frequencies (Fig. 8). The effect consists in the ability of extremely light nanofibre layers to echo with sound waves and to transfer the energy into the layer of textile fibres.

It is possible to grow human and animal cells as well as bacteria on the nanofibre layers. Therefore, the materials are used in tissue engineering and decontamination technologies (Fig. 9).

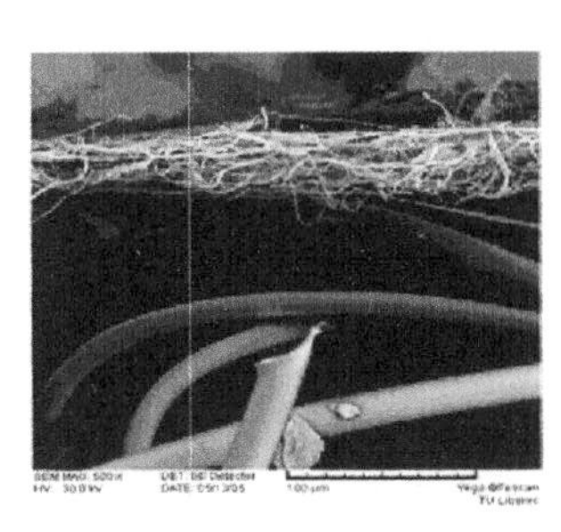

Fig. 7 Layer of nanofibres and textile fibres

Fig. 8 Sound absorption coefficientsvs. sound frequency (lower line-no nanofibres)

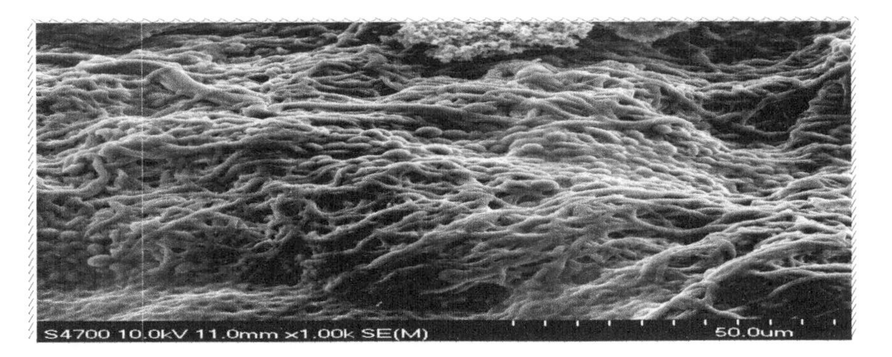

Fig. 9 A nanofibre layer covered with bacteria

References

1. Reneker, D.H., Fong, H.: Polymeric Nanofibers. American Chemical Society, Washingon (2006)
2. Ramakrishna, S., Fujihara, K., Teo, W.E., Lim, T.C., Ma, Z.: Electrospinning and Nanofibers. World Scientific Printers, Singapore (2005)
3. Brown, P.J., Stevens, K.: Nanofibers and Nanotechnology in Textiles. Woodhead Publishing Limited, Cambridge (2007)
4. Andready, A.L.: Science and Technology of Polymer Nanofibres. John Wiley and Sons, Inc., Hoboken (2008)
5. Jirsak, O., Sanetrnik, F., Chaloupek, J., Martinova, L., Lukas, D., Kotek, V. (2005) Patent WO2005024101

The Fractal Ratio as a Metric of Nanostructure Development in Hydrating Cement Paste

R.A. Livingston, W. Bumrongjaroen, and A.J. Allen

Abstract. It is necessary to have appropriate metrics to quantify the development of the nanostructure in Portland cement paste. The fractal ratio, calculated from Small Angle Neutron Scattering (SANS) data, serves as such a metric. It expresses the proportion of the volume-fractal surface area of calcium-silicate-hydrate gel (C-S-H) to the surface-fractal surface area. The volume fractal develops in the scale range from ≈ 5 nm to ≈ 100 nm, and it is associated with the formation of outer product in the capillary pore space by the through-solution mechanism. The surface fractal is attributed to the surface structure formed by colloidal particles on solid substrates such as the Portland cement grains and fly ash particles. The evolution of this ratio over time provides insight into which types of hydration processes are dominant. Applied to study of the hydration of fly ash/Portland cement mixes at later ages, the fractal ratio method showed that in every case, except two, there was a reduced hydration rate due to the dilution effect. The two exceptions involved fly ash fractions with sufficient CaO to generate significant C-S-H gel by the alkali-activated reaction. In all cases the fractal ratio increased with time, indicating the production of additional C-S-H through the topochemical reaction.

1 Introduction

The macroscopic properties of concrete are determined by the development of the nanometer scale structure in the cement paste. Therefore, in order to investigate this development it is necessary to have appropriate metrics to quantify the nanostructure. There have been few such metrics available for characterizing cement

R.A. Livingston
Department of Materials Science & Engineering, University of Maryland, College Park, MD, USA.

W. Bumrongjaroen
Vitreous State Laboratory, Catholic University of America, Washington, DC, USA.

A.J. Allen
Ceramics Division, National Institute of Standards and Technology, Gaithersburg, MD, USA.

paste. This has been due partly to the lack of analytical techniques that have nanometer spatial resolution and also to the disordered structure of the material, arising from its fractal structure. The metric that has been most used to date is the specific surface area as measured by nitrogen isothermal adsorption (BET). However, this technique is controversial [1-3]. It has at least two major disadvantages: the harsh preconditioning of the sample can induce changes to the nanostructure, and it only provides the total surface area, while as discussed below, there are at least two distinctly different surface area involved.

More recently Small Angle Neutron Scattering (SANS) has been applied to investigate cement paste microstructure. This technique uses cold neutrons to probe the paste structure on length scales from 1 nm to >100 nm. This makes it possible to observe the different types of nanostructure surface directly. In addition, SANS does not require any pre-conditioning of samples, and it is nondestructive, which enables repeated measurements on the same sample over time. Finally, it has the unique capability of being able to hide and reveal individual phases by manipulating the contrast between the paste and the porewater solution. A recent review of the literature on the application of SANS to investigate Portland cement paste microstructure is given in Allen et al. [4].

Figure 1 is an example of SANS data for a cement/ fly ash mix paste at 3 months of hydration. The horizontal axis is the magnitude of the scattering vector, Q (where $Q = (4\pi/\lambda)\sin\theta$, λ is the wavelength, and θ is half of the scattering angle),

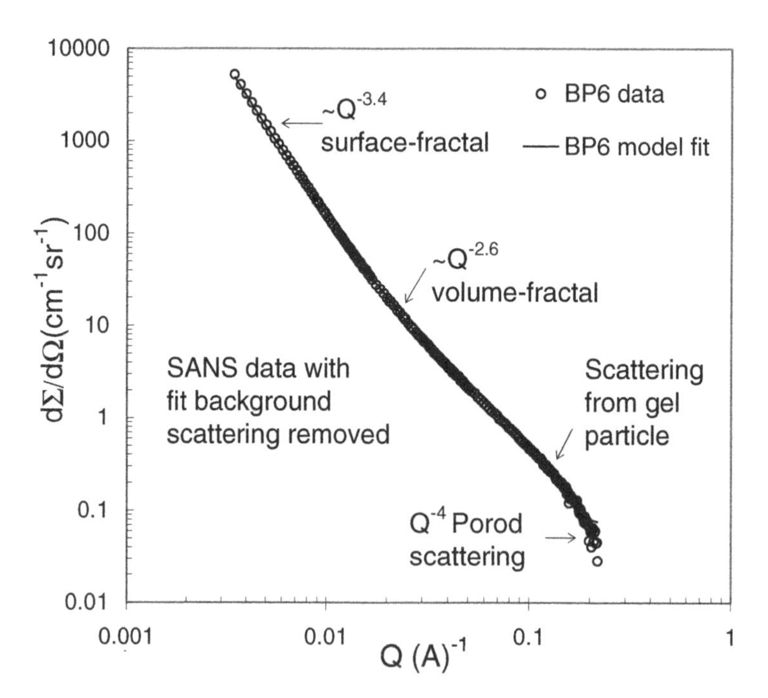

Fig. 1 Plot of SANS data for cement/fly ash paste

which is proportional to the reciprocal of the nanostructure length scale. The vertical axis is the differential macroscopic scattering cross-section, $d\Sigma/d\Omega$, which is a measure of the intensity of the scattered neutrons. It is essentially a spatial Fourier transform of the microstructure. The curve shows several distinct regions related to different features of the C-S-H gel. The plot can then be fitted with a mathematical model that provides several parameters which can be used to characterize the nanostructure.

In particular, there are two types of fractal structure that develop, defined by different fractal dimensions. A mass- (or volume) fractal develops in the scale range from $\approx$ 5 nm to $\approx$ 100 nm, and has been attributed to a disordered diffusion-limited aggregate structure composed of 5-nm colloidal C-S-H gel particles. The scattering arises because there is a neutron scattering contrast between the solid C-S-H gel particles and the pore fluid in the $\approx$ 5 nm gel pores between them. The volume-fractal component is associated with the formation of the outer product in the capillary pore space by the through-solution mechanism during the nucleation and growth period of the hydration reaction. It is quantified in terms of the surface area of the volume-fractal component per unit volume of paste, and it is symbolized by S_{vf}.

The other type of fractal is attributed to the fractally-rough surface structure produced by colloidal particles on solid substrates such as the Portland cement grains and fly ash particles. It is referred to as a surface fractal, and it is also measured in terms of its surface area per unit paste volume, S_{sf}. While this structure co-exists with the volume-fractal, the scattering associated with it is only observable at lower values of Q than the volume-fractal, because it has a steeper Q-dependence and extends to larger scale lengths, as shown in Fig 1.

Whereas the individual surface areas, S_{vf} and S_{sf}, can vary stochastically between samples, taking the ratio of S_{vf} to S_{sf}, can significantly compensate for this sample to sample random variation. The fractal ratio was first introduced by Allen and Livingston in 1995 to study the effects of silica fume on Portland cement paste microstructure during early hydration [5, 6]. More recently, it has been used to investigate the effects of fly ash morphology and chemical composition on the developing microstructure of the hydrating cement system at longer hydration times [7]. Some results of this application of the fractal ratio are described below.

2 Experimental Approach

The addition of fly ash to Portland cement paste produces competing effects: it contributes C-S-H gel through the pozzolanic and alkali-activated reactions, but dilutes the C-S-H contribution of the main Portland cement reaction. To investigate these effects, SANS was applied to several density-fractionated samples of a lignite-type (Ottumwa, IA) and a bituminous-type (Brayton Point, MA) fly ash in a Portland cement paste with a mass fraction of 20%. The experimental details are provided in Bumrongjaroen et al. [7]

The SANS measurements were performed at the National Institute of Standards and Technology (NIST). The SANS instrument, as well as the data acquisition and analysis methods have been described elsewhere [6]. To study the evolution of the cement paste microstructure over time, the SANS measurements were made after hydration for 7 d, 28 d, and 90 d. By using 3 different configurations of the instrument, an overall Q-range of 0.003 $Å^{-1}$ to 1 $Å^{-1}$, or 0.03 nm^{-1} to 10 nm^{-1}, was covered. This range is sufficient to probe features in the size range from 1 nm to 100 nm. Finally, the two-dimensional data set was reduced to one dimension by circular averaging to give $d\Sigma/d\Omega$.

The SANS intensity data were then modeled over various Q ranges using two different models. In the first model, the Q^{-4} Porod scattering law terminal slope, observed at high Q after flat-background subtraction (see Fig. 1), was used to determine the total internal surface area per unit sample volume, S_T. The value of S_T was then used with the second model which represents the full fractal system. This model combines a mass- or volume-fractal scattering term, and a surface-fractal scattering term along with several other factors related to the colloidal particles in the gel. It incorporates a significant number (9) of adjustable parameters. However, different terms dominate at different parts of the scattering curve, and hence only about 3 parameters are needed to determine the fit in any one region. A full description of the models and the fitting procedures is also given in Bumrongjaroen et al. [7].

3 Results

The fractal ratios for the pastes are plotted as a function of time in Fig. 2. For comparison the fractal ratio of the pure Portland cement paste (CCRL) is also plotted as a solid line. At a given hydration time, the fractal ratio is determined by the cumulative Portland cement hydration and pozzolanic action processes that have occurred up to that time. It can be seen that the values for the Brayton Pt mixes are all lower than the pure cement control (CCRL). For the Ottumwa samples, the values for two of the mixes are also lower than the control, but the other two, OT5 and OT6, exceed it. These two fractions have the highest potential for the alkali-activated (AA) or self-cementing reaction, based on their calcium contents. Conversely, none of the Brayton Pt fractions have significant AA potential. Therefore the AA reaction of the fly ashes appears to be an important factor in the development of the microstructure in the mixes. The reduced values of the fractal ratios of the other samples relative to the pure cement is an indication of the dilution effect caused by replacing some of the cement with fly ash.

In every case, the fractal ratio increases with time. Examination of the individual curves reveals that this trend is the result of both an increase in S_{vf} and a decrease in S_{sf}. Since the total fractal surface area, S_T, is essentially constant, this implies that the surface fractal is being transformed into the volume fractal.

By the (hydration) time these measurements were made, the formation of the outer product was essentially complete, and for both the cement and the fly ash any continuing hydration would be due primarily to the topochemical reaction. As the reaction progresses, the reacted zone becomes thicker, more closely resembling a volume fractal. However, the rate of reaction would be lower for the fly ash particles than for the Portland cement because of the requirement for dissolution of Ca from

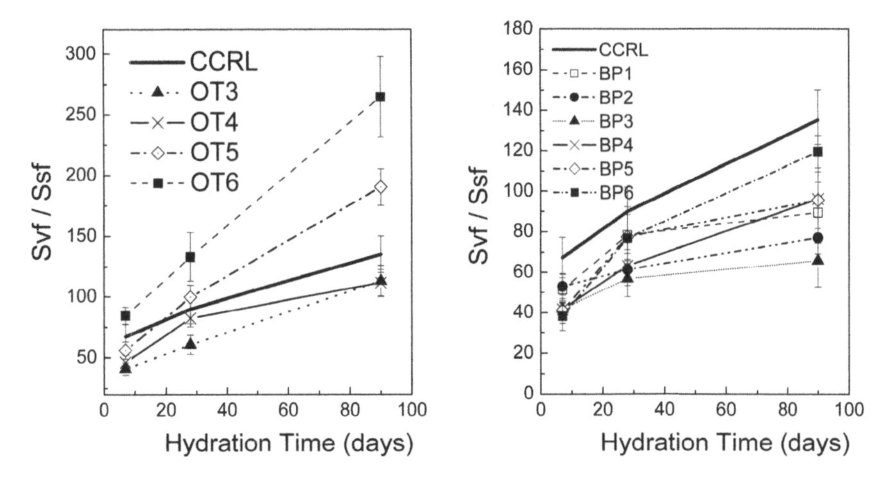

Fig. 2 Time dependence of the fractal ratio (Svf/Ssf) for Ottumwa (left) and Brayton Pt (right) fly ash cement pastes. (Vertical bars are estimated standard deviations.)

calcium hydroxide (CH) through the surrounding gel to the reactive surface of the fly ash. Thus, the overall rate of reaction of a mix would be reduced relative to the pure cement CCRL. The only exceptions are OT5 and OT6, which have enhanced rates of reaction. However, these fractions also can generate significant amounts of C-S-H by AA, which does not involve Ca diffusion.

Another consideration is the effect of the particle size distribution (PSD). Portland cement grains are typically tens of micrometers in diameter. The PSDs of all of the fly ash fractions contain many finer particles, in some cases smaller by an order of magnitude. Examination of the details of the individual SANS curves reveals that most of the fly ash/cement states start off at 7 d with S_{sf} larger than the control. This can be explained by the fact they would have larger total surface areas because of the presence of fine fly ash particle sizes. In turn, this provides more substrate for the C-S-H gel comprising the surface fractal. Conversely, the S_{vf} values for the mixes are all initially lower than the control, due to the reduction in the production of outer product because of the dilution effect. The combination of these two effects produces an initial ratio that would be lower than that of the pure cement paste.

4 Conclusions

It is necessary to have appropriate metrics to quantify the development of nanostructure in cement paste. SANS provides several advantages over nitrogen BET surface area measurements for this purpose, including the ability to discriminate between different types of surfaces, avoidance of harsh pre-conditioning of samples, and ability to make repeated measurements on the same sample over time. SANS yields several related to the nanostructure. In particular, the fractal ratio expresses the proportion of the volume fractal surface area to the surface fractal surface area. The evolution of this ratio over time provides insight into which types of hydration processes are dominant.

Applied to study of the hydration of fly ash/Portland cement mixes at later ages (7 d, onwards) the fractal ratio method showed that in every case, except two, there was a reduction in the rate of reaction due to the dilution effect. The two exceptions involved fly ash fractions with sufficient CaO to generate significant C-S-H gel by the alkali-activated reaction. In all cases the fractal ratio increased with time, indicating the production of additional C-S-H through the topochemical reaction.

Concerning industrial applications of these results, neutron scattering methods are not practical for use on a routine basis because they have to be done at specialized facilities. However, the SANS fractal ratio approach provides insights that can be used to develop reliable routine methods for characterizing fly ash reactivity.

Acknowledgments. The authors gratefully acknowledge the participation of Dan Neumann in numerous discussions about the fractal ratio. We acknowledge the support of the National Institute of Standards and Technology, U.S. Department of Commerce, in providing the neutron research facilities used in this work.

References

1. Garci Juenger, M.C., Jennings, H.M.: The Use of nitrogen adsorption to assess the microstructure of cement paste. Cem. Concr. Res. 31, 883–892 (2001)
2. Beaudoin, J.J.: A discussion on the paper, The use of nitrogen adsorption to assess the microstructure of cement paste. Cem. Concr. Res. 32, 831–832 (2002)
3. Garci Juenger, M.C., Jennings, H.M.: Reply to the discussion by J.J. Beaudoin of the paper The use of nitrogen adsorption to assess the microstructure of cement paste. Cem. Concr. Res. 32, 833–834 (2002)
4. Allen, A.J., Thomas, J.J., Jennings, H.M.: Composition and density of nanoscale calcium-silicate-hydrate in cement. Nature Mater. 6, 311–316 (2007)
5. Allen, A.J., Livingston, R.A.: Small-angle scattering study of concrete microstructure as a function of silica fume, fly ash or other pozzolanic additions. In: Malhotra, V. (ed.) Fifth Canmet/ACI International Conference on Fly Ash, Silica Fume, Slag and Natural Pozzolans in Concrete, pp. 1179–1200. American Concrete Institute, Detroit (1995)
6. Allen, A.J., Livingston, R.A.: Relationship between differences in silica fume additives and fine-scale microstructural Evolution in Cement-Based Materials. Adv. Cement. Based Mater. 8, 118–131 (1998)
7. Bumrongjaroen, W., Livingston, R.A., Neumann, D.A., Allen, A.J.: Characterization of fly ash reactivity in hydrating cement by neutron scattering. J. Mater. Res. (in press) (2009)

A Review of the Analysis of Cement Hydration Kinetics via ^{1}H Nuclear Magnetic Resonance

J.O. Ojo and B.J. Mohr

Abstract. To date, the lack of experimental data concerning entrained water transport through a cementitious microstructure during self-desiccation has limited the understanding of the mechanisms of internal curing. To improve the current knowledge state regarding the moisture transport kinetics of internal curing, novel in situ nanoscale characterization techniques, primarily ^{1}H nuclear magnetic resonance (NMR), are being applied to elucidate changes in the early age hydration effects in the porous cementitious matrix due to internal curing. Relaxation time analyses can indicate the relative intensities and percentages of free water, C-S-H interlayer (physically bound) water, and C-S-H gel (chemically bound) water. Many developments have taken place both with NMR equipment and testing technique. Consequently, this is making NMR a very useful tool in the studying permeability and moisture movement in the concrete matrix. This paper will review the current state-of-the-art regarding the application of NMR to the analysis of cementitious materials at early ages.

1 Introduction

In order to tackle the challenges of investigating the kinetics of hydration and internal curing at early ages in high-performance cementitious materials, the use of nuclear magnetic resonance (NMR) spectroscopy and magnetic resonance imaging (MRI) as novel analytical techniques has been proposed. These measurement techniques, which have been flourishing in the fields of medicine, chemistry, and biology, have more recently been utilized as an important tool for material scientists, particularly for particulate composite material studies. NMR and MRI are versatile tools due to their sensitivity to molecular movements and chemical composition as well as their ability to spatially mapping porous systems nondestructively. Though solid-state NMR is more commonly used to analyze cementitious materials, in the case of early age hydration kinetics, proton (^{1}H) analysis is

J.O. Ojo and B.J. Mohr
Department of Civil and Environmental Engineering, Tennessee Technological University, USA
e-mail: bmohr@tntech.edu

of particular interest. This review surveys the current applications of ^{1}H NMR analysis in cement-based materials research. Proposed applications to internal curing transport kinetics based on the current literature will also be discussed

2 Nuclear Magnetic Resonance Technique

Nuclear magnetic resonance (NMR) refers to the orientation of nuclei particles (e.g., protons) that occur in the nuclei of atoms when they are subjected to a static magnetic field and also exposed to another fluctuating magnetic field. During this processes, both the resonance frequencies and relaxation times for the protons are monitored and recorded. In any compound, the resonant frequency of the protons in the two atoms is not always the same. As a result, instead of a single NMR peak, two peaks are produced and the resulting images can be spatially localized. Application of gradients in orthogonal directions can lead to the reconstruction of 2-D and 3-D images using 2-D fourier transform techniques. If the applied field gradient is linear, the frequency spectrum obtained after fourier transform becomes a 1-D profile of proton (i.e., water) concentration versus distance in the chosen direction. This is known as 1-D mapping. In practice, the fourier transform experiment pulse has proven so versatile that many variations of the technique, suited to special purposes, have been devised and used effectively.

In addition, magnetic resonance imaging (MRI) is one of the more recent developments in NMR research and is a logical extension of the basic principles of magnetic resonance. The MRI scanner provides a non-invasive and non-destructive method of imaging that is sensitive to subtle differences in water distribution, structural integrity, local flow, and other measurements.

The simplest form of NMR experiment is called continuous wave (CW). A solution of the sample in a uniform 5 mm glass tube is oriented between the poles of a powerful magnet, and is spun to average any magnetic field variations, as well as tube imperfections. Radio frequency radiation of appropriate energy is broadcast into the sample from an antenna coil. A receiver coil surrounds the sample tube, and emission of absorbed radio frequency (RF) energy is monitored by dedicated electronic devices and a computer. An NMR spectrum is acquired by varying or sweeping the magnetic field over a small range while observing the RF signal from the sample. An equally effective technique is to vary the frequency of the RF radiation while holding the external field constant.

3 Application of Nuclear Magnetic Resonance to Cement-Based Materials

As stated previously, NMR is commonly used in the fields of biology and chemistry [1, 10] but is increasingly utilized for the analysis of cementitious materials due to the ability to monitor samples nondestructively and in-situ at early ages. Modern NMR spectroscopy is frequently divided into several categories: (1) high resolution mode on homogenous solutions, (2) high power mode on highly relaxing nuclei which exhibit very broad lines, (3) the study of solids, for example, magic angle spinning (MAS) techniques, (4) chromatographic separation of impurities coupled with proton detected NMR experiments, and (5) NMR 3-D imaging (i.e., MRI).

Solid-state NMR of portland cement-based materials has been used to investigate changes in the atomic structure of C-S-H and other hydration products (e.g., ^{27}Al, ^{29}Si, and ^{33}S NMR) [2, 4, 5, 11]. However, for this review, only ^{1}H (proton) NMR will be discussed, as related to the kinetics of cement hydration, and the potential application to internal curing of high-performance cementitious materials.

4 Application of ^{1}H Nuclear Magnetic Resonance to Early Age Cement Hydration

Using proton spin-spin NMR relaxation time analysis and thawing point suppression measurements, it has been shown that hydrating cement pastes have three distinct water components: (1) pore solution/capillary (unconfined/free) water, (2) water adsorbed to hydration products (gel water), and (3) water located within hydrates (chemically combined water) [3, 8, 9, 12]. The water components have been associated with T_2 relaxation times of approximately 9, 80, and 350 μs, respectively. Furthermore, a clear evolution of the relative abundance (based on normalized amplitudes) of the three water components as a function of hydration time has been observed [3]. Subsequently, it is anticipated that these results could be used to estimate the degree of hydration based on the amount of chemically bound water. There is also additional solid-like water phases associated with the conversion of ettringite to monosulfate and/or changes in C-S-H structure at an early age [6].

It has been stated [6] that these water phases are expected to play important roles in the hydration process and their direct observation and characterization offers an additional avenue for understanding details of cement hydration at standard conditions as well as at elevated temperature and pressures or with addition of chemical admixtures.

Gussoni et al. [7] used a combination of NMR relaxation spectroscopy and MRI techniques to characterize cement matrices prepared with and without organic solvents (i.e., chemical admixtures). It is also reported that residual dipolar interaction and transversal relaxation decay can be used to provide relevant information about a cementitious paste. These findings emphasized how admixtures, which are widely used in the concrete industry, can influence the cement hydration kinetics and the pore structure in concrete, particularly at very early ages.

5 Proposed Application of ^{1}H Nuclear Magnetic Resonance to Internal Curing

Among the numerous relaxation parameters that can be quantified by NMR, the spin-spin relaxation parameter (T_2) has been employed for the analysis of water in cement pastes and mortars. Previous research has shown that T_2 times for water in cement-based materials are a function of their respective locations within the microstructure. That is, T_2 relaxation times vary with the mobility of the water protons. It has been shown [3] that chemically combined water, gel (C-S-H interlayer) water, and capillary water are characterized by T_2 times of 9, 80, and 350 μs, respectively. Thus, the T_2 relaxation times are an important parameter for analyzing water in the microstructure. Furthermore, based on these relaxation times, the relative bulk percentages and evolution of the mobility types can be plotted as seen in Figure 1.

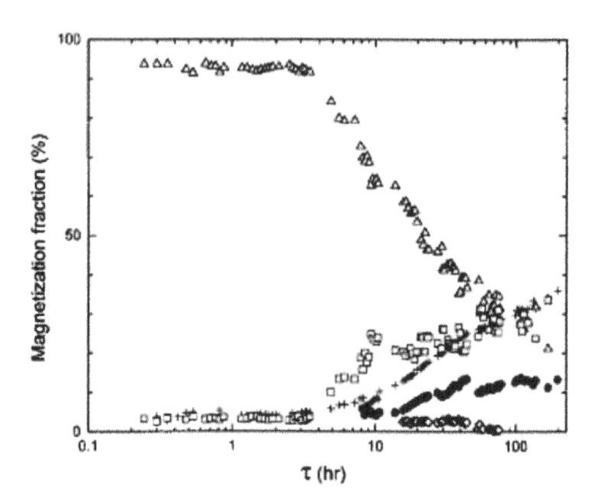

Fig. 1 Example of T_2 relaxation time analysis for determining the relative percentages of water mobility phases as a function of hydration time (● – interlayer water; □ – water in C-S-H gel pores, at early ages in capillary pores and at later ages in nanopores; • - secondary hydration water; Δ – water in large pores and intergrain space at early ages, water in C-S-H gel capillary pores at later ages; + - protons in solids environment) [6]

The purpose of these experiments is to primarily investigate changes in the amounts of free and bound water in the samples beginning as early as reasonably possible until approximately 7 days. It is anticipated that the addition of water held in the internal curing materials will delay the transition from free to bound water (compared to the w/c=0.36 control) and ultimately increase the amount of bound water in the sample (compared to the w/c=0.30 control). For the first anticipated result, the time of delay will lead to an improved understanding of the kinetics of moisture release and transport from the internal curing materials to the hydrating cement paste. As for the latter result, this would be a direct indication of an increase of the degree of hydration (amount of reaction).

Though no current literature has applied proton NMR to internal curing, it is anticipated that the NMR technique shows great promise in resolving differences in the state of water (i.e., how water is bound), as a function of hydration time for internally cured samples.

One of the drawbacks associated with this technique is that heavy water (D_2O) is generally required for NMR analysis as the broad line signal generated is more significant than that of regular water (H_2O). The partial or complete replacement of H_2O with D_2O has implications for the actual hydration kinetics, which can be decreased by a factor of almost seven with the use of D_2O. However, this disadvantage is not expected to minimize the applicability of NMR for the analysis of internal curing. For the proposed internal curing research, the relative differences between samples would be of interest. Therefore, development of NMR in situ experimental techniques to monitor the early stages of cement hydration in the presence of additional water bound within the internal curing particles would significantly contribute to the understanding of moisture transport away from the internal curing materials. Improved understanding of the moisture transport kinetics through high performance cement pastes is critical in developing an appreciation of the mechanisms responsible for self-desiccation and autogenous shrinkage mitigation.

6 Conclusion

^{1}H nuclear magnetic resonance (NMR) has been shown to be a relatively novel technique for investigating the hydration kinetics of cement-based materials, particularly at early age. T_2 relaxation time analyses can indicate the relative intensities and percentages of free water, C-S-H interlayer (physically bound) water, and C-S-H gel (chemically bound) water. In addition, potential uses of proton NMR applied to elucidate changes in the early age hydration effects in the porous cementitious matrix due to internal curing have been presented. It is anticipated that NMR will be increasingly utilized as an in-situ and non-destructive analytical technique in the studying the permeability and moisture movement in the cementitious materials.

Acknowledgments. The authors would like to acknowledge the National Science Foundation (CMS-0556015) and Tennessee Technological University for their financial support. Any opinions, findings, and conclusions or recommendations expressed in this material are those of the authors and do not necessarily reflect the views of the sponsors.

References

1. Barbotin, J., Portais, J.: NMR in microbiology: theory and applications. Horizon Scientific Press, Amiens (2000)
2. Beaudoin, J.J., Raki, L., Alizadeh, R.: A ^{29}Si MAS NMR study of modified C-S-H nanostructures. Cem. Concr. Compost. (2009) doi:10.1016/j.cemconcomp.2001.11.004
3. Bohris, A.J., Goerke, U., McDonald, P.J., Mulheron, M., Newling, B., Le Page, B.: A broad line NMR and MRI study of water and water transport in Portland cement pastes. Mag. Reson. Imaging 16, 455–461 (1998)
4. Chen, J.J., Thomas, J.J., Taylor, H.F.W., Jennings, H.M.: Solubility and structure of calcium silicate hydrate. Cem. Concr. Res. 34, 1499–1519 (2004)
5. d'Espinose de Lacaillerie, J.B., Barneron, F., Bresson, B., Fonollosa, P., Zanni, H., Fedorov, V.E., Naumov, N.G., Gan, Z.: Applicability of natural abundance ^{33}S solid-state NMR to cement chemistry. Cem. Concr. Res. 36, 1781–1783 (2006)
6. Greener, J., Peemoeller, H., Choi, C., Holly, R., Reardon, E.J., Hansson, C.M., Pintar, M.M.: Monitoring of hydration of white cement paste with proton NMR spin-spin relaxation. J. Am. Ceram. Soc. 83, 623–627 (2000)

7. Gussoni, M., Greco, F., Bonazzi, F., Vezzoli, A., Botta, D., Dotelli, G., Natali Sora, I., Pelosato, R., Zetta, L.: ^{1}H NMR spin-spin relaxation and imaging in porous systems: an application to the morphological study of white portland cement during hydration in the presence of organics. Mag. Reson. Imaging 22, 877–889 (2004)
8. Jehng, J.Y., Sprague, D.T., Halperin, W.P.: Pore structure of hydrating cement paste by magnetic resonance relaxation analysis and freezing. Mag. Reson. Imaging 14, 785–791 (1996)
9. Laganas, E., Papavassiliou, G., Fardis, M., Leventis, A., Milia, F., Chaniotakis, E., Meletiou, C.: Analysis of complex ^{1}H nuclear magnetic resonance relaxation measurements in developing porous structures: A study in hydrating cement. J. Appl. Phys. 77, 3343–3348 (1995)
10. Lamberrt, J.B., Mazzola, E.P.: Nuclear magnetic resonance spectroscopy. Pearson Education Inc., Upper Saddle River (2004)
11. Pena, P., Rivas Mercury, J.M., de Aza, A.H., Turrillas, X., Sobrados, I., Sanz, J.: Solid-state ^{27}Al and ^{29}Si NMR characterization of hydrates formed in calcium aluminate-silica fume mixtures. J. Solid State Chem. 181, 1744–1752 (2008)
12. Plassais, A., Pomiès, M.P., Lequeux, N., Boch, P., Korb, J.P., Petit, D., Barberon, F.: Micropore size analysis in hydrated cement paste by NMR. Mag. Reson. Imaging 19, 493–495 (2001)

Analysing and Manipulating the Nanostructure of Geopolymers

J.L. Provis, A. Hajimohammadi, C.A. Rees, and J.S.J. van Deventer

Abstract. Geopolymer concretes are currently being commercialised in Australia and elsewhere around the world, with a view towards enhancing the sustainability of the world's construction industry. The fundamental geopolymer binder is an aluminosilicate gel which displays key structural features on every length scale from Ångstroms up to centimetres, meaning that multiscale analysis is key to the development of a detailed understanding of geopolymer formation and performance. Here, we present results from investigations of geopolymer nanostructure, focusing on the use of infrared spectroscopy as an analytical tool. The effects of different combinations of precursors in geopolymer formation provides critical information, in particular with regard to the rate of reaction and its impact on the final distribution of elements and structures within the geopolymer binder. Formulations are designed so that the same composition is obtained by the use of precursors which release their constituent elements at very different rates under alkaline attack during geopolymerisation, and this provides essential information regarding the role of different elements in forming strong and durable geopolymer structures. Seeding the geopolymer mixture with very low doses of oxide nanoparticles presents several unexpected effects, both in terms of reaction kinetics and also in altering the nature of the zeolitic crystallites formed within the predominantly X-ray amorphous geopolymer binder.

1 Introduction

A geopolymer is a type of alkali-activated aluminosilicate cement which can have comparable or superior mechanical, chemical and thermal properties when compared to Portland-based cements, and with significantly lower CO_2 production [1]. This has led to geopolymers receiving increasing attention in the scientific literature over the past decade; however, much about these materials is still not well understood [2, 3]. Geopolymers are generally synthesised by the reaction between

J.L. Provis, A. Hajimohammadi, C.A. Rees, and J.S.J. van Deventer
Department of Chemical & Biomolecular Engineering, University of Melbourne, Australia
e-mail: jprovis@unimelb.edu.au
http://www.chemeng.unimelb.edu.au/geopolymer/

an aluminosilicate source (often fly ash, metakaolin and/or blast furnace slag) and an alkali metal hydroxide or silicate solution. The geopolymer binder structure consists of tetrahedral Si-O and Al-O bonds arranged in a predominantly X-ray amorphous gel network, where the tetrahedral Al sites are charge-balanced by an alkali cation.

The first stage of geopolymerization is the release of aluminate and silicate species from a solid source, induced by alkali attack on an aluminosilicate material. First, the surface of the solid contacts the activating solution, and hydrolysis reactions begin to occur, with the formation of oligomers and finally polycondensation to form a three-dimensional aluminosilicate network [4]. Soluble silicates are frequently used in geopolymer production to aid dissolution of the aluminosilicate starting material and enhance the mechanical properties of the binder [5, 6].

The primary aim of this paper is to analyse the role of gel nucleation in the formation of geopolymers, particularly by the use of high-surface area seed particles to modify the nucleation process [7], and also by designing geopolymer mixes with different reaction rates but the same composition. Spatially-resolved infrared spectroscopy will be used to identify the effect of the kinetics on the distribution of silicate and aluminate species within the geopolymer gel [8].

2 Materials and Methods

To synthesise geopolymers for seeding experiments, 20.8g of a 6M NaOH solution was mixed with 60g of fly ash (Gladstone Power Station, Queensland, Australia, oxide composition and detailed characterisation given in [6, 9]) and stirred mechanically for no more than 2 minutes. Additional samples were prepared with the same composition, but with 0.01g of Al_2O_3 nano-particles (NanoScale Materials, USA, mean particle size 200nm and specific surface area 275m^2/g) dispersed in the activating solution immediately before mixing with the fly ash, to act as potential nucleation sites. XRD analysis was performed after 100 days at 30ºC.

Geopolymers for infrared spectroscopy study were synthesised using a 'one-part' (just add water) procedure [10]: solid washed geothermal silica (96% SiO_2, from the Cerro Prieto geothermal power station, Mexico) and reagent-grade sodium aluminate (Aldrich) were blended to give the desired Si/Al ratios, and then mixed with water at a molar ratio of H_2O/Na_2O = 12. ATR-FTIR spectra of one-part mix geopolymers were collected using a Varian FTS 7000 FT-IR spectrometer, with a Specac MKII Golden Gate single reflectance diamond ATR attachment with KRS-5 lenses and heater top plate. Absorbance spectra were collected from 4000-400 cm^{-1} at a resolution of 2 cm^{-1} and a scanning speed of 5 kHz with 32 scans.

3 Results and Discussion

3.1 Seeded Nucleation

Fig.1 shows the X-ray diffraction data obtained from seeded and unseeded geopolymer formulations after 100 days' curing at 30°C. The most striking aspect of these diffractograms is that the presence of less than 0.01% by mass of nanosized alumina seed particles has entirely changed the nature of the zeolite product observed, from the 'normal' (widely observed) faujasite-type structure of zeolite Na-X to the unusual edingtonite-type Na-F [7]. Zeolite Na-F has previously been reported predominantly as the result of ion exchange from a product synthesised in the potassium form [11, 12].

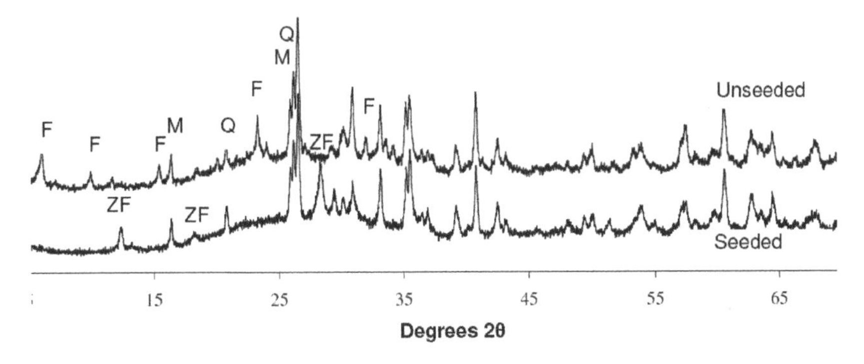

Fig. 1 X-ray diffractograms of fly ash geopolymers with and without nanoparticle seeds. Selected peaks are marked - F: zeolite Na-X (faujasite), M: mullite, Q: quartz, ZF: zeolite Na-F. Other peaks are due to unreacted mullite, quartz and iron oxides. Data from [7]

In addition to this change in the nature of the reaction product, seeding also shows a strong effect on the reaction kinetics. The induction period prior to the onset of gel formation which is observed in most hydroxide-activated fly ash geopolymer syntheses [13] was entirely absent in the seeded system [7]. This is consistent with an explanation based around nucleation behaviour; nucleation of the geopolymer gel in hydroxide-activated geopolymers usually takes place on the fly ash particle surfaces [14], but the availability of the seed particle surfaces means that the energy barrier associated with this nucleation process is greatly reduced. Nucleation therefore takes place directly around the seed particles, which direct the geopolymer structure in a different manner to the fly ash surfaces, resulting in the formation of a different type of zeolite in the geopolymer gel.

3.2 Time-Resolved Infrared Spectroscopy

Fig. 2 shows the results of infrared spectroscopy carried out at different times in the geopolymerisation process. The progression from the original powder mix to a hardened geopolymer structure is made more clearly visible by the removal of the water background from these spectra.

Bands at 1060 cm^{-1} are assigned to stretching of Si-O-Si bonds at the surface of the unreacted silica particles [15], and bands at 1100, 800 and 475 cm^{-1} relate to stretching, bending and rocking of the Si-O-Si bonds within the network of the

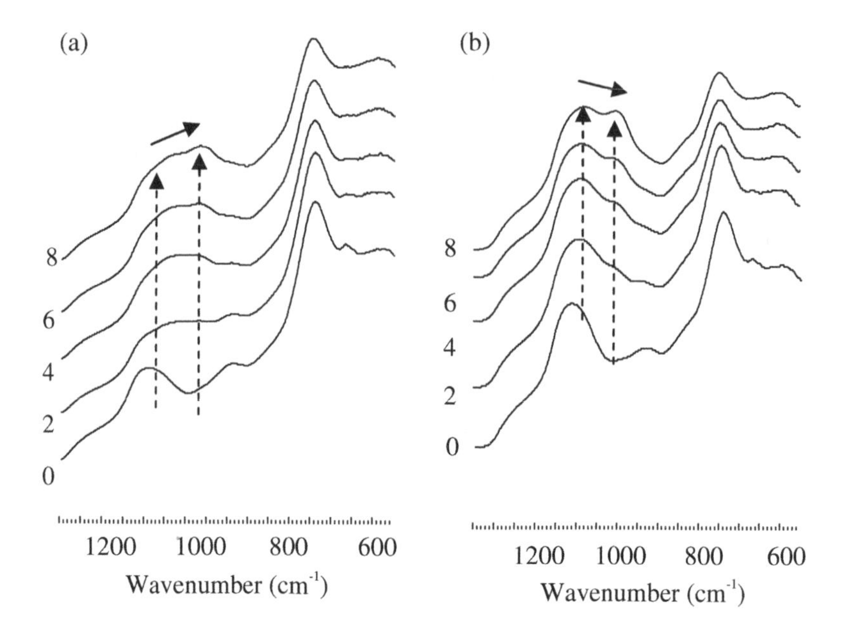

Fig. 2 ATR-FTIR spectra of one-part geopolymer samples with H_2O/Na_2O = 12 and (a) Si/Al = 1.5; (b) Si/Al = 2.5. Numbers refer to the geopolymer age in days, at a temperature of 40°C. Data from reference [10]

unreacted particles of geothermal silica [16]. Bands at 545, 625 and 700 cm^{-1} are related to vibrations in the unreacted solid aluminate [17]. The position of the main Si-O-T (T: tetrahedral Si or Al) stretching band gives an indication of the length and angle of the Si-O bonds in the silicate network [6, 13].

Initially, the main Si-O-T stretching band occurs at 1060 cm^{-1}, shifting to lower wavenumbers in both samples over time. This can indicate changes in the silicate network including an increase in non-bridging oxygen on silicate sites, charge balancing by sodium cations in the system [18] or increasing the substitution of tetrahedral Al in the silicate network [19].

Over time, there is also a reduction the in intensity of the main Si-O-Si band, indicating that the solid silica is dissolving. At the same time, a new band starts forming at about 950cm^{-1} and the intensity of this band increases over time. This particular band is associated with the Si-O-T stretch within the newly formed geopolymer network [6, 13], and it appears to grow at a similar rate in both samples depicted in Figure 2, accounting for the fact that the geothermal silica peak is larger in Fig. 2b due to the higher concentration of Si in this sample. Spectra such as these, and reaction kinetic analysis derived from in situ time resolved FTIR analysis [13], provide the opportunity to observe in detail the reaction processes taking place during geopolymerization, and the similarities and differences between 'traditional' and one-part geopolymer formulations.

4 Conclusions

The data presented in this paper show that there is clearly a strong degree of kinetic control in the formation of geopolymer gels by alkaline activation of fly ashes. The addition of seed particles and the manipulation of alumina release rates can each give control of geopolymer structure on a length scale of nanometres to microns. Such manipulation is likely to be important in the future development of geopolymer technology, particularly in optimizing mixes for specific applications including rapid controllable strength development. The application of experimental techniques such as time-resolved infrared spectroscopy is of significant value in the study of complex materials such as geopolymers, as these techniques provide the opportunity to obtain structural information which would not normally be accessible by more 'standard' methods of analysis.

Acknowledgments. This work was funded partially through a studentship awarded to Ailar Hajimohammadi by the Centre for Sustainable Resource Processing through the Geopolymer Alliance, partly through Discovery Project funding from the Australian Research Council, and partly through the Particulate Fluids Processing Centre, a Special Research Centre of the Australian Research Council.

References

1. Duxson, P., Provis, J.L., Lukey, G.C., van Deventer, J.S.J.: The role of inorganic polymer technology in the development of 'Green concrete'. Cem. Concr. Res. 37, 1590–1597 (2007)
2. Duxson, P., Fernández-Jiménez, A., Provis, J.L., et al.: Geopolymer technology: The current state of the art. J. Mater. Sci. (in press) (2007) DOI 10.1007/s10853-10006-10637-z
3. Provis, J.L., Lukey, G.C., van Deventer, J.S.J.: Do geopolymers actually contain nanocrystalline zeolites? A re-examination of existing results. Chem. Mater. 17, 3075–3085 (2005)
4. Provis, J.L., van Deventer, J.S.J.: Geopolymerisation kinetics. 2. Reaction kinetic modelling. Chem. Eng. Sci. 62, 2318–2329 (2007)
5. Rowles, M., O'Connor, B.: Chemical optimisation of the compressive strength of aluminosilicate geopolymers synthesised by sodium silicate activation of metakaolinite. J. Mater. Chem. 13, 1161–1165 (2003)
6. Rees, C.A., Provis, J.L., Lukey, G.C., van Deventer, J.S.J.: Attenuated total reflectance Fourier transform infrared analysis of fly ash geopolymer gel ageing. Langmuir 23, 8170–8179 (2007)
7. Rees, C.A., Provis, J.L., Lukey, G.C., van Deventer, J.S.J.: Geopolymer gel formation with seeded nucleation. Colloids Surf A 318, 97–105 (2008)
8. Hajimohammadi, A., Provis, J.L., Van Deventer, J.S.J.: One-part geopolymer mixes from geothermal silica and sodium aluminate. In: 2008 AIChE Annual Meeting, Philadelphia, PA (2008)
9. Keyte, L.M.: Ph.D. Thesis, University of Melbourne, Australia (2008)

10. Hajimohammadi, A., Provis, J.L., van Deventer, J.S.J.: One-part geopolymer mixes from geothermal silica and sodium aluminate. Ind. Eng. Chem. Res. 47, 9396–9405 (2008)
11. Barrer, R.M., Mainwaring, D.E.: Chemistry of soil minerals. Part XIII: Reactions of metakaolinite with single and mixed bases. J. Chem. Soc. Dalton Trans. 2534 (1972)
12. Baerlocher, C., Meier, W.M., Olson, D.H.: Atlas of Zeolite Framework Types, 5th edn. Elsevier, Amsterdam (2001)
13. Rees, C.A., Provis, J.L., Lukey, G.C., van Deventer, J.S.J.: In situ ATR-FTIR study of the early stages of fly ash geopolymer gel formation. Langmuir 23, 9076–9082 (2007)
14. Lloyd, R.R., Provis, J.L., van Deventer, J.S.J.: Microscopy and microanalysis of inorganic polymer cements. 2: The gel binder. J. Mater. Sci. (2009) (in press) DOI: 10.1007/s10853-10008-13078-z
15. Osswald, J., Fehr, K.T.: FTIR spectroscopic study on liquid silica solutions and nanoscale particle size determination. J. Mater. Sci. 41, 1335–1339 (2006)
16. Bell, R.J., Dean, P.: Atomic vibrations in vitreous silica. Discussions of the Faraday society 50, 55–61 (1970)
17. Moolenaar, R.J., Evans, J.C., McKeever, L.D.: The structure of the aluminate ion in solutions at high pH. J. Phys. Chem. 74, 3629–3638 (1970)
18. Sweet, J.R., White, W.B.: Study of sodium silicate glasses and liquids by infrared reflectance spectroscopy. Phys. Chem. Glass. 10, 246–251 (1969)
19. Roy, B.N.: Infrared spectroscopy of lead and alkaline-earth aluminosilicate glasses. J. Am. Ceram. Soc. 73, 846–855 (1990)

Nanotechnology Applications for Sustainable Cement-Based Products

L. Raki, J.J. Beaudoin, and R. Alizadeh

Abstract. Concrete is a macro-material strongly influenced by the properties of its components and hydrates at the nanoscale. Progress at this level will engender new opportunities for improvement of strength and durability of concrete materials. This article will focus on recent research work in the field of nanoscience applications to cement and concrete at the NRC-IRC. A particular attention will be given to nanoparticles and cement-based nanocomposites.

1 Introduction

Nanotechnology has been clearly identified as one of the key, cross-disciplinary areas of research for the next twenty years. Significant investments are being made in nanotechnology research in Canada and around the world. Recent studies have identified the construction industry as one of the major potential consumers of nanostructured materials [1]. Concrete is a composite material with structures ranging from nano to macro-scales. Its study represents a multidisciplinary area of research. Nanotechnology potentially offers the opportunity to further the understanding of concrete behaviour, to engineer its properties and to lower both the monetary and ecological cost of construction materials. The expected future economic and social benefits linked to the area of nanotechnology in general and in the field of concrete materials in particular will also have to include sustainability effects.

L. Raki
Institute for Research in Construction, National Research Council Canada
e-mail: laila.raki@nrc-cnrc.gc.ca

J.J. Beaudoin
Institute for Research in Construction, National Research Council Canada
e-mail: jim.beaudoin@nrc-cnrc.gc.ca

R. Alizadeh
Institute for Research in Construction, National Research Council Canada
e-mail: rouhollah.alizadeh@nrc-cnrc.gc.ca

2 Controlled Release of Admixtures

Modern concrete is more than a simple mixture of cement, water, sand and aggregates. Today, more is demanded from concrete than ever before as it is being used in different forms and special applications such as high performance (HPC), self-levelling concrete (SLC), self-compacting concrete (SCC), ultra high performance concrete (UHPC) etc. These very high demand developments require the use of chemical admixtures in order to modify/control one or more properties of the fresh or hardened properties of concrete. The most used admixtures in cement and concrete include accelerators, set retarders, air entraining agents, superplasticizers and others. Today, chemical admixtures have become an integral part of the concrete technology and practice. Like drugs, chemical admixtures can produce side effects that can be beneficial or detrimental depending on the situation. A particular challenge of interest to concrete scientists is to optimize and maximize the use of supplementary cementing materials in high performance concretes. Dispersing agents such as superplasticizers are commonly used in these special concretes. There are, however, practical problems such as loss of workability with time that are controlled by interactions with cement components. Controlling the timing of the availability of an admixture in cement systems is essential for its optimal performance.

There have been a number of applications in cement and concrete technology where different means of controlling the effect of admixtures via a controlled release technique were used. A number of patents and research articles describe "encapsulation" procedures for delivery of liquids and solids. Encapsulation often relies on a soluble coating (gelatine or wax) to effect control. Mechanisms involve dissolution (coating), diffusion (membrane), desorption (porous materials), or mechanical removal (during mixing). A corrosion inhibitor, such as calcium nitrite, was dispersed by encapsulation in coated hollow polypropylene fibers [2] via an automated activation. Porous aggregates were also used to encapsulate antifreezing agents [3]. Porous solid materials were used to encapsulate oil well treating fluids [4]. Another method to control the release of chemicals in cement-based materials is by "intercalation-deintercalation". A cement additive for inhibiting concrete deterioration was formulated as a mixture of an inorganic cationic exchanger: zeolite, and an inorganic anionic exchanger: hydrocalumite [5].

Admixtures are most often added at time of mixing, which is not necessarily optimum for the desired chemical effects. For instance, it is sometimes required to delay release of compounds such as superplasticizers, and other additives. Development of new materials for programmed delivery and control of admixtures in concrete and other materials presents a significant technological advance. Recent work from the authors [6, 7] examined means to control the timing of the release of chemical admixtures through their incorporation in nanoscale composite materials. More specifically, a series of new hydrocalumite derivatives were prepared by an anion exchange reaction of a synthetic precursor, hydrocalumite, and different model organic molecules used as admixtures in concrete technology. The technique consisted of intercalating an admixture into a hydrocalumite-like material and adding this composite to a cement-based mix. De-intercalation of the admixture could be actively programmed through controlled chemistry involving, for example, type of layered inorganic material, charge density, concentration of the admixture, and/or pH. A sulphonated naphthalene formaldehyde-based superplasticizer, Disal™, was used to produce the controlled release formulation (CaDisal) [7]. The effectiveness of Disal™ alone in controlling the slump-loss versus time characteristic was compared to that of the controlled release formulation CaDisal. Mini-slump measurements [8] indicated that the controlled release formulation provided a longer time for the superplasticizer to keep cement workability at a reasonable level after mixing [7]. Compressive strength measurements on mortar cubes at w/c=065 are shown in Figure 1. A general increase in strength with time for all mixes was observed. The mix with higher dosage of Disal (0.6% Disal) exhibited the lowest strength value in the series. The addition of the controlled release formulation 2.4% CaAl_Disal (2.4% CaDisal by mass of cement contains an equivalent amount of 0.3% Disal™) showed an increase with time and a higher value at 28 days (35.9Mpa) compared to other mixes including the control (33.4 Mpa). This later strength development could be due to slow hydration kinetics as a result of a slow release of the Disal, as shown by results from a separate study on hydration kinetics of C_3S systems with and without the controlled release formulation. The results obtained with this new composite additive confirmed that a slow release of the superplasticizer not only maintained the workability of the fresh mix but also improved the strength development of mortar samples. Further studies on long-term durability of concrete samples containing the controlled release composite are ongoing.

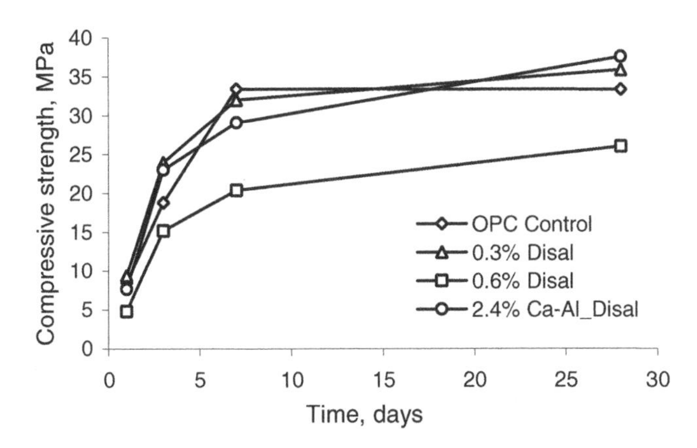

Fig. 1 Compressive strength for different mortar cubes (W/C=0.5)

3 C-S-H and C-S-H Composites

The primary binding agent in the hydrated Portland cement is a nearly amorphous material called Calcium-Silicate-Hydrate (C-S-H). It forms about 60% of the hydration products and is mainly responsible for important properties of hardened cement such as strength and volume stability. The nanostructure of C-S-H has been the subject of extensive debate for more than fifty years [9]. Improvement in our understanding of the nature of these hydration products would eventually result in better indentifying the underlying mechanisms in the mechanical and durability performance of concrete structures. It would also assist in engineering the nanostructure of cement-based systems and producing novel materials having superior characteristics.

A clearer image of the C-S-H nanostructure has emerged through the recent advancements in the analytical methods and experimental techniques [10-12]. Understanding of the relation between the chemical composition of C-S-H and its physical properties has certainly improved. The state of water in C-S-H is one of the most controversial issues in cement nanoscience. Although the pioneering model by Feldman and Sereda [13] is generally accepted for describing the microstructural behavior of hydrated Portland cement, further compelling evidence was recently presented in support of their layered model for C-S-H in cement paste [14]. Water molecules adsorbed on the surface of C-S-H or those located in the interlayer regions can significantly contribute to the mechanical behavior of hardened cement systems. Changes in the silicate structure or the interaction of calcium ions between the C-S-H sheets upon the removal of water may be responsible for variations in the mechanical response of C-S-H [15]. Cement paste and synthetic calcium silicate hydrate show analogous behavior in this regard.

As a part of the global efforts to improve the sustainability of concrete structures, investigations on the fabrication of organic/inorganic cementitious nanocomposites have shown early promise in achieving nanostructurally modified systems [16]. C-S-H based materials can be tailored at the nano level using several types of organic molecules in order to enhance the mechanical performance and volume stability. The organic moieties are interacted within the nanostructure of C-S-H during or after its preparation. There are indications that the produced nanocomposites have improved characteristics. It has been postulated that the organic guest molecules may possibly graft at defect sites of the silicate chain where bridging tetraherda is missing [17-19], intercalate into the interlayer space of the C-S-H [16], or form a covalent boding with the silicate structure [20, 21].

The extent of the interaction of organic molecules with the C-S-H nanostructure largely depends on the preparation procedure. X-ray diffraction and nuclear magnetic resonance spectroscopy are the primary tools that have resulted in obtaining more insight about the nature of the C-S-H based nano-hybrids. It has been shown using XRD that in some cases the organic substances can result in an increase of the 002 basal spacing of C-S-H up to as much as two times [22]. The ^{29}Si MAS NMR has revealed more details about the structural positions on the guest molecules with respect to the disordered crystal structure of C-S-H. The increase in the Q^2/Q^1 ratio determined from the NMR spectra following the interaction of several organic molecules (e.g. hexadecylmethylammonium, poly(ethylene glycol) and methylene blue) was attributed to the increased shielding of Q^1 silicate sites by the organic substances grafted in the defect locations on the silicate structure [17-19]. The effectiveness of the grafting process is dependent on the C/S ratio of the

C-S-H as the number of defect sites generally increases with C/S ratio >1.0. A schematic representation of the grafting mechanism is shown in Figure 2. It is speculated that the chemical stability of the polymer-modified C-S-H in aggressive environments is improved as the access to the interlayer region and the CH sheet is restricted by the organic molecules. True nano-hybrid systems can be prepared by the covalent bonding of the modified organic groups to the inorganic C-S-H base [11, 20, 21]. Polymers having silylated functions are introduced during the precipitation of C-S-H. The silylated group is incorporated into the silicate chain of the C-S-H and forms T silicon sites.

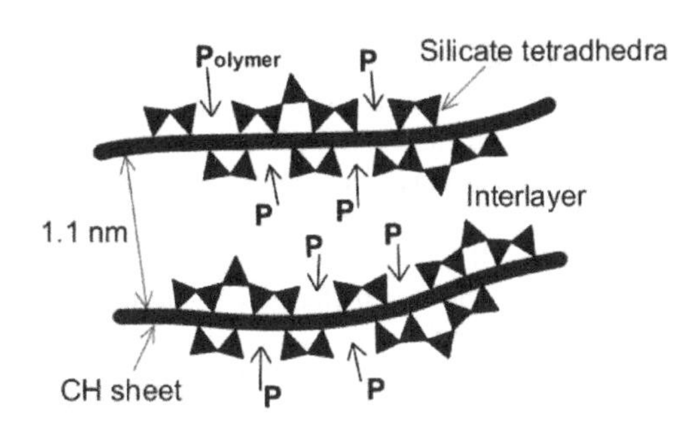

Fig. 2 Schematic of polymer-modified C-S-H nanostructure

4 Concluding Remarks

Significant advancements have been made in the application of nanotechnology in cement science. The molecular structure of cement-based materials can be modified using organic molecules in order to achieve a certain characteristics. The engineering performance of polymer-modified C-S-H nano-strucutures remains to be investigated. Novel approaches are currently taken at NRC to estimate the extent of ingress of foreign molecules into the modified C-S-H. Strength and elasticity of polymer-modified C-S-H materials can also be measured on compacted powder specimens. Producing a new range of construction materials is promising.

References

1. Bartos, P.J.M.: Nanotechnology in construction: A roadmap for development. In: Proceedings of the 2nd International Symposium on Nanotechnology in Construction, Bilbao, Spain, November 13-16, 2005, pp. 27–36 (2005)
2. Dry, C.M., Corsaw, M.J.T.: A time-release technique for corrosion prevention. Cem. Concr. Res. 28(8), 1133–1140 (1998)
3. Dry, C.M.: Alteration of matrix permeability and associated pore and crack structure by timed release of internal chemicals. Ceram. Trans. 16, 729–768 (1991)
4. Reddy, B.R., Crook, R.J., Chatterji, J., King, B.J., Gray, D.W., Fitzgerald, R.M., Pwell, R.J., Todd, B.L.: Controlling the release of chemical additives in well treating fluids, US. Patent 6,209, 646 (2001)

5. Tatematsu, H., Nakamura, T., Koshimuzu, H., Morishita, T., Kotaki, H.: Cement additive for inhibiting concrete deterioration, US. Patent 5,435, 848 (1995)
6. Raki, L., Beaudoin, J.J., Mitchell, L.D.: Layered double hyroxides-like materials: nanocomposites for use in concrete. Cem. Concr. Res. 34, 1717–1724 (2004)
7. Raki, L., Beaudoin, J.J.: Controlled release of chemical admixtures. Canadian Patent # CA 2554347, US patent Application US 2007/0022916 A1 (2007)
8. Kantro, D.L.: Influence of water-reducing admixtures on properties of cement pastes-a miniature slump test. Cem. Concr. and Aggre. 2(2), 95–102 (1980)
9. Taylor, H.F.W.: Cement Chemistry, 2nd edn., p. 459. Thomas Telford, London (1997)
10. Alizadeh, R., Beaudoin, J.J., Ramachandran, V.S., Raki, L.: Applicability of Hedvall effect to study the reactivity of calcium silicate hydrates. Journal of Advances in Cement Research, 1–8 (2009) DOI: 10.1680/adcr.2008.00008
11. Pellenq, R.J.-M., Lequeux, N., Van Damme, H.: Engineering the bonding scheme in C–S–H: The iono-covalent framework. Cem. Conc. Res. 38, 159–174 (2008)
12. Constantinides, G., Ulm, F.-J.: The effect of two types of C-S-H on the elasticity of cement-based materials: Results from nanoindentation and micromechanical modeling. Cem. Conc. Res. 34, 67–80 (2004)
13. Feldman, R.F., Sereda, P.J.: A model for hydrated Portland cement paste as deduced from sorption-length change and mechanical properties. Matériaux et Construction 1, 509–520 (1968)
14. Alizadeh, R., Beaudoin, J.J., Raki, L.: C-S-H (I) - A Nanostructural model for the removal of water from hydrated cement paste. J. Am. Ceram. Soc. 90, 670–672 (2007)
15. Alizadeh, R., Beaudoin, J.J., Raki, L.: Dynamic mechanical response of Calcium-Silicate-Hydrate systems, 47 (2009) (under preparation)
16. Matsuyama, H., Young, J.F.: Intercalation of Polymers in Calcium Silicate Hydrate: A New Synthetic Approach to Biocomposites? Chem. Mat. 11, 16–19 (1999)
17. Beaudoin, J.J., Drame, H., Raki, L., Alizadeh, R.: Formation and properties of C-S-H - HDTMA nano-hybrids. J. Mat. Res. 23, 2804–2815 (2008)
18. Beaudoin, J.J., Patarachao, B., Raki, L., Alizadeh, R.: The interaction of methylene blue dye with calcium-silicate-hydrate. J. Am. Ceram. Soc. 92, 204–208 (2009)
19. Beaudoin, J.J., Drame, H., Raki, L., Alizadeh, R.: Formation and properties of C-S-H - PEG nano-hybrids. Mat. Struct., 1–6 (2009) DOI: 10.1617/s11527-008-9439-x
20. Minet, J., Abramson, S., Bresson, B., Franceschini, A., Van Damme, H., Lequeux, N.: Organic calcium silicate hydrate hybrids: a new approach to cement based nanocomposites. J. Mater. Chem. 16, 1379–1383 (2006)
21. Franceschini, A., Abramson, S., Mancini, V., Bresson, B., Chassenieux, C., Lequeux, N.: New covalent bonded polymer-calcium silicate hydrate composites. J. Mater. Chem. 17, 913–922 (2007)
22. Matsuyama, H., Young, J.F.: Synthesis of calcium silicate hydrate/polymer complexes: Part I. J. Mater. Res. 14, 3379–3388 (1999)

Nanoscale Modification of Cementitious Materials

S.P. Shah, M.S. Konsta-Gdoutos, Z.S. Metaxa, and P. Mondal

Abstract. This research investigates changes in the nanostructure and the nanoscale local mechanical properties of cement paste with micro- and nano-modifiers. Silica fume and multiwall carbon nanotubes (MWCNTs) were used as micro- and nano-modifiers. An effective method of dispersing CNTs in cement matrix was developed. A detailed study on the effects of CNTs concentration and aspect ratio on the fracture properties, nanoscale properties and microstructure of nanocomposite materials, was conducted. Significant improvements on the macro and nanomechanical properties of cement paste were observed with the incorporation of CNTs. Results suggest that CNTs can strongly modify and reinforce the cement paste matrix at the nanoscale.

1 Introduction

The fundamental properties of cementitious materials, such as concrete, are affected by the material properties at the nanoscale [1]. The ultimate goal of this research is to modify the properties at the nanoscale and develop nano-engineered materials with improved macroscopic properties.

In general, cement based materials are typically characterized as quasi-brittle materials that exhibit low tensile strength. Typical reinforcement of cementitious materials is usually done at the millimeter scale and/or at the micro scale using macrofibers and microfibers, respectively. However, cement matrix exhibits flaws which are at the nanoscale. The development of new nanosized fibers, such as carbon nanotubes (CNTs), has opened a new field for nanosized reinforcement within concrete [2-8]. The remarkable mechanical properties of CNTs [9-10] suggest that are ideal candidates for high performance cementitious composites.

S.P. Shah and Z.S. Metaxa
Center for Advanced Cement Based Materials, Department of Civil and Environmental Engineering, Northwestern University, IL, USA

M.S. Konsta-Gdoutos and Z.S. Metaxa
Department of Civil Engineering, School of Engineering, Democritus University of Thrace, Greece

P. Mondal
Department of Civil and Environmental Engineering, University of Illinois at Urbana Champaign, IL, USA

The major drawback however, associated with the incorporation of CNTs in cement based materials is poor dispersion [11]. To achieve good reinforcement in a composite, it is critical to have uniform dispersion of CNTs within the matrix [12]. Few attempts have been made to add CNTs in cementitious matrices at an amount ranging from 0.5 to 2.0% by weight of cement. Previous studies have focused on the dispersion of CNTs in liquids by pre-treatment of the nanotube's surface via chemical modification [2-8]. Preliminary research has shown that small amounts of CNTs can be effectively dispersed in cementitious matrix [13].

In this study, the effect of multiwall carbon nanotubes (MWCNTs) on the macro and nanoscale mechanical properties of cement paste was investigated. An effective method of dispersing carbon nanotubes in cement paste matrix, by applying ultrasonic energy and using a surfactant, was developed. A detailed study on the effects of CNTs concentration and aspect ratio, was conducted. The nano-mechanical properties of CNTs nanocomposites were compared to cement paste with silica fume, which is commonly used as an additive for high strength concrete.

2 Experimental Details

For the preparation of the CNTs nanocomposites ordinary Portland cement and multiwall carbon nanotubes (MWCNTs) were used. Prior to their addition to cement MWCNTs were dispersed in water using a surfactant and by applying ultrasonic energy. A 500 W cup-horn high intensity ultrasonic processor was used to apply constant energy (1900-2100 J/min) to the CNT dispersions. After the sonication, cement was added into the CNT dispersions at a water to cement ratio of 0.5. To make the samples with the silica fume ordinary Portland cement and silica slurry (1000D from W.R. Grace) was used. 15wt% silica fume was used to replace cement. The percentage of water present in the slurry was considered so as to maintain water to binder ratio of 0.5. All materials were mixed according to ASTM C 305. Following mixing, the paste was cast in plastic molds. After demolding the specimens were cured in water saturated with lime until testing.

The morphology and the microstructure of the fracture surface of CNTs nanocomposites were investigated using an ultra-high resolution field emission SEM (Hitachi S5500). Specimens were tested after 18 hours of curing. Prior to their observation, the fracture surface of the specimens was sputter-coated at a 20nm thick layer of gold-palladium (Au/Pd).

The mechanical performance of the CNTs nanocomposites was evaluated by fracture mechanics tests. Notched specimens of 20×20×80 mm were tested at the age of 3, 7 and 28 days, by three-point bending. The tests were performed with a closed-loop testing machine with a 89 kN capacity. The feedback control signal for running the test was the crack mouth opening displacement (CMOD) at the notch, which was advanced at a rate of 0.12 mm/min. The load and the CMOD were recorded during the test. The Young's modulus was calculated from the load versus CMOD results using the two-parameter fracture model by Jenq and Shah [14].

The nanomechanical properties of the CNTs nanocomposites were investigated using a Hysitron Triboindenter following the method described in Ref. 15. Before testing, thin slides of approximately 5 mm were cut out of the specimens. The surfaces were polished with silicon carbide paper discs and diamond lapping films in order to obtain a very smooth and flat surface. Nanoindentation was performed in a 12×12 grid (10 μm between adjacent grid points). This procedure was repeated in at least two different areas on each sample.

3 Results and Discussion

In order to investigate the effectiveness of the dispersing method nanoimaging of the fracture surfaces of samples reinforced with 0.08% by weight of cement CNTs, were performed. Results from SEM images of cement paste samples reinforced with CNTs that were added to cement as received (without dispersion) and CNTs that were dispersed following the method described previously are presented in Fig. 1. As expected, in the samples where no dispersing technique was used [Fig. 1 (a)] CNTs appear poorly dispersed, forming large agglomerates and bundles. On the other hand, in the samples where dispersion was achieved by applying ultrasonic energy and using a surfactant [Fig. 1 (b)] only individual CNTs were identified on the fracture surface. The results indicate that the application of ultrasonic energy and the use of surfactant can be employed to effectively disperse CNTs in cementitious matrix.

Fig. 1 SEM images of cement paste reinforced with CNTs dispersed with (FIG. 1(b)) and without (FIG. 1(a)) the application of ultrasonic energy and the use of surfactant

To evaluate the reinforcing effect of CNTs fracture mechanics tests were performed using MWCNTs with aspect ratios of 700 and 1600 for short and long CNTs, respectively. Additionally, to investigate the effect of CNTs concentration cement paste samples reinforced with lower and higher amounts of CNTs (0.048wt% and 0.08wt%, respectively) were tested. The fracture mechanics test results of the average Young's modulus of the nanocomposites which demonstrated the best mechanical performance are illustrated in Fig. 2. In all cases, the samples reinforced with CNTs exhibit much higher Young's modulus than plain cement paste. More specifically, it is observed that the specimens reinforced with either short CNTs at an amount of 0.08wt% or long CNTs at an amount of 0.048wt% provide the same level of mechanical performance. Generally, it can be concluded that the optimum amount of CNTs depends on the aspect ratio of

CNTs. When CNTs with low aspect ratio are used a higher amount close to 0.08wt% by weight of cement is needed to achieve effective reinforcement. However, when CNTs with high aspect ratio are used less amount of CNTs close to 0.048 wt% is needed to achieve the same level of mechanical performance.

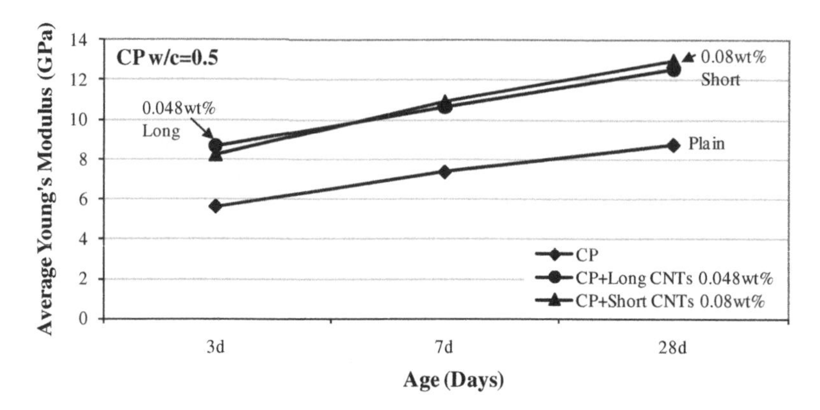

Fig. 2 Fracture mechanics test results of the Young's modulus of CNTs nanocomposites which exhibit the best mechanical performance among the different mixes tested

Comparing the 28 days Young's modulus of the nanocomposites with that of the plain cement paste, a 50% increase is observed. Based on the parallel model [16] the predicted Young's modulus of cement paste nanocomposites reinforced with either 0.048wt% or 0.08 wt%CNTs at the age of 28 days (~9.1 GPa) is much lower than the experimental values obtained (~13 GPa). To further investigate the increase of the Young's modulus, nanoindentation tests on 28 days cement paste samples reinforced with CNTs were performed. The results were compared with cement paste samples with silica fume.

Fig. 3 illustrates the probability plot of the Young's modulus of plain cement paste (w/c=0.5), cement paste reinforced with 0.08wt% short CNTs, cement paste reinforced with 0.048wt% long CNTs and cement paste with silica fume. A peak analyzing protocol was used to fit four normal distributions to the probability plot of the Young's modulus corresponding to the porous phase, low stiffness C-S-H, high stiffness C-S-H and calcium hydroxide [1, 17]. It is observed that the peak of the distribution of the nanoindentation modulus of the plain cement paste and cement paste with silica fume falls in the area of 15 to 20 GPa which corresponds to the low stiffness C-S-H. However, the peaks of the probability plot of the Young's modulus of the CNTs nanocomposites were found to be in the area of 20 to 25 GPa which corresponds to the high stiffness C-S-H gel. These results indicate that the incorporation of CNTs increased the amount of high stiffness C-S-H gel resulting in a stronger material. Additionally, it is observed that the probability of the Young's modulus below 10 GPa, which represents the porous phase, is reduced for the samples with silica fume and CNTs. These results suggest that CNTs similarly with silica fume reduce the nanoporosity of cement paste by filling the gaps between the C-S-H gel. Furthermore, comparing the probability plots of the

Young's modulus of the CNTs nanocomposites it is observed that the probability of high-stiffness C-S-H is higher for the samples with short CNTs and lower for the samples with long CNTs. This indicates that the samples reinforced with 0.08wt% short CNTs exhibit more improved properties at the nanoscale than the samples reinforced with 0.048wt% long CNTs. This response comes into agreement with the macromechanical properties of the samples where the samples with 0.08wt% short CNTs exhibit slightly higher Young's modulus than the samples reinforced with 0.048wt% long CNTs.

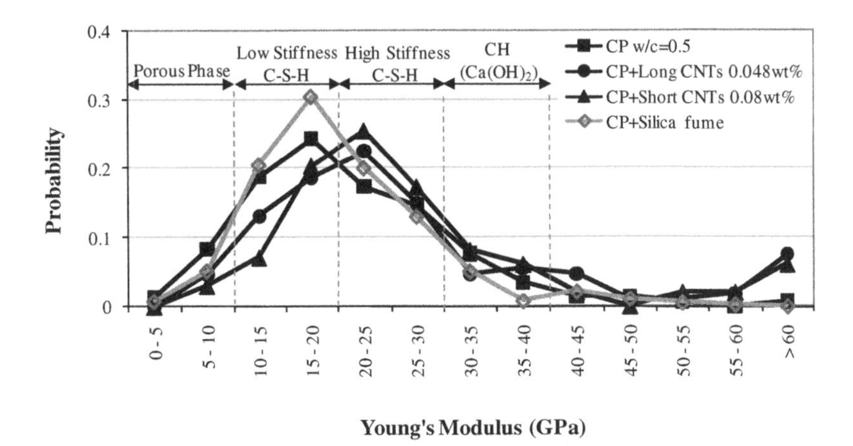

Fig. 3 Probability plot of the calculated Young's modulus of 28 days cement paste (w/c=0.5), cement paste reinforced with 0.048wt% long CNTs, cement paste reinforced with 0.08wt% short CNTs and cement paste with silica fume

4 Conclusions

The effect of multiwall carbon nanotubes (MWCNTs) on the nanostructure as well as the macro and nanoscale mechanical properties of cement paste has been investigated. An effective method of dispersing carbon nanotubes in cement paste matrix by applying ultrasonic energy and with the use of a surfactant has been developed. The fracture mechanics test results indicate that the fracture properties of cement matrix are increased through proper dispersion of small amounts of CNTs (0.048wt% and 0.08wt%). In particular, when short CNTs are used, higher amounts of CNTs (0.08wt%) are required to achieve effective reinforcement, while when longer CNTs are incorporated, lower amounts of CNTs (0.048wt%) are needed to achieve the same level of mechanical performance. The nanoindentation results suggest that CNTs can strongly modify and reinforce the cement paste matrix at the nanoscale by increasing the amount of high stiffness C-S-H and decreasing the porosity. A comparison of the nano-mechanical properties of bulk paste, cement paste reinforced with CNTs and cement paste with silica fume has shown that CNTs substantially enhance the Young's modulus of the C-S-H phase.

References

1. Mondal, P.: Nanomechanical properties of cementitious materials. PhD thesis, Northwestern University (2008)
2. Makar, J.M., Beaudoin, J.J.: Carbon nanotubes and their applications in the construction industry. In: Nanotechnology in construction. Proceedings of the 1st International Symposium on Nanotechnology in Construction, pp. 331–341. Royal Society of Chemistry (2004)
3. Makar, J.M., Margeson, J., Luh, J.: Carbon nanotube/cement composites – Early results and potential applications. In: Proceedings of the 3rd International Conference on Construction Materials: Performance, Innovations and Structural Implications, Vancouver, Canada, pp. 1–10 (2005)
4. Li, G.Y., Wang, P.M., Zhao, X.: Mechanical behavior and microstructure of cement composites incorporating surface-treated multi-walled carbon nanotubes. Carbon 43, 1239–1245 (2005)
5. Li, G.Y., Wang, P.M., Zhao, X.: Pressure-sensitive and microstructure of carbon nanotube reinforced cement composites. Cem. Concr. Comp. 29, 377–382 (2007)
6. Saez de Ibarra, Y., Gaitero, J.J., Erkizia, E., Campillo, I.: Atomic force microscopy and nanoindentation of cement pastes with nanotube dispersions. Physica Status Solidi (a) 203, 1076–1081 (2006)
7. Wansom, S., Kidner, N.J., Woo, L.Y., Mason, T.O.: AC-impedance response of multi-walled carbon nanotube/cement composites. Cem. Concr. Comp. 28, 509–519 (2006)
8. Cwirzen, A., Habermehl-Chirzen, K., Penttala, V.: Surface decoration of carbon nanotubes and mechanical properties of cement/carbon nanotube composites. Adv. Cem. Res. 20, 65–73 (2008)
9. Salvetat, J.P., Bonard, J.M., Thomson, N.H., Kulik, A.J., Forro, L., Benoit, W., et al.: Mechanical properties of carbon nanotubes. Appl. Phys. A 69, 255–260 (1999)
10. Belytschko, T., Xiao, S.P., Schatz, G.C., Ruoff, R.: Atomistic simulations of nanotube fracture. Phys. Rev. B 65, 235430–235437 (2002)
11. Groert, N.: Carbon nanotubes becoming clean. Mater. Today 10, 28–35 (2007)
12. Xie, X.L., Mai, Y.W., Zhou, X.P.: Dispersion and alignment of carbon nanotubes in polymer matrix: A review. Mater. Sci. Eng. Rep. 49, 89–112 (2005)
13. Konsta-Gdoutos, M.S., Metaxa, Z.S., Shah, S.P.: Nanoimaging of highly dispersed carbon nanotube reinforced cement based materials. In: Seventh Intl. RILEM Symp. on Fiber Reinforced Concrete: Design and Applications, Chennai, India, pp. 125–131 (2008)
14. Shah, S.P., Swartz, S.E., Ouyang, C.: Fracture mechanics of concrete: application of fracture mechanics to concrete, rock and other quasi-brittle materials. John Willey and Sons, New York (1995)
15. Mondal, P., Shah, S.P., Marks, L.D.: Nanoscale characterization of cementitious materials. ACI Mater. J. 105, 174–179 (2008)
16. Mindess, S., Young, J.F., Darwin, D.: Concrete. Prentice Hall, Upper Saddle River (2003)
17. Constantinides, G., Ulm, F.-J.: The nanogranular nature of C-S-H. J. Mech. Phys. Solids 55, 64–90 (2007)

Progress in Nanoscale Studies of Hydrogen Reactions in Construction Materials

J.S. Schweitzer, R.A. Livingston, J. Cheung, C. Rolfs, H.-W. Becker, S. Kubsky, T. Spillane, J. Zickefoose, M. Castellote, N. Bengtsson, I. Galan, P.G. de Viedma, S. Brendle, W. Bumrongjaroen, and I. Muller

Abstract. Nuclear resonance reaction analysis (NRRA) has been applied to measure the nanoscale distribution of hydrogen with depth in the hydration of cementitious phases. This has provided a better understanding of the mechanisms and kinetics of cement hydration during the induction period that is critical to improved concrete technology. NRRA was also applied to measure the hydrogen depth profiles in other materials used in concrete construction such as fly ash and steel. By varying the incident beam energy one measures a profile with a depth resolution of a few nanometers. Time-resolved measurements are achieved by stopping the chemical reactions at specific times. Effects of temperature, sulfate concentration, accelerators and retarders, and superplasticizers have been investigated. Hydration of fly ashes has been studied with synthetic glass specimens whose chemical compositions are modeled on those of actual fly ashes. A combinatorial chemistry approach was used where glasses of different compositions are hydrated in various solutions for a fixed time. The resulting hydrogen depth profiles show significant differences in hydrated phases, rates of depth penetration and amount of surface etching. Hydrogen embrittlement of steel was studied on slow strain rate specimens under different corrosion potentials.

J.S. Schweitzer, T. Spillane, and J. Zickefoose
University of Connecticut, Storrs, CT, USA

R.A. Livingston
University of Maryland, College Park, MD, USA

J. Cheung
W.R. Grace, Cambridge, MA, USA

C. Rolfs and H.-W. Becker
Ruhr-Universität Bochum, Bochum, Germany

S. Kubsky
Synchrotron SOLEIL, Saint-Aubin, Gif-sur-Yvette CEDEX, France

M. Castellote, N. Bengtsson, I. Galan, and P.G. de Viedma
Institute of Construction Science "Eduardo Torroja" (CSIC), Madrid, Spain

S. Brendle
Delft University of Technology, Delft, The Netherlands

W. Bumrongjaroen and I. Muller
Catholic University, Washington, DC, USA

1 Introduction

At the previous International Symposium on Nanotechnology in Construction (NICOM2) in 2005 in Bilbao, the use of nuclear resonant reaction analysis (NRRA) to investigate the induction period in the hydration of cementitious phases was described [1]. This method gives in-situ measurements of hydrogen concentration with depth at a depth resolution of a few nanometers [2]. It uses the E_R = 6.400 MeV resonance in the $^{1}H(^{15}N,\alpha\gamma)^{12}C$ reaction [3]. A nitrogen ion beam with precisely regulated energies and good energy resolution is produced by a 4 MeV Dynamitron tandem accelerator at Ruhr-Universität Bochum, Germany which provides an H-detection sensitivity of about 10 ppm and an H-depth resolution of a few nm at the surface [4]. This has enabled the detailed investigation of the effect of temperature and other factors on the induction period [4-6].

Since NICOM2 the application of NRRA has gone beyond the silicate phases in Portland to look at the hydration of the calcium aluminate phase. It has also been applied to study the pozzolanic reactions of fly ash glasses and the process of hydrogen embrittlement of steel. These studies are reported here.

2 Experimental Approach

The experimental procedure for studying hydration has been described in detail elsewhere [2]. A major difference between this method and others for studying cement, like calorimetry, is the material being studied is not in powdered form, but is rather a solid pellet, that presents a smooth surface to the ion beam. Cementitious phases such as tricalcium silicate (C_3S) and tricalcium aluminate (C_3A) are molded into cylindrical pellets of 12.7 mm diameter and fused. A typical experiment involves 8-12 samples. They are hydrated in a common solution bath of specific composition and temperature; individual samples are removed at specific times. Samples are stored and handled under inert atmosphere both before and after the chemical reaction. Reacted samples are kept in vacuum until analysis.

Each sample is a single point in the material's hydration time history. To obtain an H depth profile, the beam energy is increased stepwise from just below the resonance energy of 6.400 MeV. As this is an isolated resonance in the H cross section the reaction only occurs when the ^{15}N ion energy is at the resonance energy. If its energy is greater, no reactions occur until the beam loses enough energy by scattering to get down to the resonance energy. At each energy step, the ^{15}N ion will reach the resonance energy at a particular sample depth, and the hydrogen concentration at that depth is measured. For each energy step, a gamma-ray spectrum is acquired, typically 10,000 cts per minute. The beam energy is increased in 10 keV steps to 7 MeV to resolve thin surface layers, and then in coarser steps (100-500 keV) as the profile typically changes more slowly in this region. The maximum beam energy is limited to 12 MeV to avoid interference from the next higher energy resonance.

A plot of the H signal as a function of incident beam energy allows a visualization of the H depth profile, as shown in Fig. 1 for a C_3S sample during the induction period. It shows the typical Gaussian peak associated with a surface layer, on the left edge of the figure, followed by a diffusion-type region at greater depths.

The NRRA coordinates of beam energy and counts per charge have been converted to depth and H concentrations on the upper and right axes, respectively. Profiles for three times are shown to illustrate the diffusion region growth with the Gaussian peak unchanged. The induction period ends with the surface layer breakdown. This is easily recognized by the absence of the Gaussian peak and a change in the shape of the diffusion region curve that allows the time for the induction period to be determined to a relative precision better than 5%.

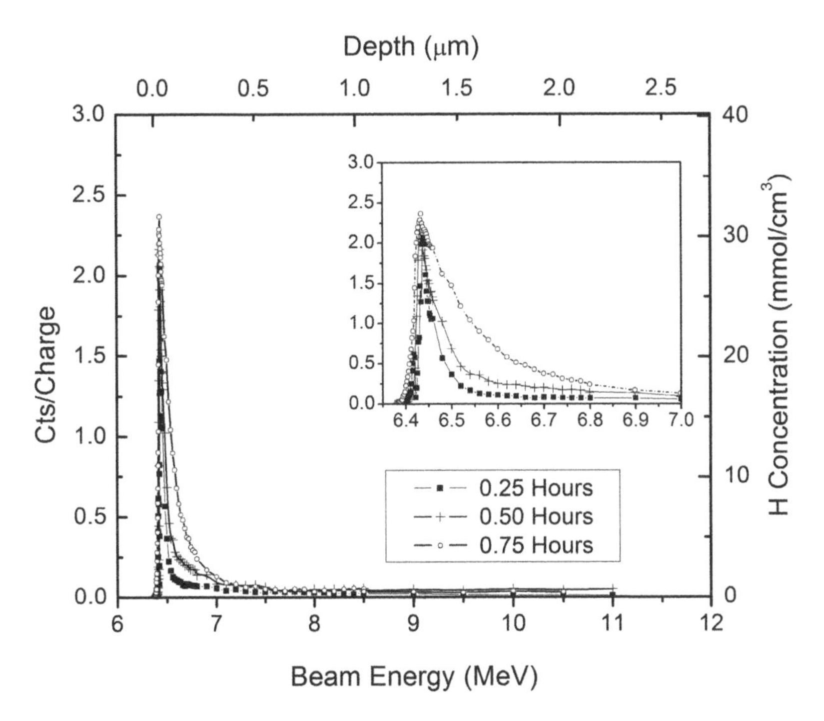

Fig. 1 Evolution of hydrogen depth profile for triclinic C_3S hydrated at 30 °C. The inset expands the left portion of the figure for clarity

3 Cement Measurements

NRRA hydration studies have been reported on several of the silicate cement phases including the effects of retarders, absorbers, and temperature. From the temperature dependence it has been possible to determine the activation energy. The studies have been extended to investigate the hydration properties of calcium aluminate (C_3A). As is well known, C_3A appears to have rapid early hydration reactions. Therefore, to obtain better sensitivity for seeing changes in the hydration properties, we have performed studies at temperatures of typically 5-10 °C. Factors that have been studied for C_3A include the effects of gypsum, retarders, and superplasticizers. Figure 2 shows a comparison of the hydration profiles after 40 minutes of hydration at a temperature of 10 °C. This figure shows the results for three different conditions. All samples are hydrated in a fully saturated calcium hydroxide water solution. In one case, gypsum was also added to the solution. In the third case, both gypsum and a superplasticizer were added.

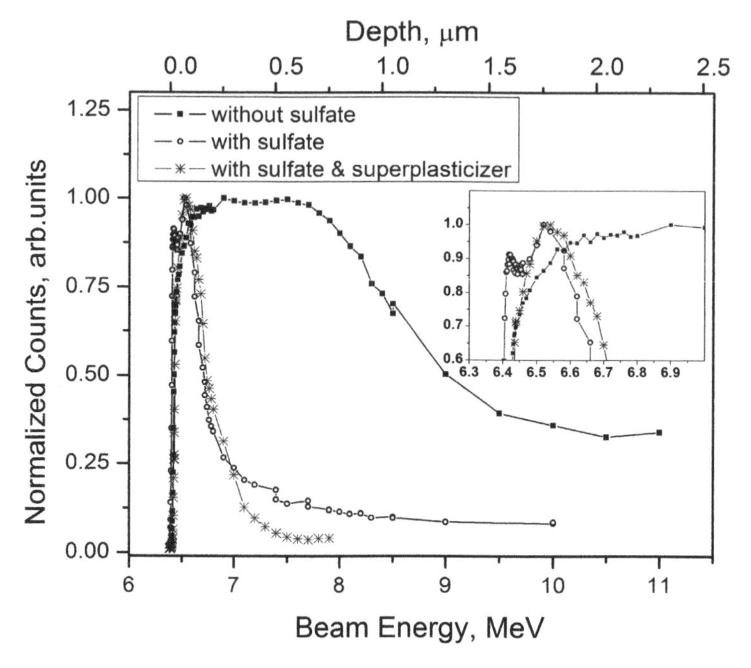

Fig. 2 C_3A hydration after 40 min. at 10 °C for three different solutions

All three profiles are very different from the hydration profile for tricalcium silicate (Fig. 1). When no gypsum is present the hydration profile has saturated down to about ¾ of a micron with a diffusion tail to greater depths. This indicates rapid early reaction and corresponds to flash set of Portland cement. When gypsum is added to the hydrating solution, hydration has only occurred to about a ¼ of a micron. As shown in the inset in Fig. 2, a very sharp rise and flat plateau appears at the leading edge. We believe this is due to a crystalline surface layer, presumably ettringite, that has impeded the hydration rate. Examination of the profiles in pellets taken at other times in this series show essentially no difference. The implication is that the formation of the ettringite layer stops the hydration reaction in less than five minutes. When a superplasticizer is also added to the gypsum and calcium hydroxide in solution, the leading edge has a different shape that resembles a Gaussian curve that would be typical of a disordered noncrystalline layer. Nevertheless, the rest of hydration curve falls off even more rapidly than the one with the ettringite layer, suggesting that the superplasticizer interferes with the formation of the surface layer or alters its character, but similarly stops the hydration reaction at a very early time.

4 Fly Ash Measurements

A simple way to study fly ash hydration is with glasses whose chemical compositions are identical to those of a particular fly ash. The glass samples are made with smooth flat surfaces for NRRA study, so we can study the hydration changes as a function of small changes in the fly ash chemical composition.

NRRA provides direct observation of the depth profile of hydrogen diffusing into the fly ash in exchange for the alkalis and Ca^{2+} as shown in Fig. 3. These H depth profiles are for three calcium aluminate silicate glasses that have chemical compositions based on data from actual fly ash glass samples [7]. The major difference among these specimens is the Ca/(Na+K) ratio. The specimens were hydrated for 72 hrs in simulated concrete pore solution (0.4 M KOH + 0.0215 M $Ca(OH)_2$, pH=13.5). The profile for the low Ca specimen appears to be much shallower than the others, which suggests that it reacts more slowly. Another possibility however, as shown by analysis of the dissolved ions in the solution, is that the rate of etching of this glass is so rapid that a full depth gradient cannot develop, as initial hydrated portions are etched leaving a fresh surface. The profiles for the other two glasses show hydrogen profiles that reach a plateau that extends over the 1-micron range of the measurement. This indicates the presence of a saturated phase. Given the higher Ca content of these fly ashes, this is likely to be a C-S-H gel formed through the alkali-activated, or self-cementing reaction.

Repeating these measurements at different hydration times makes possible the determination of parameters such as kinetic rate constants and diffusion coefficients. However, these profiles generally cannot be fitted to a simple Fick's Law model based on the erfc function. Instead, more complicated mathematical relations would be required, possibly involving concentration dependent diffusion coefficients [8] . This diffusion process within the unreacted core is not the same as the one that is associated with the topochemical reaction that involves diffusion through the surrounding C-S-H gel layer. Both diffusion processes affect the overall rate of growth of the gel. However, the reaction at the core/gel interface is affected by the fly ash glass composition, whereas the diffusion to the core's reaction surface is determined by the properties of the gel. By studying the hydration

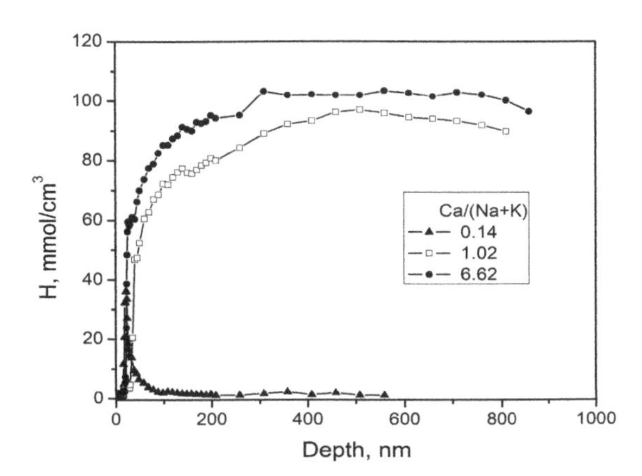

Fig. 3 NRRA hydrogen depth profiles of three synthetic fly ash glass specimens

reactions of the fly ash in isolation and in a cement paste, it is possible to determine the contributions of the individual diffusion processes.

5 Hydrogen Embrittlement Measurements

The application of NRRA to investigate hydrogen embrittlement is illustrated by Fig. 4 [9]. This presents the H profiles for three samples of a commercial cold drawn pearlitic steel (0.77% C) that were exposed to different corrosion conditions designed to accelerate hydrogen embrittlement. The samples were subjected to the combined action of stress and aggressive environment through the Slow Strain Rate Test (SSRT) in which a potential is applied so that H is generated within cracks and pits in the specimen. The aggressive environment was a naturally aerated 0.05 M aqueous solution of $NaHCO_3$ (pH = 8.5) previously shown to be capable of promoting SCC in cold drawn steels [10, 11]. An anodic potential of -300 mV (SCE) was applied to sample N1. This condition promotes pitting corrosion, a localized attack in which hydrogen is produced by acid hydrolysis of the corrosion products in the interior of the pit[10, 11]. In contrast, sample N14 was subjected to a cathodic potential of -1200 mV (SCE). At this potential, water is not thermodynamically stable [12]. This produces molecular hydrogen at the metal surface some of which diffuses into the metal and promotes embrittlement.

For the NRRA measurement, the sample was mounted so that the beam hit it normal to the fracture surface. It can be observed in Fig. 4 that the resulting shape of the H profile is different depending on the treatment. The -1200 mV (N14) sample has a higher concentration near the fracture surface, but it diminishes to values very close to that of the inert sample at about 0.3 μm. In between it exhibits a convex shape, which indicates a departure from a simple Fick's Law diffusion process. The N1 sample shows a more Fickian profile but exhibits a flat profile of hydrogen deeper in the sample at depths greater than 0.5 μm. The total amount of hydrogen in each sample can be determined by integrating the area under the curve. The peak at the surface in each curve may be due to H adsorption, the first step in the HE process. However, surface contamination such as grease from handling or exposure to atmospheric humidity may also be present. Therefore, to eliminate this peak the integration of the hydrogen concentration was restricted to the range between 0.020 μm and 0.48 μm, the maximum depth of the NRRA technique. The resulting area densities are 0.13, 0.67 and 0.46 $\mu mol/cm^2$ for the N4 (inert), N1 (-300 mV) and N14 (-1200 mV) samples, respectively. Therefore, the sample tested under pitting corrosion conditions took up a greater amount of hydrogen than the one tested at the more negative potential.

These results can also be used to estimate key parameters in the hydrogen embrittlement process. By the nature of the SSRT procedure, the H concentration in the sample is the critical value for fracture. The average of the two samples N1 and N14 is 0.56 $\mu mol/cm^2$. Taken over a depth of 0.46 μm, this is an average concentration of 12 $mmol/cm^3$ or 7.3 x 1021 $atoms/cm^3$. Steel has a number density of 8.53×10^{22} $atoms/cm^3$. Hence the critical H value is 8.6 atomic percent.

Concerning the diffusion constant, as noted above it is not possible to fit the profiles with Fick's Law models. But it is possible to make a rough estimate, given the time of exposure to solution of ~ 48 hrs, and the effective depth of 0.46

μm. The resulting effective diffusion constant is on the order of 10^{-15} cm^2/sec. For comparison, the values in the literature are in the range of 10^{-10}-10^{-11} cm^2/sec [20, 21]. However, these were determined indirectly, whereas these NRRA values are measured directly. A major advantage of NRRA is that it is nondestructive and requires no special treatment of the samples. Thus it is possible to measure the same steel sample repeatedly over time to observe how the depth profile evolves. Hence, this time-resolved approach can be used to sort out the different diffusion mechanisms that control the hydrogen embrittlement process.

6 Conclusions

We have used NRRA with the $^{1}H(^{15}N,\alpha\gamma)^{12}C$ reaction to study physical and chemical processes at a nanometer scale involved in the hydration of various components of Portland cement. This has given new insights into the mechanisms that control the setting and curing of concrete and has helped to resolve some long-standing controversies. For the calcium silicate phases, there is generally the development of a semi-permeable surface layer that controls the hydration rate of reaction during the induction period. The induction period length has a classic Arrhenius-type dependence on temperature.

Tricalcium aluminate shows a very different spatial pattern of hydration than the calcium silicates. There is no semi-permeable surface layer. Instead a crystalline layer develops rapidly when gypsum is present and apparently slows further reaction with water.

Initial experiments with accelerators and retarders have shown that they significantly affect the development the hydration profile, either by changing the permeability of the surface layer or the diffusion coefficients in the substrate. For a commercially available retarder like sodium gluconate, the rate of hydration is strongly dependent on the dosage. These studies have demonstrated the great promise of nuclear resonance reaction analysis for better understanding the hydration properties of cements and that the hydration of Portland cement is a valid topic for nanoscience research.

The NRRA technique has also been used to study hydration reactions of fly ash and hydrogen concentrations resulting in hydrogen embrittlement of steels.

Acknowledgments. The authors are indebted to the support of the National Science Foundation under contract CMS-0600532 that made this research possible.

References

1. Schweitzer, J.S., Livingston, R.A., Rolfs, C., Becker, H.-W., Kubsky, S., Spillane, T., Castellote, M., Garcia de Viedma, P.: Nanoscale Measurements Of Cement Hydration During The Induction Period. In: Bartos, P.J.M., de Miguel, Y., Porro, A. (eds.) NICOM: 2nd International Symposium on Nanotechnology for Construction, pp. 125–132. RILEM Publications SARL Bilbao, Spain (2006)

2. Livingston, R.A., Schweitzer, J.S., Rolfs, C., Becker, H.-W., Kubsky, S.: Characterization of the Induction Period in Tricalcium Silicate Hydration by Nuclear Resonance Reaction Analysis. J. Mat. Res. 16, 687–693 (2001)
3. Lanford, W.A.: Nuclear Reactions for Hydrogen Analysis. In: Tesmer, J., Nastasi, M. (eds.) Handbook of Modern Ion Beam Analysis, pp. 194–204. Materials Research Society, Pittsburgh (1995)
4. Schweitzer, J.S., Livingston, R.A., Rolfs, C., Becker, H.-W., Kubsky, S.: Study of Cement Chemistry with Nuclear Resonant Reaction Analysis. In: Duggan, J.L., Morgan, I.L. (eds.) Proceedings of the Sixteenth International Conference, Applications of Accelerators in Research and Industry, pp. 1077–1080. American Institute of Physics, Denton (2001)
5. Schweitzer, J.S., Livingston, R.A., Rolfs, C., Becker, H.-W., Kubsky, S.: Ion beam analysis of the hydration of tricalcium silicate. Nucl. Instr. Meth. B 207, 80–84 (2003)
6. Schweitzer, J.S., Livingston, R.A., Rolfs, C., Becker, H.-W., Kubsky, S., Spillane, T., Castellote, M., Garcia de Viedma, P.: Nanoscale studies of cement chemistry with 15N Resonance Reaction Analysis. Nucl. Instr. Meth. B 241, 441–445 (2005)
7. Bumrongjaroen, W., Muller, I.S., Pegg, I.L., McKeown, D., Viragh, C., Davis, J.: Characterization of Glassy Phase in Fly Ash from Iowa State University, VSL-07R520X -1, Vitreous State Laboratory, Washington, DC (2008)
8. Crank, J.: The Mathematics of Diffusion. Oxford University Press, Oxford (1967)
9. Castellote, M., Fullea, J., de Viedma, P.G., Andrade, C., Alonso, C., Lorente, I., Turrillas, X., Campo, J., Schweitzer, J.S., Spillane, T., Livingston, R.A., Rolfs, C., Becker, H.-W.: Hydrogen Embrittlement of High Strength Steel studied by Nuclear Resonance Reaction Analysis and Neutron Diffraction. Nucl. Instr. Meth. B (in review) (2007)
10. Acha, M.: Univ. Complutense de Madrid, Madrid (1993)
11. Alonso, M.C., Procter, R.P.M., Andrade, C., Saenz de Santa Maria, M.: Susceptibility to stress-corrosion cracking of a prestressing steel in $NaHCO_3$ solutions. Corros. Sci. 34, 961 (1993)
12. Pourbaix, M., de Zoubov, N.: Iron. In: Pourbaix, M. (ed.) Atlas of Electrochemical Equilibria in Aqueous Solutions, pp. 307–321. Pergamon Press, London (1966)

Engineering of SiO_2 Nanoparticles for Optimal Performance in Nano Cement-Based Materials

K. Sobolev, I. Flores, L.M. Torres- Martinez, P.L. Valdez, E. Zarazua, and E.L. Cuellar

Abstract. The reported research examined the effect of 5-70 nm SiO_2 nanoparticles on the mechanical properties of nano-cement materials. The strength development of portland cement with nano-SiO_2 and superplasticizing admixture was investigated. Experimental results demonstrate an increase in the compressive and flexural strength of mortars with developed nanoparticles. The distribution of nano-SiO_2 particles within the cement paste plays an essential role and governs the overall performance of these products. Therefore, the addition of a superplasticizer was proposed to facilitate the distribution of nano-SiO_2 particles. Superplasticized mortars with 0.25% of selected nano-SiO_2 demonstrated a 16% increase of 1-day compressive strength, reaching 63.9 MPa; the 28-day strength of these mortars was 95.9 MPa (vs. strength of reference superplasticized mortars of 92.1 MPa). Increase of 28-day flexural strength of superplasticized mortars with selected nano-SiO_2 was 18%, reaching 27.1 MPa. It is concluded that the effective dispersion of nanoparticles is essential to obtain composite materials with improved performance.

1 Introduction

Recent nano-research in cement and concrete has focused on the investigation of the structure of cement-based materials and their fracture mechanisms [1-5]. With new advanced equipment it is possible to observe the structure at its atomic level and even measure the strength, hardness and other basic properties of the micro- and nano-scopic phases of materials [2]. The application of Atomic Force Microscopy (AFM) for the investigation of the "amorphous" C-S-H gel revealed that at the nanoscale this product has a highly ordered structure [6]. Better understanding the nano-structure of cement based materials helps to control the processes related to hydration, strength development, fracture, and corrosion. For instance, the

---

K. Sobolev
Department of Civil Engineering, CEAS, University of Wisconsin-Milwaukee, USA

I. Flores, L.M. Torres- Martinez, P.L. Valdez, and E. Zarazua
Facultad de Ingenieria Civil, Universidad Autonoma de Nuevo Leon, Mexico

E.L. Cuellar
Facultad de Ingeniería Mecánica y Eléctrica, Universidad Autonoma de Nuevo Leon, Mexico

development of materials with new properties such as self-cleaning, discoloration resistance, anti-graffiti protection, and high scratch and wear resistance, is extremely important for many construction applications [7-11]. A self-cleaning effect (i.e., decomposition of organic pollutants and gases) is achieved when TiO_2 based photocatalyst thin film is deposited on the material's surface and exposed to UV light [7-9]. Other examples are related to the development of new superplasticizers for concrete, namely, Sky, which is based on polycarboxylic ether (PCE) polymer. This product was recently developed by BASF with a nano-design approach targeting the extended slump retention of concrete mixtures [12]. For decades, major developments in concrete performance were achieved by the application of super-fine particles: fly ash, silica fume, metakaolin, and now nanosilica. The optimal performance of these systems was attributed to the high-density continuous packings of the binder constituents (Fig. 1) that are realized at high fluidity levels with the help of the effective superplasticizers [3-5].

For example, silicon dioxide nanoparticles (nanosilica, nano-SiO_2) proved to be a very effective additive to polymers for improving strength, flexibility, and durability. Nano-SiO_2 can be used as an additive to improve workability and the strength of high-performance and self-compacting concrete [3-5, 13-15]. Nano-binders composed of nano-sized cementitious material, pozzolanic nanoparticles, and a finely ground mineral additive-portland cement mixture were proposed [5]. The nano-sized cementitous component is obtained by the colloidal milling of a conventional or low temperature sintered portland cement clinker (the top-down approach), or is manufactured by the self-assembly route using sol-gel method or mechano-chemically induced topo-chemical reactions (the bottom-up approach) [3, 4].

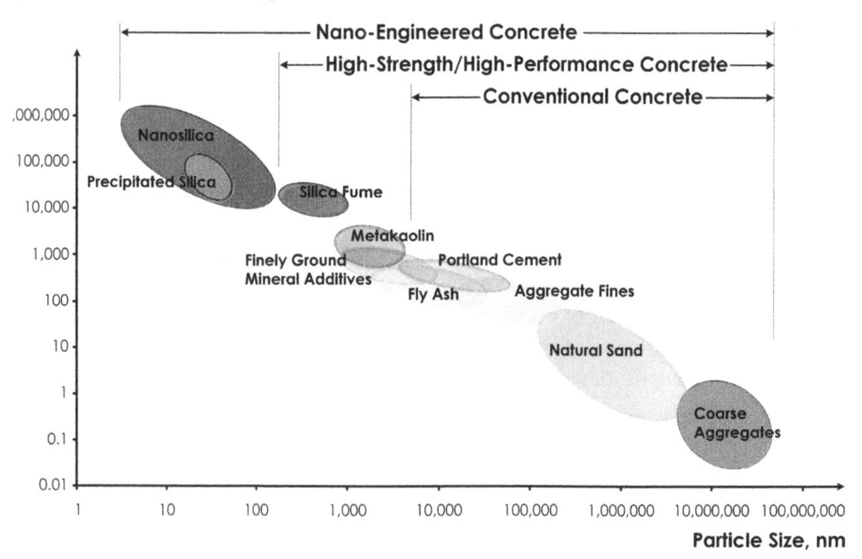

Fig. 1 The particle size and specific surface area scale related to concrete materials [3]

2 Methods to Produce Nanomaterials

The relatively small quantities, less than 1% of nano-sized materials are sufficient to improve the performance of nanocomposites [3-5, 16]. Yet, the commercial success of nanomaterials depends on the ability to manufacture these materials in large quantities and at a reasonable cost relative to the overall effect of the nano-product. The nanomaterials technologies, which could lead to the industrial outputs, involve plasma arching, flame pyrolysis, chemical vapour deposition, electrodeposition, sol-gel synthesis, mechanical attrition and the use of natural nanosystems [17]. Among chemical technologies, sol-gel synthesis (Fig. 2) is one of the widely used "bottom-up" production methods for nano-sized materials, such as nano-silica. The process involves the formation of a colloidal suspension (sol) and gelation of the sol to form a network in a continuous liquid phase (gel). Usually, trymethylethoxysilane or tetraethoxysilane (TMOS/TEOS) is applied as a precursor for synthesizing nanosilica. The sol-gel formation process can be simplified to few stages [3, 17]:

1. Hydrolysis of the precursor;
2. Condensation and polymerization of monomers to form the particles;
3. Growth of particles;
4. Agglomeration of particles, followed by the formation of networks and, subsequently, gel structure;
5. Drying (optional) to remove the solvents and thermal treatment (optional) to remove the surface functional groups and obtain the desired crystal structure.

The chemical reaction of nanosilica synthesis can be summarized as:

$$n\mathrm{Si(OC_2H_5)_4} + 2n\mathrm{H_2O} \xrightarrow[\mathrm{NH_3}]{\mathrm{C_2H_5OH}} n\mathrm{SiO_2} + 4n\mathrm{C_2H_5OH}$$

There are a number of parameters that affect the process, including pH, temperature, concentration of reagents, H_2O/Si molar ratio (between 7 and 25), and type of catalyst [17, 18]. When precisely executed, this process is capable of producing perfectly spherical nanoparticles of SiO_2 within the size range of 1–100 nm (Figs. 2, 3). Processing natural nanosystems, such as clays, is a promising technique used for manufacturing nanoporous materials (with pores in the order of 10 nm) [18]. Phyllosilicates consisting of alternating layers of silicate tetrahedra and aluminium octahedra are normally used. The technology involves the application of a pillaring solution (usually containing polycation $[Al_{13}O_4(OH)_{24}]^{7+}$) to widen the distance between the platelets of phyllosilicate. The resulting intercalated clay is converted to oxide by thermal treatment [19]. Due to their high pozzolanic activity, nanoclays have a good potential for improvement of portland cement properties. Mechanical attrition or mechano-chemical activation of powder particles is another method capable of synthesizing of nanomaterials at an industrial scale. It was successfully applied to manufacture new alloys of metals with differentiating melting points and metal-ceramic composites of improved strength and corrosion resistance. With mechanical attrition, the intermixing of individual components at an atomic scale occurs due to solid-state amorphization. The process is governed by both the mechanical alloying [17, 19] and the incorporation of lattice defects into the crystal structure. Mechano-chemistry, which is realized by high energy ball milling, helps activate the solid state displacement reactions. This technology has been used to manufacture ceramic nanoparticles such as Al_2O_3 and ZrO_2 [17]. Mechano-chemical activation is also a very effective method to attach the organic functional groups to the surface of inorganic powders, such as portland cement (i.e., organo-mineral hybridization) [3, 20, 21]. Intergrinding cement and dry modifiers in a ball mill results in a binders of improved performance [20, 21].

3 Concrete with Nanoparticles

The mechanical properties of cement mortars with nano-Fe_2O_3 and nano-SiO_2 were studied [15, 22]. Experimental results demonstrated an increase in the compressive and flexural strength of mortars containing nanoparticles. It was found that increasing the nano-SiO_2 dosage improves the strength of mortars, which was higher than that for mortars with silica fume. M. Collepardi et al. investigated self-compacting concretes with low-heat development [13-14]. Mineral additives such as ground limestone, fly ash and ground fly ash were used to control the heat of hydration. Nanosilica (of the size of 5–50 nm) was used as viscosity modifying agent at a dosage of 1-2% of cementitious materials. The best performance was demonstrated by concrete with ground fly ash, 2% nanosilica and 1.5% of superplasticizer. This concrete had the highest compressive strength of 55 MPa at the age of 28 days along with the desired behavior in a fresh state: low bleeding, perfect cohesiveness, better slump flow, and very little slump loss. The beneficial action of the nanoparticles on the microstructure and performance of cement-based materials can be explained by the following factors [3-5, 13-15]:

- Well-dispersed nanoparticles increase the viscosity of the liquid phase, helping to suspend the cement grains and aggregates, improving the segregation resistance and workability of the system;
- Nanoparticles fill the voids between cement grains, resulting in the immobilization of “free” water (“filler” effect);
- Well-dispersed nanoparticles act as centers of crystallization of cement hydrates, therefore accelerating the hydration;
- Nanoparticles favour the formation of small-sized crystals (such as $Ca(OH)_2$ and AF_m) and small-sized uniform clusters of C-S-H;
- Nano-SiO_2 participates in the pozzolanic reactions, resulting in the consumption of $Ca(OH)_2$ and formation of an “additional” C-S-H;
- Nanoparticles improve the structure of the aggregates’ contact zone, resulting in a better bond between aggregates and cement paste;
- Nanoparticles provide crack arrest and interlocking effects between the slip planes, which improve the toughness, shear, tensile and flexural strength of cement-based materials.

4 Design and Performance of Nano-SiO_2

The details on the "bottom-up" synthesis of nano-SiO_2 particles (nano-SiO_2) and the effect of this material on the performance of cement systems were reported [3]. The nanoparticles of SiO_2 (with the size range of 5–100 nm) were synthesized using the sol-gel method. Tetraethoxysilane (98% TEOS, supplied by Aldrich) was used as a precursor and the reaction was realized in a base reaction media with ammonia as a catalyst (ammonia solution at pH = 9). The specimens of nano-SiO_2 were obtained at different experimental conditions (ethanol-to-TEOS molar ratios 6 or 24 and water-to-TEOS ratios 6 or 24), as presented in Table 1. Portland cement (NPC, conforming to ASTM Type I, supplied by CEMEX) and silica fume (SF, supplied by Norchem) were used in the experimental program. In addition to synthesized nano-silica materials, commercially available nano-SiO_2 admixture Cembinder-8 (CB8, available in a form of 50% suspension in water, supplied by Eka Chemicals) was used as a reference material. Commercially available polyacrilate/polycarboxylate superplasticizer (PAE/SP, 31% concentra-tion, supplied by Handy Chemicals) was used as a modifying admixture. Prior to its application in mortars, SP was premixed with 18% (by weight) of tributyl-phosphate (99% TBP, supplied by Aldrich) in order to compensate for air-entraining effect of PAE. Graded Ottawa sand (ASTM C778) was used as a fine aggregate in all tested mortars. Distilled water was used for the preparation of mortars.

The synthesized nano-SiO_2 particles were characterized by the x-ray diffraction (XRD, Bruker AXS D8), transmission electron microscopy (TEM, JEOL-2010), scanning electron microscopy (SEM, JEOL), and nitrogen absorption (according to Brunauer-Emmett-Teller theory, BET; Quantachrome Nova E2000), as reported in Table 1.

The performance of nano-SiO_2 based mortars (at nano-SiO_2 dosage of 0.25% by the weight of the binder, water-to-cement ratio, W/C of 0.3 and sand-to-cement ratio, S/C of 1) was compared with the properties of two reference mixtures: plain mortar (with water-to-cement ratio of 0.3 and sand-to-cement ratio of 1) and superplasticized mortar (at SP dosage of 0.1%). Relevant ASTM standards were used for evaluation of mortars flow (ASTM C 1437), compressive (ASTM C 109) and flexural strength (ASTM C 348).

Table 1 Design and properties of nano-SiO_2

Specimen Type*	Molar Ratio TEOS/Etanol/H2O	Reaction Time, hours	Particle Size (TEM), nm	Surface Area (BET), m^2/kg
1B3	1/ 24 / 6	3	15-65	116,000
2B3	1/ 6 / 6	3	30, 60-70	145,000
3B3	1/ 6 / 24	3	15-20	133,000
4B3	1/ 24 / 24	3	5	163,000

* Sample coding: ABC
- First number denotes molar ratio combination as per as experimental matrix
- Last number corresponds to the reaction time in hours
- Letter – denotes reaction media: A- for acid and B- for base

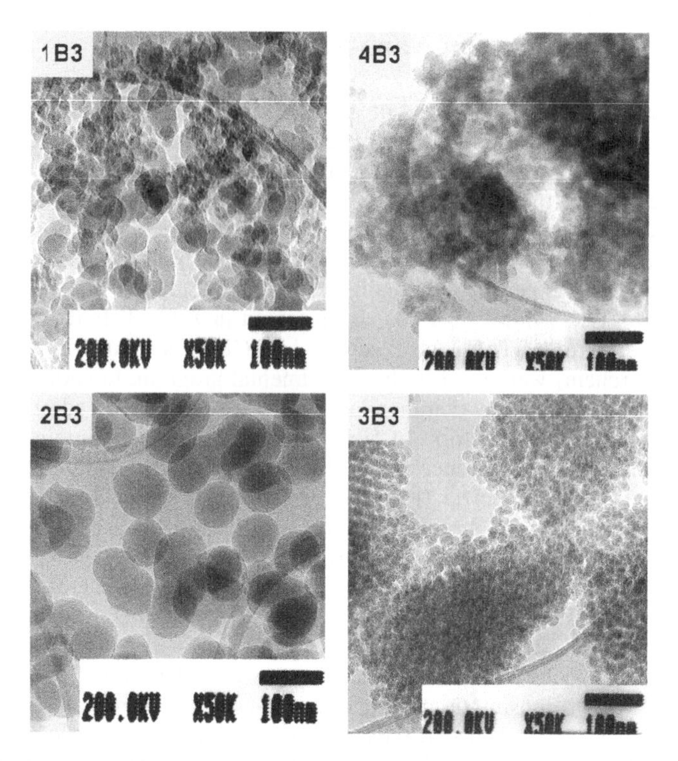

Fig. 2 The morphology and particle size distribution of developed nano-SiO_2 (TEM) [3]

Fig. 3 Effect of ultrasonification on xerogel agglomerates of nano-SiO_2 [3]: after drying at 70°C (left), the same particles, after ultrasonification for 5 minutes (right)

According to the results of x-ray diffraction, the obtained nano-SiO_2 is a highly amorphous material with predominant crystallite size of 1-2.5 nm [3]. TEM was used to characterize the morphology and particle size distribution of nano-SiO_2 (Fig. 2). It can be observed that nano-SiO_2 particles are represented by highly agglomerated xerogel clusters with the size of 0.5 - 10 μm (Fig. 3). The particles within the clusters are of the size of 5 - 70 nm and obtained xerogels are characterized by the BET surface area of 116,000-163,000 m^2/kg.

The majority of obtained nano-SiO_2 particles (used at a dosage of 0.25%) reduce the flow of plain mortars to some extent [3]. In cases when superplasticizer (PAE) was applied at a dosage of 0.1%, the major part of the obtained nano-SiO_2 did not affect the fluidity of reference superplasticized mortar (Table 2). The performance of developed nano-SiO_2 depends on the conditions of synthesis (i.e., molar ratios of the reagents, type of the reaction media and of the duration of the reaction) [3]. The best nano-SiO_2 products with particle size ranging from 5 to 20 nm are synthesized at highest molar concentrations of water (specimens 3B and 4B). The addition of nano-SiO_2 to plain portland cement mortars at a dosage of 0.25% (by the weight of cementitious materials) improves the 1-day strength by up to 17%. Early strength (up to three days) was also improved by the majority of obtained nano-SiO_2. The addition of nano-SiO_2 improves the 28-day compressive and flexural strengths of plain mortars by 10% and 25%, respectively [3].

The distribution of nano-SiO_2 particles within the cement paste is an important factor governing the performance of these products [3]. Therefore, the disagglomeration of nanoparticles is essential to obtain the composite materials with improved properties. The application of PAE superplasticizer, ultrasonification and/or high-speed mixing was proven to distribute nano-SiO_2 effectively [3]. According to the results of the experiment, the strength of superplasticized cement mortars with developed nano-SiO_2 was improved (Table 2). The addition of 0.1% of superplasticizer and 0.25% of nano-SiO_2 (specimen 4B3, manufactured at TEOS / Ethanol / H_2O ratio of 1:24:24 using base medium) provides the 1-day strength of 63.9 MPa, about 20% increase vs. plain cement mortar (Table 2); this value also represents 16% strength improvement vs. superplasticized cement mortar. The strength of mortars with 4B3 at the age of 3 to 7 days was similar to the strength of superplasticized mortar. Only a slight improvement in strength, about 4% (vs. strength of superplasticized mortar), was observed for this specimen at the 28-day age, reaching 95.9 MPa. Remarkable, 18% improvement of 28-day flexural strength was demonstrated by nano-SiO_2 specimen 4B7 (manufactured at the same reaction conditions as 4B3, but at a longer reaction time of 7 hours). Application of SF and commercial nano-SiO_2 (CB8) in superplasticized mortars demonstrated similar behaviour; however, the increase in the 1-day strength was only 8–9% (vs. the strength of superplasticized mortar). These materials slightly reduced the 28-day strength of superplasticized mortars.

It can be expected that the application of wider range of particle sizes and better dispersion of nano-SiO_2 can further improve the performance of engineered nano-particles [3].

5 Concluding Remarks

Nanotechnology has changed and will continue to change our vision, expectations and ability to control the materials world. Major progress in cement and concrete technology is expected in the next several years due to adaptation of new knowledge generated by this fast-growing field of science. Mechano-chemistry and nano-catalysts could change the face of the modern cement industry by significantly reducing the clinkering temperature and even realizing the possibility of low-temperature sintering of clinker minerals in mechano-chemical reactors. Development of the following concrete-related nanoproducts can be anticipated:

- Catalysts and processes for the low-temperature synthesis of clinker including nano-crystalline clinker;
- Grinding aids for superfine grinding and admixtures for mechano-chemical activation of cements;
- Binders with enhanced/nano-engineered internal bond between the hydration products;
- Binders reinforced with nano-particles, nano-rods, nano-tubes (including SWNTs), nanofibers, nano-dampers, nano-nets, or nano-springs;
- Binders modified by nano-sized polymer particles, their emulsions or polymeric nano-films;
- Bio- and bio-mimicking materials (including those imitating the structure and performance of mollusc shells);
- Cement-based composites reinforced with new fibers containing nanotubes, including fibers covered by nano-layers (to enhance the bond, corrosion resistance, electrical conductivity);
- Next-generation superplasticizers for "total workability control" and supreme water reduction;
- Cement-based materials with supreme strength, ductility and toughness;
- Cement-based materials with engineered nano- and micro- structures exhibiting supreme durability;

Table 2 Performance of nano-SiO_2 in mortars

Specimen Type	Dosage, %		Flow, %	Compressive Strength, MPa at the age of				Flexural Strength,* MPa
	SiO_2-based Additive	PAE		1 day	3 days	7 days	28 days	
NPC	-	-	74	53.3	65.8	74.7	86.1	19.0
NPC-SP	-	0.1	111	55.0	70.6	78.3	92.1	23.0
1B3	0.25	0.1	108	63.8	66.6	74.7	93.3	19.3
1B5	0.25	0.1	112	56.5	72.7	76.8	91.2	25.3
2B3	0.25	0.1	103	60.9	71.6	80.0	93.0	23.4
2B5	0.25	0.1	109	59.6	76.0	78.7	92.7	24.4
3B3	0.25	0.1	107	61.2	74.2	79.3	95.0	20.1
3B5	0.25	0.1	108	62.3	65.3	76.8	98.6	24.9
3B7	0.25	0.1	112	54.4	69.1	80.1	95.0	24.7
4B3	0.25	0.1	108	63.9	71.3	77.1	95.9	21.2
4B7	0.25	0.1	107	59.9	74.4	78.3	92.9	27.1
CB8	0.25	0.1	114	59.2	75.1	80.2	91.3	23.3
SF	0.25	0.1	107	59.9	75.4	81.7	91.6	22.7

* at the age of 28 days

- Eco-binders modified by nanoparticles and produced with substantially reduced volume of portland cement component (down to 10-15%) or binders based on the alternative systems (MgO, phosphate, geopolymers, gypsum);
- Self-healing materials and repair technologies utilizing fullerenes, nanoparticles, nanotubes and chemical admixtures;
- Materials with self-cleaning/air-purifying features based on photocatalyst technology;
- Materials with controlled electrical conductivity, deformative properties, non-shrinking and low thermal expansion;
- Smart materials, such as temperature-, moisture-, stress-sensing or responding materials.

It was observed that the efficiency of nanoparticles such as nano-SiO_2 depends on their morphology and genesis, as well as on the application of superplasticizer and additional treatment options such as thermal treatment and ultrasonification. With the sol-gel method, it is possible to manufacture a wide range of nanoparticles with engineered parameters such as particle size, porosity and surface conditions. It was demonstrated that all synthesized nano-SiO_2 at a small dosage, 0.25% can improve the 1-day and 28-day compressive strength of portland cement mortars. The major problem related to application of nano-SiO_2 is related to the agglomeration of nanoparticles and their even distribution within the cement matrix. The application of effective superplasticizer, ultrasonification and high-speed mixing were found to be very effective disagglomeration techniques that improve the strength of superplasticized portland cement mortars, reaching the compressive strength of 63.9 MPa and 95.9 MPa at the age of 1 day and 28 days, respectively. Importantly, the application of selected nano-SiO_2 resulted the improvement of 28-day flexural strength by 18%, reaching 27.1 MPa.

Further research is required to modify the sol-gel method to avoid the formation of dense xerogel agglomerates (such as the development of nano-SiO_2 products in liquid state, application of surfactants, ultrasonification and microwave drying), and to achieve better dispersion of developed nano-SiO_2. Further investigation is necessary to quantify the effect of synthesized nano-SiO_2 on the hydration of portland cement-based systems.

Acknowledgments. The reported study was performed under the research grants of CONACYT 46371 and PROMEP 103.5/07/0319. Support of CEMEX with cement materials, sample preparation and characterization is gratefully acknowledged. Special appreciation is conveyed to Handy Chemicals and Eka Chemicals for supply of superplasticizers and nano-products. The financial support of CONACYT, PROMEP and PAICYT (Mexico) is gratefully acknowledged.

References

1. Gann, D.: A Review of Nanotechnology and its Potential Applications for Construction. SPRU, University of Sussex (2002)
2. Trtik, P., Bartos, P.J.M.: Nanotechnology and concrete: what can we utilise from the upcoming technologies? In: Proceeding of the 2nd Anna Maria Workshop: Cement & Concrete: Trends & Challenges, pp. 109–120 (2001)

3. Sobolev, K., et al.: Development of nano-SiO_2 based admixtures for high-performance cement-based materials. Progress report, CONACYT, Mexico (2006)
4. Sobolev, K., Ferrada-Gutiérrez, M.: How nanotechnology can change the concrete world: Part 1. Am. Ceram. Soc. Bull. 10, 14–17 (2005)
5. Sobolev, K., Ferrada-Gutiérrez, M.: How nanotechnology can change the concrete world: Part 2. Am. Ceram. Soc. Bull. 11, 16–19 (2005)
6. Plassard, C., et al.: Investigation of the surface structure and elastic properties of calcium silicate hydrates at the nanoscale. Ultramicroscopy 100, 331–338 (2004)
7. Super-hydrophilic photocatalyst and its applications (accessed June 15, 2008), `http://www.toto.co.jp`
8. Pilkington Activ™ - Self-cleaning glass (accessed June 15, 2008), `http://www.pilkington.com`
9. Watanabe, T., et al.: Multi-functional material with photocatalytic functions and method of manufacturing same. US patent 6294247 (2001)
10. Roco, M.C., Williams, R.S., Alivisatos, P.: Vision for nanotechnology R&D in the next decade. IWGN Report on Nanotechnology Research Directions, National Science and Technology Council, Committee on Technology (1999)
11. Scottish Centre for Nanotechnology in Construction Materials (accessed June 15, 2008), `http://www.nanocom.org`
12. Corradi, M., et al.: Controlling performance in ready mixed concrete. Concrete Int. 26, 123–126 (2004)
13. Collepardi, M., et al.: Influence of amorphous colloidal silica on the properties of self-compacting concretes. In: Proceedings of the International Conference - Challenges in Concrete Construction - Innovations and Developments in Concrete Materials and Construction, Dundee, UK, pp. 473–483 (2002)
14. Collepardi, M., et al.: Optimization of silica fume, fly ash and amorphous nanosilica in superplasticized high-performance concretes. In: Proceedings of 8th CANMET/ ACI International Conference on Fly Ash, Silica Fume, Slag and Natural Pozzolans in Concrete, SP-221, Las Vegas, USA, pp. 495–506 (2004)
15. Li, H., et al.: Microstructure of cement mortar with nano-particles. Compos Part B-Eng. 35, 185–189 (2004)
16. Bhushan, B.: Handbook of nanotechnology. Springer, Heidelberg (2004)
17. Wilson, M., et al.: Nanotechnology - Basic science and emerging technologies. Chapman & Hall/CRC, Boca Raton (2000)
18. Kang, S., et al.: Preparation and characterization of epoxy composites filled with functionalized nanosilica particles obtained via sol-gel process. Polymer 42, 879–887 (2001)
19. Edelstein, A.S., Cammarata, R.C.: Nanomaterials: Synthesis, properties and applications. Institute of Physics Publishing, Bristol (1996)
20. Sobolev, K.: Mechano-chemical modification of cement with high volumes of blast furnace slag. Cem. Concr. Compos. 27(7-8), 848–853 (2005)
21. Sobolev, K.: The effect of complex admixtures on cement properties and development of a test procedure for the evaluation of high-strength cements. Adv. Cem. Res. AC15, 65–75 (2003)
22. Li, G.: Properties of high-volume fly ash concrete incorporating nano-SiO_2. Cement and Concrete Res. 34, 1043–1049 (2004)

Improving the Performance of Heat Insulation Polyurethane Foams by Silica Nanoparticles

M.M. Alavi Nikje, A. Bagheri Garmarudi, M. Haghshenas, and Z. Mazaheri

Abstract. Heat insulation polyurethane foam materials were doped by silica nano particles, to investigate the probable improving effects. In order to achieve the best dispersion condition and compatibility of silica nanoparticles in the polymer matrix a modification step was performed by 3-aminopropyltriethoxysilane (APTS) as coupling agent. Then, thermal and mechanical properties of polyurethane rigid foam were investigated. Thermal and mechanical properties were studied by tensile machine, thermogravimetric analysis and dynamic mechanical analysis.

1 Introduction

Polyurethanes are one of the most versatile groups of plastic materials. The variety of polyurethane types reaches from flexible and rigid foams over thermoplastic elastomers to adhesives, paints and varnishes. Rigid polyurethane foams are highly applied in construction industry as heat insulator and also shock-noise absorber. The thermal characteristics of these polymeric materials had made them as a very useful product for decrement in energy loss in buildings and constructions [1,2]. Their low thermal conductivity is due to a unique combination of blowing agent properties, cell size, and closed cell morphology. However, PUs also have some disadvantages, such as low thermal stability and low mechanical strength, etc. There is an interest to improve the physical and thermal properties of these polymers. To overcome these disadvantages, great deals of effort have been devoted to the development of nanocomposites in recent years [1]. One route is to utilize nano technology to modify them. In this research it has been tried to introduce silica nano particles in polyurethane rigid foam formulation in order to investigate its probable improving effect on foam samples. This would produce high performance composites. There are also many reports, indicating the physical mixing of nano material and polyol as the synthesis route. The main problem in this

M.M. Alavi Nikje, A. Bagheri Garmarudi, and Z. Mazaheri
Chemistry Department, Faculty of Science, IKIU, Qazvin, Iran

M.M. Alavi Nikje, A. Bagheri Garmarudi, and M. Haghshenas
Department of Chemistry & Polymer Laboratories, Engineering Research Institute, Tehran, Iran

kind of prepared nanocomposites is the heterogeneous dispersion of nano particles [3-7]. While it can be expected that the nano silica, with extremely large surface area would affect the PU properties much more than regular fillers, it is noticeable that how the dispersion of nano filler in polyol matrix is homogenized. The main objective of this research was to examine the effect of nano silica on the properties of rigid PU foams. In addition, silane based agent was applied to play the role of coupling agent between polyol media and nano phase surface. Effect of well dispersed nano filled on thermal and physical properties of PU was monitored.

2 Experimental

2.1 Materials and Apparatus

DatloFoam TA® 14066 polyether polyol containing all of required additives, MDI (Suprasec®5005) for rigid polyurethane foam were from HUNTSMAN Co. Technical data are listed in tables 1 and 2. Nano silica (Particle size 12 nm, AEROSIL®, surface area 200 m^2g^{-1}) from Degussa, 3-aminopropyltriethoxysilane (Amino A-100) from Silquest® chemicals (wetting area 353 m^2g^{-1}) and toluene from MERCK.

Table 1 Technical data for MDI

Suprasec®5005	
Appearance	Dark brown liquid
Viskosity	220 cps @ 25°C
Specific gravity	1.23 $g.cm^{-3}$@ 25 °C
Flash point	233 °C
Fire point	245 °C

Table 2 Technical data for polyether polyol

DatoFoamTA®14066	
Appearance	Viscous yellow liquid
Viskosity	5260 cps @ 25°C
Specific gravity	1.06 $g.cm^{-3}$@ 25 °C
Water kontent	2.3%
pH	12

Infrared spectra were obtained by a Tensor 27 Brucker Infrared spectrometer. Thermal properties of nanocomposites were analyzed by using Dupont 2000 instrument under nitrogen flows from room temperature to 700 °C at heating rate of 10°C/min. tensile strength and elongation at break of these foams were measured using by INSTRON 1122 tensometer at a test speed 5 mm/min. Dynamic mechanical analysis was carried out by Dupont 2000 instrument at a frequency 1Hz and heating rate of 5°C/min from -50 to 150°C.

2.2 Sample Preparation

Five grams of nano-silica in 200 ml toluene was refluxed with mechanical stirring (1000 rpm) at 80-90°C for 2h. Then 2.83 g 3-aminopropyltriethoxysilane (APTS) was added to this mixture. The mixture was refluxed for 12h under same condition. After isolation and washing with methanol the free APTS was removed. Finally, modified nano-SiO_2 powder was dried in an oven at 70°C for 24h and resulted powder was dried and sieved. Figure 1 shows the schemoatic process of modification. Then polyether polyol (200 g) was mixed with modified nano silica in acetone media. High shear mechanical stirrer (15 min) and ultrasonic homogenizer (20 min) were used to perform the mixing process. In the next step, acetone was removed under vacuum at 20 °C. MDI portion was then added to the mixture, being well homogenized during the fast curing step.

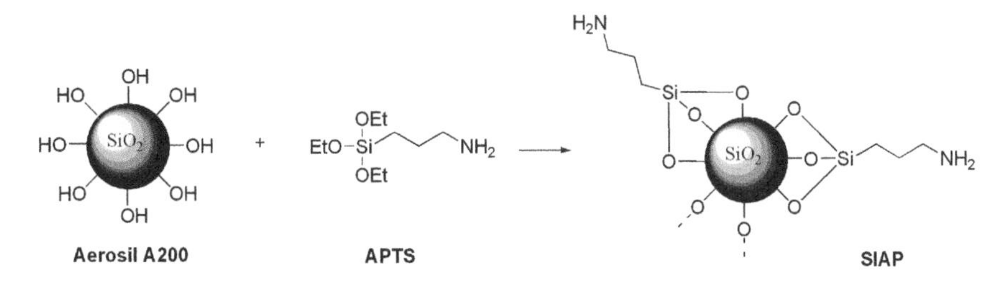

Fig. 1 Schematic structure of modified nanosilica

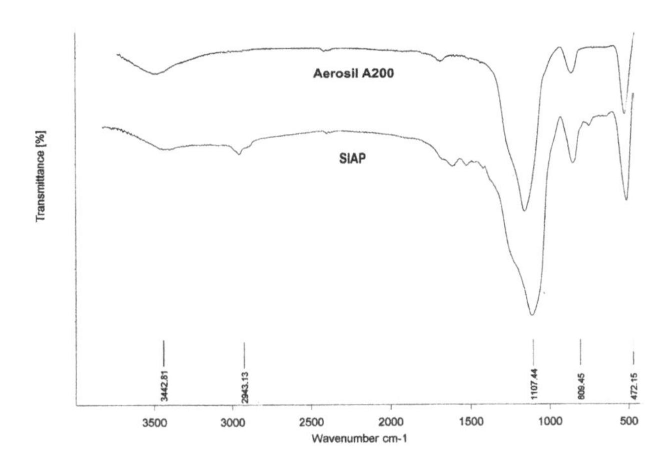

Fig. 2 FT-IR spectra

3 Result and Discussions

3.1 Monotoring of Modification Process

Comparison of FTIR spectra obtained form nono silica and modification's product made some useful information. Aerosil A200 spectrum absorption peak at 3442 cm^{-1} is attributed to silanol group on to Aerosil surface and the signal due to Si-O band would appeare at 1107 cm^{-1}. In the modification product's spectrum it was observed that the absorbance peaks of the C-H stretching vibrations would appeare below 3000 cm^{-1}. This would indicate the covalent band between silanol group on the nano-silica surface and APTS (Figure 2).

3.2 Thermomechanical Studies

The results of tensile strength, elongation at break and elasticity modulus are shown in table 3.

Table 3 The results of mechanical and thermal tests and glass transition temperature

Nano silica content (%w)	Elasticity Modulus (MPa)	Tensile strength (MPa)	Elongation at break (%)
0.0	3.7	0.5	31.2
0.5	3.6	0.50	22.1
1.0	4.3	0.45	17.9
1.5	3.4	0.4	24.2
2.0	3.5	0.6	24.5
2.5	5.1	0.5	27.4
3.0	2.5	0.3	21.7

Tensile strength in 2 wt% loaded sample 2.5% sample was the highest of all. This result indicated that with increasing of nano silica up to 2% loading, an interfacial interaction between the functionalized silica surface and the nearby polymer chains is strong. Silica nano particles inherently possess high module and would strengthen the PU matrix when dispersed in the nano scale. In the other hand, surface modification of nano silica by coupling agent would reduce the possible heterogeneity of network which is also effective as a negative effective parameter. Finally in 3 wt% sample it was reduced to 0.3 MPa. Decrement in the tensile strength at high concentration 3 wt% was attributed to the aggregation nanosilica due to additional hydrogen bonding between silica surface, that result in increasing number of voids in the polymer layer next to the filler surface. The elongation at break at nanocomposites was decreased comparing to those of pure PU foam. Because the functionalized silica surface would act as crosslinker and not as a chain extender which leads to decrease of elongation at break. But whitin the nanofilled PU, in 2.5% sample, the highest elongation at break was seen. Perhaps it could be assumed that in 2.5% loaded sample, a part of amino group on nanosilica surface acted as chain extender to some extent.

Also with infusion of nano silica in polymer matrix, the glass transition temperatures (T_g) were increased to 2% loading and in higher nano silica content were decreased. The increasing in T_g up to 2% nanocomposites may be attributed to decrease in phase separation between soft and hard segment in presence of nano silica. Increment in the T_g of soft segments, indicates that silica nanoparticles have been in a very tight correlation with polymer network because of their high surface area. In the other hand, as shown in figure 1 the surface activity of nano silica would interact with crosslinking polymer network in presence of silane coupling agent strongly which leads to hindered relaxational mobility in the polymer segments near the interface, increasing the T_g. TGA data are also shown in Table 3. It is observed that temperature corresponding to 50% decomposition with increasing of nano silica would shift to higher temperatures. This means that the incorporation of SIAP in to PU foam offers a stabilizing effect against decomposition.

Table 4 TGA data

Nano silica content (%w)	T_g (°C)	T_d at 50%w (°C)
0	85	307
0.5	95	311
1	85	313
1.5	95	310
2	95	320
2.5	80	337
3	57	315

References

1. Hsu-Chiang, K., Hsun-Yu, S., Chen-Chi, M.: Synthesis and characterization of polysilicic acid nanoparticles/waterborne polyurethane nanocomposite. J. Mater. Sci. 40, 6063 (2005)
2. Petrovic, Z.S., Cho, Y.J., Javni, I., Magonov, S., Yerina, N., Schaefer, D.W., Waddon, J.I.A.: Effect of silica nanoparticles on morphology of segmented polyurethanes. Polymer 45, 4285 (2004)
3. Xiong, M.N., You, B., Zhou, S.X., Wu, L.M.: Study on acrylic resin/titania organic-inorganic hybrid materials prepared by the sol-gel process. Polymer 45, 2967 (2004)
4. Hsiue, G.H., Kuo, W.J., Huang, Y.P., Jeng, R.J.: Microstructural and morphological characteristics of PS-SiO2 nanocomposites. Polymer 41, 2813 (2000)

5. Chen, X.C., You, B., Zhou, S.X., Wu, L.M.: Surface and interface characterization of polyester-based polyurethane/nano-silica composites. Surf Interface Anal. 35, 369 (2003)
6. Gao, Y., Choudhury, N.R., Dutta, N., Matisons, J., Reading, M., Delmotte, L.: Organic-inorganic hybrid from ionomer via sol-gel reaction. Chem. Mater. 13, 3644 (2001)
7. Chan, C.K., Peng, S.L., Chu, I.M., Ni, S.C.: Effects of heat treatment on the properties of poly(methyl methacrylate)/silica hybrid materials prepared by sol-gel process. Polymer 42, 4189 (2001)

Eco-innovation Strategies in the Construction Sector: Impacts on Nanotech Innovation in the Window Chain

M.M. Andersen and M. Molin

Abstract. In this paper we examine the strategic response of the construction companies to the climate agenda and the emerging nanotechnologies. A case is brought on the Danish window industry. The construction industry has for a long time been considered little innovative and this also goes for the window section. It seems that a combination of an intensified focus on climate issues and the potential of nanotechnology is invoking a new innovation potential and pressure in the window industry. Specifically we identify a strategic shift among the major players towards more systemic innovations that places the window as a part of the wider energy system of the house, a trend that isn't driven by but could be reinforced by the new nanotech opportunities that so far only play a limited strategic role.

1 Introduction

The climate agenda is becoming increasingly important to the construction industry, none the least because of the strong focus on energy efficiency as a core means to reach climate goals. As around 40 percent of energy is used in buildings the construction sector is becoming a main target for climate policies world wide. As the financial crisis is hitting hard on construction activities, climate issues are becoming a rare potential growth area. Eco-innovation strategies, therefore, are on the rise among construction companies.

Nanotechnology is only slowly emerging in construction, despite the fact that the construction sector was among the first to be identified as a promising application area for nanotechnology back in the beginning of the 1990s [10]. Nanotechnology is essentially a new processes technology, enabling new functionalities of

M.M. Andersen
DTU Management
e-mail: mmua@man.dtu.dk
http://www.man.dtu.dk

M. Molin
DTU Management
e-mail: mmol@risoe.dtu.dk
http://www.man.dtu.dk

materials in the range of 1-100 nanometres, and can as such be particularly interesting for renewing traditional material industries such as construction and window production/glass-processing. As a general purpose technology it is expected to have a pervasive impact on the economy but little is known on the effects in different industrial sectors. Until now nanotechnology has mainly been developed for 'high-technology' areas such as semiconductors and medicine while more traditional sectors, such as construction, have gained less attention [3].

This study contributes to a better understanding of the industrial dynamics involved in the uptake of nanotech in construction. It appears that the generally low tech-tech and conservative construction sector is falling behind other sectors in applying nanotechnology [10]. There are widespread claims of major climate benefits to nanotechnologies (see e.g. [6, 8, 12, 13, 15-17]), but also risk issues related to health and environment prevail [1, 2, 4, 5, 7, 9, 11, 14, 16]. It is therefore interesting to look into the eco-innovation strategies related to nanotech more in detail in order to investigate if and how the climate claims translate into activities on the market.

The paper investigates shortly the interrelationship between the eco-innovation and nanotech strategies among construction companies. In other words, does the rise of the climate agenda lead to a greater interest into nanotech opportunities - or vice versa? A case is brought from the Danish window chain focusing on the core Danish companies and their nanotech strategies and activities. The window chain is interesting because it represents a relative early application area of nanotech in construction as well as entailing a strong emphasis on energy efficiency issues.

The paper focuses on the three largest window producers in Denmark – Velux, Velfac and Rationel[1].

2 Competitive Conditions and Eco-innovation Strategies in the Danish Window Industry

In Denmark there is a range of small window producers but only relatively few big ones. There is no glass production in Denmark. The glass in bought in from major foreign companies. The biggest Danish window producers all belong to the same group, the VKR Group. The VKR Group has approx. 16,000 employees in more than 40 countries and activities in the following five business areas: Roof windows and skylights, vertical windows, thermal solar energy, decoration and sunscreening as well as natural ventilation. The group's 17.3 billion DKR revenue comes mainly from the Velux entity (60% of it's employees) producing and selling roof windows and skylights where they hold the a well known brand. Velfac and Rationel form part of the Dovista company specializing in vertical windows.

The Danish window producers have traditionally competed primarily on price, quality and design and the ability to make customized solutions. The building regulation plays a key role in setting very direct framework conditions; hence the window companies strategies have been very much oriented towards new policy

[1] We would like to thank Torben Hundevad (Dovista), Carl Hammer (former Velfac), Erik Bjørn (Dovista), Ellen K. Hansen (Velfac), Jens Winther (Velfac), Pia Tønder (Velux), Peter Sønderkær (Velux), Claus Holm (Velux) for valuable information and interviews.

trends more than the market. There has generally been limited incentive to undertake more radical innovations. While energy efficiency has been an issue since the mid 1970s it has not been a major innovation driver among these incumbent companies. This seems to be changing quite substantially the last couple of years with the rise of the climate agenda. New more proactive and radical eco-innovation strategies are formed.

Windows as part of the building has changed character considerable during the last 30 years. In the 80th the aim was to minimize the amount of windows a house because of the loss of heat that they caused. This was to a large extent driven by the building laws of the late 70th and beginning of 80th regulating energy consumption of a house by specifying the maximum area of windows that could be placed in a house. This has created incentives for developing more narrow window panes. The contemporary building regulations in Denmark have quite a flexible formulation. The total energy consumption of a house is regulated, but how you solve this is up to the construction of the house. The total area of windows, its energy value or the amount of insulation of a house is not regulated as long as the house as a whole lives up to an energy budget.

The current climate focus has placed a much stronger demand for energy efficient windows which represent new market opportunities for the window producers; but it also represents a threat to the window industry because of the trend towards reducing the amount of windows in the low-energy/-passiv energy/plus energy houses.

The Danish companies analyzed have all reacted to these new competitive conditions in reformulating their strategies. An overall trend by these companies (in the same group) is that the window increasingly is seen as an integrated part of the house and none the least the energy system of the house. Instead of just focusing on the light that a window provides they focus increasingly on the functions that a window can provide as an integral part of the house. More advanced window innovations, including nanosolutions, such as new coatings enable new effects on heating and indoor climate of the house. These new features of the window is expanding the strategic focus of the Danish companies. These strategic considerations can be seen in a joint Velux and Velfac project called "bolig for livet" (housing for life), where they take a broader and more integrated understanding on energy efficiency, architecture and indoor climate than they have done earlier. During the project they will build eight houses around Europe in the coming years with a specific vision on energy, comfort and aesthetics.

According to Velfac The "bolig for livet" project is a different way to deal with the same issues that the concept passive house does[2], in emhpåasizing not only

---

[2] Velfac in cooperation with Velux, Aart (architects) and Esbensen Rådgivende Ingeniører A/S is building a future house called "bolig for Livet". This house is to be: 1) CO2 neutral: it gives more than it takes and uses only sustainable energy. 2) Comfort and Wellbeing: there is harmony between light/shadow, heat/cold and in-door/outdoor conditions. 3) Good indoor climate: Light, air and sound materials create comfort and good health. 4) Architectural wholeness: The house is formed in interplay between energy, com-fort, and aesthetics. 5) Good economy: energy costs = 0 DKK.

energy efficiency issues but also the wellbeing of the people living in the house. So Velfac together with Velux seeks to show that it is possible to build a house that lives up to the standards of future energy efficient housing without compromising on light and indoor climate. Though still on the drawing table, the project aims to show that it is possible to live up to the energy standards of tomorrow with 40 percent windows in the house.

This has though demanded innovations on the system level of the house, where they are working with an active heat/ventilation system. Included in this cooperation is another VKR company called Window Master that develops computer systems to control the opening and closing of windows. To this system has also been added different sunscreens and shutters that close in front of the window during the night. Sunscreens both help to control the solar heat of the house, but also reduce the airflow around the house, thus reducing heat loss. Also the shutters fulfill two functions in the system: first, they keep more of the heat inside the house during the night, and second, they reduce the problem of condensation on the windows.

Overall, the project illustrates a shift in competition from the window to the system level (the house). Velfac and Velux seek to launch an "active house" concept building on the principles of energy, comfort and aesthetics as a contrast to the passiv house concept. Strategically they are seeking to form a market standard thatfits their own products better. Traditionally product manufacturers in the construction industry has not taken a great interest in the role as system integrator [18]. However with this project Velfac and Velux is changing their role in the value chain where they now play a more active role as systems integrator which is related to a greater interest into systemic innovation.

3 Green Nanoinnovation in the Danish Window Industry

In this section we will shortly illustrate the main nanorelated innovation activities of the involved companies. Generally, nano innovation activities are as yet limited but with an increasing interest. Velux is more active involved than Velfac and Rationel (Dovista). Dovista handles the R&D for Velfac and Rationel, while Velux has its own R&D department.

3.1 Velux

Velux, as a producer of roof windows which are difficult to clean, naturally has a strong interest into self-cleaning windows. Their sun tunnel has "easy to clean" coating, which is a TiO2 coating, as standard in some countries[3]. In some countries you can also get the "easy to clean" coating as optional choice on all Velux windows. It is, however, not a standard on all windows because the discussion on TiO2 and its health and environmental effects still has not settled on any definite

[3] See http://www.velux.dk/Produkter/Lystunnel/Product/Hovedkomponenter/Main_components.htm
http://www.velux.co.nz/products/SunTunnels/Products/FlexibleSunTunnel/default.htm

conclusion. At the start Velux used Saint Gobains' self-cleaning glass bioclean (www.selfcleaningglass.com) but they considered that the layer of TiO2 was too thick which affected the view through the glass. Now they have "easy to clean" which is a glass they have developed together with an American company. Velux persistently look for other solutions to the self-cleaning effect on windows and they test different products that reach the market. They have for example tested hydrophobic surfaces (the so called "lotus effect", see www.lotus-effekt.de), but without any promising results. The structures get clogged after a while reducing the lotus effect. There are other ways of solving the self-cleaning effect.[4]

The other parts of a window where Velux has an interest into nanoinnovations are wood, metal and plastic. Here the different areas have their own explicit focus. Although wood is one of the best materials to use in a window, because of longevity and its structural specifications, to improve it by different treatments (paint and impregnation) has always been a focus. Metal is a significant environmental focus area because of the use of Chrome VI to extend the longevity of ironworks. Most industries have been obligated to phase out Chrome VI but the construction industry has been given an extended time frame because of the longevity requirement of building materials. Plastic is not a main focus today, but Velux expects this to change relatively soon because of the improvements in bio-composites and bio-plastics.

As part of their general technological and environmental strategies Velux participated in NanoPaint (www.nanopaint.dk)[5]; a project that aims to find different nanotechnological solutions for wood, plastic and metal protection. Velux has had some promising results with a new anti-fungus treatment of wood, but they still need to develop a process to get it into the wood. One possible solution to this is another company in the VKR group called Superwood (www.superwood.dk). Superwood is also interesting in itself since they use supercritical conditions to impregnate wood. This method does not use any heavy metals in the impregnation process, which thus has some very nice environmental advantages. Other participants of NanoPaint work with sol-gel to create a Teflon surface on metals. This can also have an effect on the windows industry, since a solution can present an alternative to the usage of Chrome VI in the ironwork of the window frames.

Other nanorelated issues considered are new glazing's that will improve the energy efficiency of windows, new impregnation materials and processes and the integration of solar cells with windows.

3.2 Dovista

Dovista (Velfac and Rationel) is so far more hesitant on nanotechnology, but they are scanning their suppliers for new advanced solutions to old problems. These include nano solutions but there is no targeted search into nanotech innovations. One such nanorelated area is for example paint, but they have not yet found any

---

[4] See Parkin and Palgrave, 2005, Journal of Materials Chemistry, vol 15, pp 1689-1695., `http://www.toto.co.jp/hydro_e/index.htm`, and `http://www.freepatentsonline.com/6789906.html`.

[5] Participants of NanoPaint where Velux, Dyrup, Danish Technological institute, SP Group, Ciba, and Aarhus University.

nanotech application that improves the factors of paint they are interested in. Another area has been researching in the use of thin film solar cells to generate aesthetic effects and energy saving solution in houses, in part related to the above mentioned project "bolig for livet" (houses for life). Velfac and Rationel have not used any self-cleaning classes in their products, the reason being that they believe their (vertical) windows can be easily cleaned anyway.

An interesting new development is Velfac's attempt to develop better, more energy efficient window panes with the help of composite materials. The first product that has reached the market is a window pane where a profile in composite material separates the aluminium from the wood. This design lowers a windows U-value by 0,1 to 0,2. Hither too energy innovations in the pane have been neglected among Danish producers where the emphasis has been on the energy performance on the glass only. But the rising use of aluminum in window panes, for longevity and design reasons, remains a problem with regard to thermal bridges.

Velfac's activities may be seen as a reaction to increasing competition from new players. Composite materials, also nanobased, are starting more generally to enter the windows industry. In Denmark it was, however, an upstart company, Pro Tec windows (www.protecwindows.com), who succedeed in developing the first composite window in Denmark at a tiem when the incumbeont cmpanies showed little interest. Together with the leading composite material producer of Denmark, Fiberline, they developed a window frame where composite materials replace the use of aluminum displaying a high energy efficiency compared to traditional window panes. It is likely that composite materials will continue to play a rising role in window production as more emphasis is being placed on the energy performance of the entire window and not only the glass fraction.

4 Conclusions

A combination an intensified demand for climate solutions, energy policies that are open towards new solutions together with the emergence of new advanced solutions including nanotechnologies, are invoking a new innovation potential and a reframing of strategies among the Danish window industry. Specifically we identify a strategic shift among the major players towards more systemic innovations that places the window as a part of the wider energy system of the house. The window companies new roles as system integrators is likely to lead to more systemic and radical innovations, a trend that isn't driven by but could be reinforced by the new nanotech opportunities. Clearly, eco-innovation strategies function as catalysts for nanotech solutions which so far only play a limited but rising role amongst Danish window producers.

References

1. Aitken, R.J., Chaudhry, M.Q., Boxall, A.B.A., Hull, M.: Manufacture and use of nanomaterials: current status in the UK and global trends. Occupational Medicine 56, 300–306 (2006)
2. Andersen, M.M., Rasmussen, B.: Environmental opportunities and risks from nanotechnology. Risoe-R-1550-EN Risø National Laboratory, Roskilde (2006)

3. Andersen, M.M., Molin, M.: NanoByg: A survey of nanoinnovation in Danish construction. Risoe-R-1234(EN) Risø National laboratory, Roskilde (2007) (accessed December 15, 2008), http://www.risoe.dk/rispubl/reports/ris-r-1602.pdf
4. Arnall, A.H.: Future Technologies, Today's Choices: Nanotechnology, Artificial Intelligence and Robotics - A technical, political and institutional map of emerging technologies. Greenpeace Environmental Trust, London (2003), http://www.greenpeace.org.uk/MultimediaFiles/Live/FullReport/5886.pdf
5. Colvin, V.: Nanotechnology: environmental impact. Presentation at National Center for Environmental Research (NCER). EPA (2002)
6. EC, Towards a European Strategy for Nanotechnology. European Commission (2004) (accessed December 15, 2008), http://cordis.europa.eu/nanotechnology/actionplan.htm
7. EC SANCO, Nanotechnologies: A Preliminary Risks Analysis, report on the basis of a workshop organized in Bruxelles on 1-2 March by the Health and Consumer Protection Directorate General of the European Commission (SANCO), European Communities, Bruxelles [20] (2004)
8. European Parliament Scientific Technology Options Assessment Committee, The Role of Nanotechnology in Chemical Substitution (2007) (accessed December 15, 2008), http://www.nanowerk.com/spotlight/spotid=2212.php
9. Friends of the Earth Germany - BUND, For the Responsible Management of Nanotechnology (2007) (accessed December 15, 2008), http://www.bund.net/lab/ reddot2/pdf/bundposition_nano_03_07.pdf
10. Gann, D.: A Review of Nanotechnology and its Potential Applications for Construction. SPRU/CRISP (2003) (accessed December 15, 2008), http://www.crispuk.org.uk/REPORTS/LongNanotech240203.pdf
11. Hansen, S.F., Larsen, B.H., Olsen, S.B., Baun, A.: Categorization framework to aid hazard identification of nanomaterials. Nanotoxicology 1–8 (2007)
12. Luther, W., Zweck, A.: Anwendungen der Nanotechnologie in Architektur und Bauwesen. VDI Technologiezentrum, Düsseldorf (2006)
13. Nanoforum, Nanotechnologies help solve the world's energy problems. Nanoforum (2003) (accessed December 15, 2008), http://www.nanoforum.org
14. Nanoforum, Benefits, risks, ethical, legal and social aspects of nanotechnology. Nanoforum (2004) (accessed December 15, 2008), http://www.nanoforum.org
15. NSET, The National Nanotechnology Initiative: Research and Development Leading to a Revolution in Technology and Industry: Supplement to the Presidents FY 2004, Budget. National Science and Technology Council, Washington D.C (2003)
16. Royal Society, Nanoscience and nanotechnologies: opportunities and uncertainties. The Royal Society & The Royal Academy of Engineering (2004) (accessed December 15, 2008), http://www.nanotec.org.uk/finalReport.htm
17. Schmidt, K.F.: Green Nanotechnology. Woodrow Wilson International Center for Scholars - Project on Emerging Nanotechnologies (2007) (accessed December 15, 2008), http://www.nanotechproject.org/process/assets/files/2701/187_greennano_pen8.pdf
18. Winch, G.: Zephyrs of creative destruction: understanding the management of innovation in construction. Build Res. Inf. 26, 268–279 (1998)

Interpretation of Mechanical and Thermal Properties of Heavy Duty Epoxy Based Floor Coating Doped by Nanosilica

M.M. Alavi Nikje, M. Khanmohammadi, and A. Bagheri Garmarudi

Abstract. Epoxy-nano silica composites were prepared using Bisphenol-A epoxy resin (Araldite® GY 6010) resin obtained from in situ polymerization or blending method. SiO_2 nanoparticles were pretreated by a silan based coupling agent. Surface treated nano silica was dispersed excellently by mechanical and ultrasonic homogenizers. A dramatic increase in the interfacial area between fillers and polymer can significantly improve the properties of the epoxy coating product such as tensile, elongation, abrasion resistance, etc.

1 Introduction

Some of the main advantages of epoxy based coatings are thermal stability, mechanical resistance, low density and minimal shrinkage. The main factor influencing their performance is the molecular architecture [1-3]. Owing to their densely cross-linked structure, they exhibit a number of superior qualities such as high glass transition temperature, high modulus and high creep resistance. Developments in the synthesis of nanoparticles have made it possible to process nanocomposites [4,5]. Investigations have been reported during the past few years for the development of the high-performance nanocomposites, which consist the incorporation of a nanometer-size inorganic component into the organic resin matrix. Industrial application of polymer–inorganic nanocomposites has attracted a lot of interests during the last 5 years. The main reason is that nanocomposites demonstrate combined properties of polymers e.g. flexibility, and also those of inorganic chemicals e.g., rigidity, thermal stability, hardness, etc [6,7]. Application of epoxy compounds in a variety of products for construction industry has been emerged since their properties, such as thermal stability, mechanical response, low density and electrical resistance. Floor coating, humidity resistant

M.M. Alavi Nikje, M. Khanmohammadi, and A. Bagheri Garmarudi
Chemistry Department, Faculty of Science, IKIU, Qazvin, Iran

M.M. Alavi Nikje and A. Bagheri Garmarudi
Department of Chemistry & Polymer Laboratories, Engineering Research Institute, Tehran, Iran

paints, polymer blended concrete and epoxy grout are some of the main products applied in constructions. Molecular architecture, curing conditions and the ratio of the epoxy: hardener are the main factors influencing the performance of these chemicals.

2 Experimental

2.1 Sample Preparation

Bisphenol-A epoxy resin (Araldite® GY 6010) was from JANA Resin Manufacturing Co. (epoxy value: 0.5208-0.5498 eq per 100 g; weight per epoxide: 182-192 g per eq; residual epichlorohydrin: below 100 ppm). Cycloaliphatic polyamine (Aradur 43) from HUNTSMAN® Co. (amine value: 260-280 mg KOH per g). The resin:hardener stoichiometric ratio was 100:60 pbw. Coupling agent is γ-aminopropyltriethoxysilane (Amino A-100) manufactured by Silquest® Chemicals which was added 5 pbw to the epoxy resin. Nano-SiO_2 was AEROSIL® 200 with specific surface area of 200 m^2g^{-1} and average particle size of 12 nm, from Degussa. Water free acetone was from Merck®. One of the main issues in preparation of nanocomposites is to disperse the nanoparticles in resin media which has been reported to increase the resin's viscosity. SiO_2 nanoparticles were pretreated by coupling agent in acetone. Using a "high shear" laboratory-mixing device for mechanical mixing and an ultrasonic homogenizer, dispersion process was conducted. Acetone content of the sample was removed by vacuum at 40 °C , homogenizing by ultrasonic apparatus. The hardener was added to the formulation, being mixed and degasse. Cure process was in a chamber with room temperature. Table 1 shows the content of nano silica in flooring samples

Table 1 Prepared nano SiO_2–epoxy flooring samples

Flooring Sample	Nano Silica (%)
EP-0	0.0
EP-1	0.5
EP-2	1.0
EP-3	1.5
EP-4	2.0
EP-5	2.5
EP-6	3.0

2.2 Thermophysical Analysis

The tensile strength of cured flooring samples was determined, using an Instron testing machine at a crosshead speed of 5 mm min^{-1} at room temperature, according to ASTM D638. The thermomechanical properties of samples were investigated by a DuPont Instrument operating in the three-point bending mode under nitrogen atmosphere. Data were collected in -20 to 200 °C temperature range at a scanning rate of 5 °C min^{-1}, using 10 Hz frequency. The friction resistance test was performed by an abrasion machine. Hardness of flooring samples was determined in Shore D scale. Thermal gravimetric analyzer and scanning electron microscope were also utilized the characteristics of the nano-composites.

3 Results and Discussion

It was observed that reinforcement of epoxy flooring by SiO_2 nanoparticles would dramatically improve the tensile properties of these coating. However improvement from 3.22 MPa (EP-0) to 12.59 MPa (EP-5) is really excellent as much more interfacial surfaces can be generated between polymer and nanoparticles, which assists in absorbing the physical stress. Generally, nanoparticles inherently possess high module and would strengthen the polymeric matrix when dispersed in the nano scale level. The maximum tensile strength and elongation was in EP-5 which would drop in EP-6 sample with higher amount of SiO_2 nanoparticles. There are several possible reasons for this decrement in tensile strength. One would be the weak boundaries between nanoparticles and probable micronized trapped bubbles. The other responsible reason may be the effect of high amounts of nanoparticles on homogeneity in crosslinking of the epoxy network. One of the main problems in floor coating by epoxy based materials is the existence of expansion joints on concrete surface, which their dimensional variations would cause the flooring to crack. The improvement in elongation at break of flooring would result in higher resistance against concrete dimensional variations. Interestingly the hardness and abrasion resistance of epoxy flooring would be increased by addition of SiO_2 nanoparticles (Table 2).

DMA experiments showed that δ_{tan} would decrease with addition of nanoparticles while glass transition tempreature is vise versa. the Chemical bonding at the interface of the nanoparticles and polymer matrix could lead to hindered relaxational mobility in the polymer segments near the interface, which

Table 2 Physical properties of prepared nano silica - epoxy floor coating

	Tensile Strength (MPa)	Standard Deviation (%)	Elongation (%)	Abrasion (mm^3)
EP-0	3.22	1.8	5.94	490
EP-1	3.36	2.1	7.17	426
EP-2	5.08	1.7	6.99	416
EP-3	5.14	2.5	8.47	397
EP-4	8.72	2.8	14.29	368
EP-5	12.59	2.2	31.58	318
EP-6	11.19	2.1	33.04	256

leads to increase of Tg. The loss in the mobility of epoxy chain segments according to nanoparticle: matrix interaction would result in restricted chain mobility by improving the homogenized dispersion. Better dispersion would reduce the distance between nanoparticles providing better interaction with each other and also with epoxy matrix. In the other hand, dynamic modulus would be benefited by relative hindering of epoxy structure motion (Figures 1 and 2).

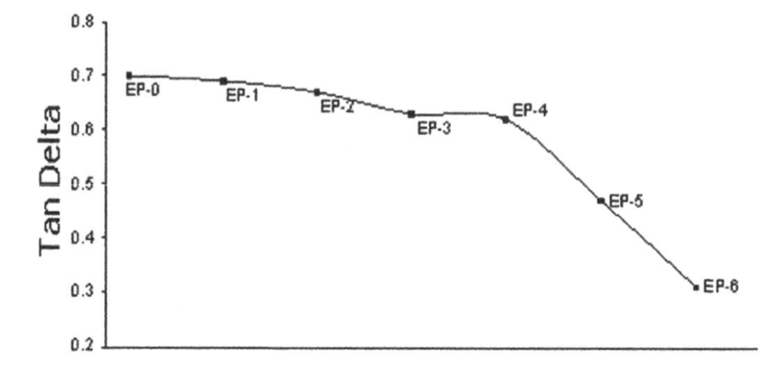

Fig. 1 Variations in δ_{tan} of flooring samples

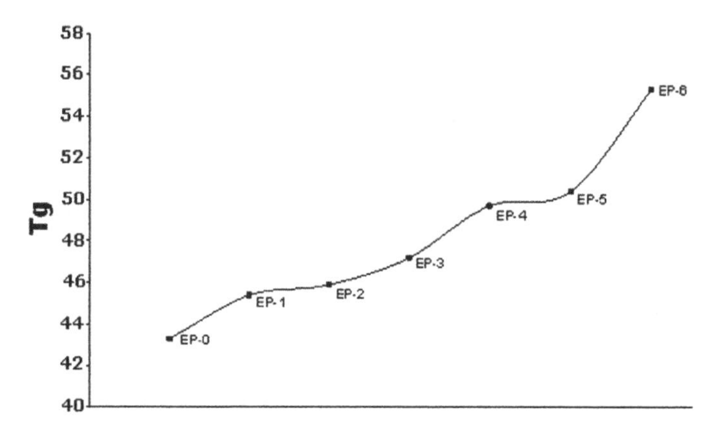

Fig. 2 Effect of nano SiO_2 on glass transition temperature of epoxy flooring

Achieving a homogenous dispersion is considered as a difficult goal according to their strong tendency in agglomeration. According to the SEM images form the dispersion of silica nano particles in the epoxy flooring matrix, it is concluded that good dispersion may occur by surface modification of the nanoparticles under an appropriate processing condition. Of course the homogenizing steps (sonification and high speed mechanical mixing) would be so effective [8]. Investigating the breaking surface of nano composites, it was observed that nano silica would affect the surface and breaking direction. As seen in figure 3, the breaking surface is more homogenous and edge of separations are in a same direction in comparison with neat flooring or low nano silica content samples.

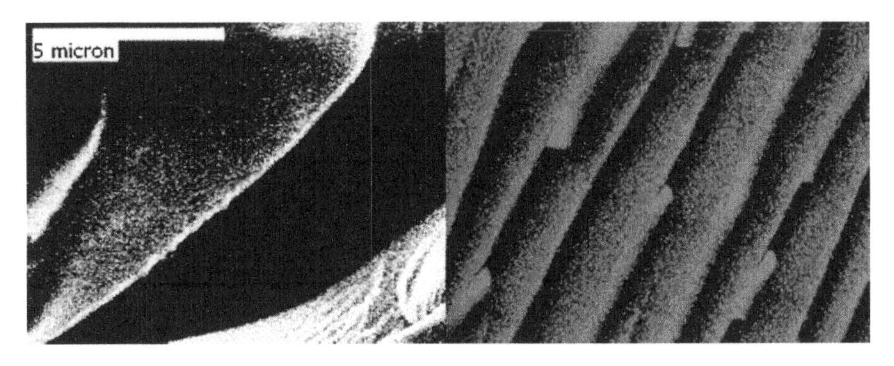

Fig. 3 SEM image of EP-0 (left) and EP-5 (right)

4 Conclusions

It was concluded that nano silica is industrially applicable in epoxy based coatings for metal structures, factories, ships, and buildings. Nano silica particles fill up the weak micro-regions of epoxy resin to boost the interaction forces at the polymer–filler interfaces.

References

1. May, C.A.: Epoxy Resins: Chemistry and Technology, 2nd edn. Marcel Dekker, New York (1988)
2. Lee, H., Nevellie, K.: Handbook of Epoxy Resin. McGraw-Hill, New York (1967)
3. Wong, C.P.: Application of polymer in encapsulation of electronic parts. Adv. Polym. Sci. 84, 63 (1988)
4. Saegusa, T., Chujo, Y.: Macromol Symp., vol. 46, p. 1 (1992)
5. Matejka, L., Dukh, O., Kolarik, J.: Block-copolymer organic-inorganic networks. Structure, morphology and thermomechanical properties. Polymer 41, 1449 (2000)
6. Cheng, K.C.: Kinetic model of diepoxides with reactive diluents cured with amines. J. Polym. Sci. Polym. Phys. Ed. 36, 2339 (1998)
7. Queiroz, S.M., Machado, J.C., Porto, A.O., et al.: Positron annihilation and differential scanning calorimetry studies of plasticized polyethylene oxide. Polymer 42, 3095 (2001)
8. Chan, C.M., Wu, J.S., Li, J.X.: Polypropylene/calcium carbonate nanocomposites. Polymer 43, 2981 (2002)

Nanoindentation Study of Na-Geopolymers Exposed to High Temperatures

I. Beleña and W. Zhu

Abstract. This paper reports the usefulness of nanoindentation as a characterization and monitoring tool for studying thermal behaviour of Geopolymer materials. The influence of the manufacturing process of Na-Geopolymers in their micromechanical properties and thermal behaviour has been studied. Two types of metakaolin-based geopolymer panels with almost identical composition were prepared by injection and pouring methods. Micro-mechanical properties of the two samples exposed to high temperatures up to 1000 ºC were studied by nanoindentation technique, supplemented by X-ray diffraction (XRD), Nuclear magnetic resonance (NMR), Thermogravimetric analysis (TGA) and Microscopy. Remarkable differences in micro-mechanical properties and thermal behaviour between the two samples were found. Statistical nanoindentation has been successfully used to provide information about the micro-mechanical properties of different phases in the material and their volume distributions.

1 Introduction

Geopolymers are ceramic-like materials obtained by alkali activation of aluminosilicates raw materials at high pH environment, atmospheric pressure and temperatures below 100 ºC. These materials are inorganic polymers consisting in 3D amorphous networks of SiO_4 and AlO_4 tetrahedron alternately linked by sharing all the oxygen atoms. The Al^{3+} forth-coordinated is charge balance achieved by the presence of alkali cations (usually Na^+ or K^+). Their properties are much influenced by the Si/Al ratio in the final network. The starting aluminosilicate materials are disaggregated in a high alkali media and the SiO_4 and AlO_4 oligomers (dimers, trimers…) condense giving as a result an amorphous aluminosilicate polymer and water [1].

I. Beleña
Technological Institute of Construction of Valencia (AIDICO), Technological Park, Valencia, Spain
e-mail: irene.belenya@aidico.es

W. Zhu
Advanced Concrete and Masonry Centre, University of the West of Scotland, Scotland, UK
e-mail: Wenzhong.Zhu@uws.ac.uk

Due to the low energy requirements in the production and good mechanical performance, thermal behaviour and durability, Geopolymer has attracted increasing interest as an ecologically friendly fireproof building material. Many studies about the thermal behaviour of geopolymers with sodium or potassium silicate have been reported [1-4]. The metakaolin-based geopolymers (with a potassium-containing activator) have shown to achieve thermal stability up to 1200-1400ºC. The chemical composition, microstructure and macro-mechanical properties have been widely studied. No attempt, however, has been made to determine the micro-mechanical properties and their changes in different conditions for these materials.

The aim of the present work was to study the micro-mechanical properties of Na-geopolymers using nanoindentation technique, which has been used to study properties of nano/micro-scale features (e.g. Elastic modulus and Hardness) in many different materials, including composites or multiphase materials [5-8]. In this study, geopolymers with the same starting materials and composition were made using two different processes, i.e. direct pouring and injection. Micro-mechanical properties of the two samples and their changes due to high temperature exposure up to 1000 ºC were studied by Nanoindentation and supplemented by Microscopy, XRD, MAS-NMR and TGA.

2 Experimental

2.1 Materials and Specimens

Geopolymers tested were synthesised using Metakaolin (Sigma & Aldrich Chemistry, S.A.), Sodium Silica (Massó y Carol, S.A.) and Sodium Hydroxide (Prolabo, S.A.) (Table 1). The specimens were prepared using a previous optimized composition [9] and different manufacturing processes (Table 2).

Table 1 Raw material composition and physical properties

% oxides (wth)	SiO_2	Al_2O_3	Fe_2O_3	Na_2O	K_2O	P.C*	d (g/cm^3)	d_{50} (μm)	BET (m^2/g)
Metakaolin	52.4	43.8	1.4	0.0	0.5	0.0	2.78	5.78	9.0
$Na_2SiO_4.xH_2O$	27	0.0	0.0	10.6	0.0	62.4	1.6	---	---

Table 2 Details of specimens used

	Manufacture process			Chemical composition			
Sample	Moulding	Curing	Drying	SiO_2/Al_2O_3	Na_2O/Al_2O_3	H_2O/Na_2O	%Solid
NaMk-I	Injection	~25°C Closed mould Two weeks	~25°C In the air	3.5	0.8	13	34
NaMk-P	Pouring	~25°C Open mould 24h	~25°C In absorbent material	3.5	0.8	15	36

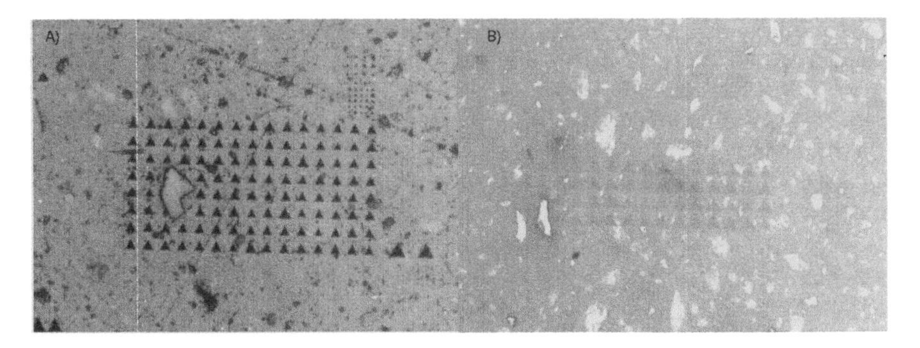

Fig. 1 Optical images of the typical tested specimens (intent spacing being 30 μm): a) NaMk-I unheated, b) NaMk-P unheated

2.2 Testing Details

The methodology and operating principle for the nanoindentation technique have been reviewed and presented in detail elsewhere [5-8]. Briefly, the test consists of making contact between a sample surface and a diamond indenter of known geometry, followed by a loading-unloading cycle while continuously recording the load, P, and indentation depth, h. The P-h curve obtained is a finger print of the mechanical properties of the test area. Most commonly, the elastic modulus (E) and hardness (H) of the test area are determined by analysing the initial part of the unloading data according to a model for the elastic contact problem. For studying multiphase composite materials, a refined statistical indentation method has been used, which involves testing and statistically analyzing a large number of indentation points within a representative sample area [7-8]

The nanoindentation apparatus used in this study was Nanoindenter XP with a Berkovich indenter. In this study, all testing was programmed in such a way that the loading started when the indenter came into contact with the test surface and the load maintained for 30 seconds at the pre-specified maximum value before unloading. In order to provide statistical analysis of the micro-mechanical properties of different phases in the specimen, a grid of 8x15 indentation points with a indent spacing of 30 μm was selected, as shown in Fig.1. For studying the effect of high temperature exposure on properties of the geopolymers, selected samples were placed in an electrical furnace at various temperatures up to 1000°C. The temperature was set to increase at 1°C/min until 100°C and 2°C/min until the specified temperature and then maintained for 24 hours before natural cooling.

3 Results and Discussions

3.1 Micromechanical Properties of the Geopolymers before Heating

The results obtained from the statistical indentation test of the two geopolymer samples are presented in Table 3. Generally, for both samples, three peaks were observed in the frequency-mechanical property (i.e. E and H) plots, which led to the determination of E and H values and volume fractions for the three different phases in each sample. Microscopic (optical and electron) analysis of the indented area has been used to identify the nature of these phases. In Fig.1 the grey background mass is the geopolymer matrix, the brightest irregular shaped pieces are the quartz (present in metakaolin as an impurity) and the slightly bright/white pieces are those incompletely reacted metakaolinite raw material.

Table 3 Micromechanical properties and volume fractions of phases present in test samples

Phase	NaMk-I			NaMk-P		
	E, GPa	H, GPa	V%	E, GPa	H, GPa	V%
Geopolymer (GP)	14	0.5	89	7	0.2	83
Metakaolinite (Mk)	25	1.0	7	12	0.5	13
Quartz (Q)	99	15	4	96	14	4

Results in Table 3 appear to indicate that the micromechanical properties of the geopolymeric matrix are significantly higher in NaMk-I than in NaMk-P sample, as well as there is a higher amount of unreacted metakaolinite in NaMk-P than in NaMk-I. This is probably due to the difference in curing and drying of the two processes. For NaMk-P the curing time was short (24 h) and in open mould with water likely evaporating from the top surface, while the NaMk-I was cured in closed mould for two weeks, thus likely leading to more complete polymerization.

3.2 Thermal Behaviour Study of Geopolymer Samples

Nanoindentation tests were carried out on selected geopolymer samples which had undergone exposure at various high temperatures (Table 4). They show a moderate increase in the mechanical properties of the geopolymer matrix and partially reacted metakaolin phases in both specimens with the rising exposure temperature up to 400ºC. It is believed that such changes are probably due to the dehydration of both phases, together with a progressive increase in the development of new bonds (as suggested by the thermal analysis results). The quartz phase remained unchanged until 400°C but disappeared at 700°C. There seemed to be little change in the GP and the MK phases between 400ºC and 700ºC. Dramatic changes in the sample were observed at high temperature exposure between 700 to 1000ºC. The mechanical properties of the GP phase in both specimens were more than tripled, likely due to a partial softening of geopolymer structure from 788ºC and forming a vitreous phase upon cooling, as suggested by the TDA results. In the 1000°C sample, the MK phase was no longer present, but a new phase, Mullite was observed in both samples. Nepheline was also found in the NaMk-I..

Optical images of the 1000°C specimens (Fig.2) revealed that no MK and quartz phases were present in both samples, and many pores (up to 150 um) were developed in NaMk-I sample. Different phase identification was support by XRD. It is

Table 4 Properties of different phases in geopolymer samples heated to high temperatures

T[a]	NaMk-I				NaMk-P			
°C	Phase	E	H	%	Phase	E	H	%
					GP	15	0.7	82
200					Mk	27	1.6	14
					Q	95	12	4
	GP	23	1.2	89	GP	16	0.7	92
400	Mk	31	1.8	7	Mk	35	1.6	4
	Q	99	15	4	Q	96	14	4
700					GP	17	0.7	95
					Mk	35	1.4	5
	GP	75	7.4	94	GP	83	7.6	92
1000	Nepheline	85	9.0	5	Mullite	107	12	8
	Mullite	115	10.6	1				

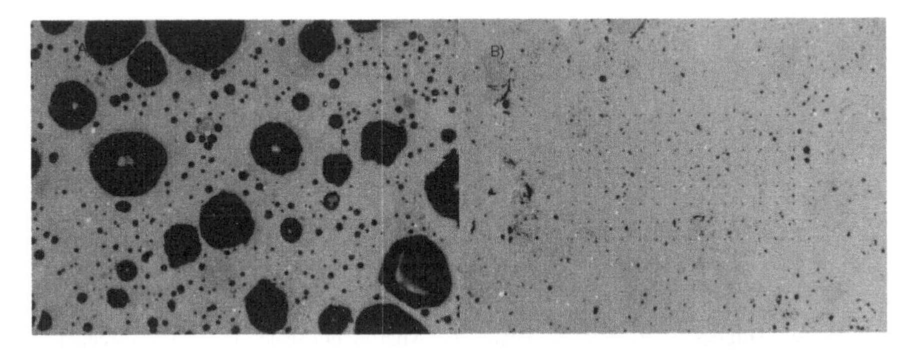

Fig. 2 Optical images of specimens after the 1000°C exposure: a) NaMk-I, b) NaMk-P

believed that in the injection process, migration and accumulation of Na^+ in pores occurred due to the long and closed curing, leading to the formation of nepheline at about 1000°C accompanied with a volume reduction and resulting voids.

4 Conclusions

Statistical nanoindentation has been successfully used to provide information about the micro-mechanical properties of different phases and their volume distributions in the geopolymer samples. The manufacturing process was found to significantly affect the micro-mechanical properties and thermal stability of geopolymer materials produced. A long curing period in closed mould associated with the injection process appeared to be responsible for a higher degree of reaction and better mechanical properties of the geopolymeric phase. The long and closed curing, however, seemed to result in higher Na^+ concentration in pore solutions, which led to formation of nepheline and large porosity at 700 – 1000 °C. On the other hand, the sample produced by direct pouring process using open mould and short curing period was found to lead to an incomplete reaction of the starting materials and, thus, lower mechanical properties of geopolymers. Generally, the mechanical properties of the geopolymer matrix showed a moderate increase when the sample was exposed to temperature up to 400ºC, no significant change at 400 – 700 °C, and then an increase more than three times at 1000 °C.

Acknowledgments. We are indebted to the Valencian Institut of Small and Medium Size Industries (IMPIVA) and European Social Fund for the financial support through the Program High Specialization in Industrial Technologies with the project High Specialization in Composed Nanomaterials with application in the Construction Sector. IMAETA /2005/1.

References

1. Davidovits, J.: Geopolymer, Chemistry and Applications. Institute Géopolymère (Geopolymer Institute) (2008)
2. Hammell, J.A., Balaguru, P.N., Lyon, K.E.: Strength retention of fire resistant aluminosilicate-carbon composites under wet-dry conditions. Compos. Part B-Eng. 31, 107–111 (2000)
3. Barbosa, V.F.F., McKencie, K.J.D.: Thermal behaviour of inorganic geopolymers and composites derived from sodium polysialate. Mater. Res. Bull. 38, 319–331 (2003)
4. Barbosa, V.F.F., McKencie, K.J.D.: Synthesis and thermal behaviour of potassium sialate geopolymers. Matter Lett. 57, 1477–1482 (2003)
5. Oliver, W.C., Pharr, G.M.: Measurement of hardness and elastic modulus by instrumented indentation: Advances in understanding and refinements to methodology. J. Mater. Res. 19, 3–20 (2004)
6. Ficher-Cripps, A.C.: Nanoindentation. Springer, Heidelberg (2002)
7. Ulm, F., Vandamme, M., Bobko, C., Ortega, J.A.: Statistical indentation techniques for hydrated nanocomposites: concrete, bones and shale. J. Am. Ceram. Soc. 90, 2677–2692 (2007)
8. Zhu, W., Hughes, J., Bicanic, N., Pearce, C.: Micro/nano-scale mapping of mechanical properties of cement paste and natural rocks by nanoindentation. Mater. Charact. 11, 1189–1198 (2007)
9. Beleña, I., Tendero, M.J.L., Tamayo, E.M., Vie, D.: Estudio y optimización de los parámetros de reacción para la obtención de material geopolimérico mediante ^{29}Si y ^{27}Al RMN y DRX. Boletín de la Sociedad Española de Cerámica y Vidrio 43 (2004)

Nanoscale Agent Based Modelling for Nanostructure Development of Cement

E. Cerro-Prada, M.J. Vázquez-Gallo, J. Alonso-Trigueros, and A.L. Romera-Zarza

Abstract. Most of macroscopic properties of materials are consequences of processes taking place at the nanoscale. Investigation of phenomena at the sub-micro level to improve the performance of construction materials is one of the main applications of Nanotechnology in Construction. The modelling and simulation of nanostructures is then essential to provide a better understanding of the behaviour of construction materials. We present in this paper an agent-based modelling approach in which a set of interacting agents, capable of organizing themselves dynamically and of adapting to the environment, encapsulate the behaviour of the whole system. This new approach is used for studying the nanostructure development of cement, considering that each individual particle develops its own C-S-H shell critical thickness which triggers the agent for shifting the algorithm to next stage. The algorithm results were compared and found in good agreement with reported experimental work.

1 Modelling the Hydration Process of Cement

As computational capabilities have evolved rapidly over the last ten years, the ability of mathematical modelling and computation to contribute to the field of materials science has been clearly demonstrated. For porous materials such as hardened cement paste and concrete, major emphasis has been placed on the topic of this work, mathematical and computational modelling of the nanostructure and its potential for predicting mechanical properties and durability. The degree of hydration is one of the key points when studying properties of cement-based materials. For engineering practical purposes, it is very useful to provide calculation procedures or macro models that help to predict mechanical properties as stress and strength, in hardening concrete systems. A computational simulation can be used in producing the hydration curve in terms of initial parameters, as chemical composition of cement, particle size distribution or water/cement ratio.

Computational modelling of the hydration process of cement is found to be a difficult issue due to the large number of factors that should be considered within

E. Cerro-Prada, M.J. Vázquez-Gallo, J. Alonso-Trigueros, and A.L. Romera-Zarza
E. U. I. T. de Obras Públicas, Universidad Politécnica de Madrid, Madrid, Spain

a three-dimensional spatial domain. Based on our previous work [1], we present in this paper an agent-based modelling approach in which a set of agents encapsulate the behaviour of the whole system while its hydration process. This computational modelling technique has been used for studying microstructural models of cement system at two scales: hydrated cement pasted at the micrometer level and calcium silicate hydrated gel at the nanolevel.

1.1 Agents Encapsulating the Behavior

In order to construct models involving multiple types of particles and different timing scales, one may choose between two different classes of models: equation-based approaches that evaluate or integrate sets of equations relating the system variables and, in the other hand, agent-based modelling which consists of a set of agents reflecting the behaviour of the various individuals that makes up the behavior of the whole system. In this paper, the latter approach has been adopted. Therefore, a meta-algorithm is developed as a framework for integrating heterogeneous units *–agents-* to form a complex model, as the cement system is indeed. With this style of scheme integration, the whole system is divided into *subsystems*, which means that a degree of modularity can be assumed for each subsystem simulated by using a particular algorithm. If this requirement is satisfied, it can be expected an established and well-studied model to be useful when simulating the hydration process of cement.

In the work presented here, this new modelling approach is employed for studying the nanostructure development of cement. A combination of stochastic, deterministic and adaptive rules allows the cement system to evolve from an anhydrous separated state to a bond state with enhanced mechanical properties. The model simulates the two main chemical reactions producing C-H-S gel and portlandite, considering the related anhydrous cement components, C_3S and C_2S, as two autonomous agents, while water and hydration time are viewed as activator agents.

Initially, each agent chooses, at random, one of the two types of particles: type A for C_3S, or type B for C_2S. During the simulation, each agent can take one of the following actions: (i) keep inactive, (ii) react with water, or (iii) stop to reacting

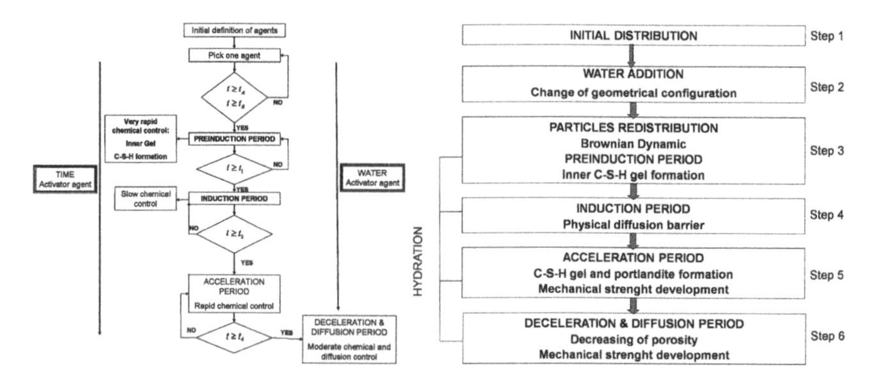

Fig. 1 Left: Flowchart representation of the agent-based model. Right: Algorithm Flowchart

with water. The rules for keeping chemically inactive are deterministic as they depend on the type of particle along with the time lasted. Reacting with water is also deterministic by the amount of water available. During the induction period, there is a physical diffusion barrier formed by the reaction products and the hydration process speeds down. The acceleration period follows and hydration process continues faster, until products layer for each particle reaches the critical thickness, according to van Breugel's work [2,3]. When that fact occurs, the deceleration period controls the hydration process at moderate rate. The overall agent-based model algorithm takes the form outlined in Fig. 1, left.

1.2 C-S-H Critical Thickness

The principal aim of this work is to measure the degree of hydration as well as reproduce the nanostructure of the C-S-H phase, resulting from hydration of alite and belite in ordinary Portland cement, through a self-controlled algorithm following the agent-based model described in section 1.1. The algorithm comprises a set of rules based on a systematic *ab-initio* study of hydration process of Portland cement. The main steps of the algorithm reproduce five hydration stages: preinduction, induction, acceleration, deceleration and diffusion [4].

The hydration kinetics are based on Van Breugel's model [2,3], consisting of an initial phase-boundary reaction in which a layer of C-S-H gel is formed over the surfaces of cement particles followed by a diffusion process, starting at a certain *critical time* when the product layer around each cement particle reaches the so called *transition thickness*. The novelty presented here with respect to previous works, has been when considering different rate factors for each individual particle. When hydration occurs, the algorithm randomly assigns to each cement particle its rhythm for decreasing, being this rhythm into a specific time interval that depends on the size and composition of the particle, coherently built from experimental data. Another new ingredient is how the transition to the diffusion period has been modelled. The critical thickness for each particle depends on its size and its chemical type. The values given by the algorithm have shown to be in good agreement with experimental work [2].

Regarding the kinetics of diffusion period, while most previous models presenting the exponential rule for governing the process simplifies the exponential law to a lineal rate, in our model the exponential character on the kinetic law has been preserved. These new approaches could be seen as an indirect but simple way of reflecting interactions between particles which, in general, appear to be difficult to simulate.

2 Experimental Work

In this study, the material representation is based on the Stroeven's work [5], which simulates the granular structure of cement at the microlevel as a partial distribution of spherical particles (anhydrous cement) diluted in an aqueous matrix – water-. This paper presents a non-investigated system configuration based on a *two-dimensional granular fluid*. Initially, anhydrous cement particles are randomly located throughout a thin film with rectangular base and very small height. When water is added the spatial restriction changes to a cylinder-shaped thin film whose thickness coincides with the previous height. The cement particles are not allowed to overlap any previously placed particle. Immediately after water addition, the hydration process starts and the algorithm simulates the corresponding preinduction, induction, acceleration, deceleration and diffusion periods, as it is shown in Fig. 1, right.

The implementation of the algorithm has been performed by using the algebraic computation software Maple [6].The degree of hydration and the thickness of C-H-S gel both can be obtained in terms of the initial size and proportion of the anhydrous particles, as well as the relative consumption of water, that can be written as a function of the water/cement ratio.

Particle size distribution of the cement is one of the factors which determine the progress of the hydration process. This distribution has been often described in terms of the Rosin-Rammler function [2]. Here, we consider just two *chemical types* of anhydrous particles: C_3S, and C_2S, corresponding to agent-A and agent-B respectively. Spheres have been chosen for modelling the *particle shape* (as broadly chosen by most of models, see [2] for a discussion about the suitability of this choice). Regarding *size particle*, the radius for type A and type B spheres are also set as initial parameters. The *water/cement ratio*, while most of previous studies consider a set of discrete possible values for this ratio, this algorithm allows this ratio to be an independent variable in terms of which the relative consumption of water can be written. For the rate factor controlling the decreasing radius of cement particles, each one of them has its own rate, taken at random within some interval according to the size and type of particle considered. The exponential character of diffusion period kinetic used here, allows each particle to have a randomly assigned exponent between 1 and 2, in agreement with experimental knowledge.

According to van Breugel's work [2], once the lost cement volume in some interval of time is computed, the volume of C-H-S produced is proportional to that volume by a factor ${}_1$= 2.2, being temperature constant at 20ºC. In the same way, the lost water volume is proportional to the lost cement volume, by a factor equal to ${}_2$= 0.4 times the density of cement. The values of these two constants are given by experimental works. Other elements involved in the hydration process of cement, like temperature and different curing conditions, have not been taken into account in this modelling approach.

3 Results and Discussion

The algorithm effectiveness has been quantified by building a fineness function to measure the *degree of hydration*, defined as the ratio of reacted cement volume to initial cement volume, and the *relative consumption of water*, defined as the ratio of lost water volume to initial water volume, at certain stages of the process. In Table 1, the degree of hydration for two different particle sizes, is shown for three different periods of the hydration process. Interpolating these data, it is possible to plot the degree of hydration versus time in these two cases –as it is shown in Fig. 2.A - and discuss whether the curves obtained are in good agreement with experimental

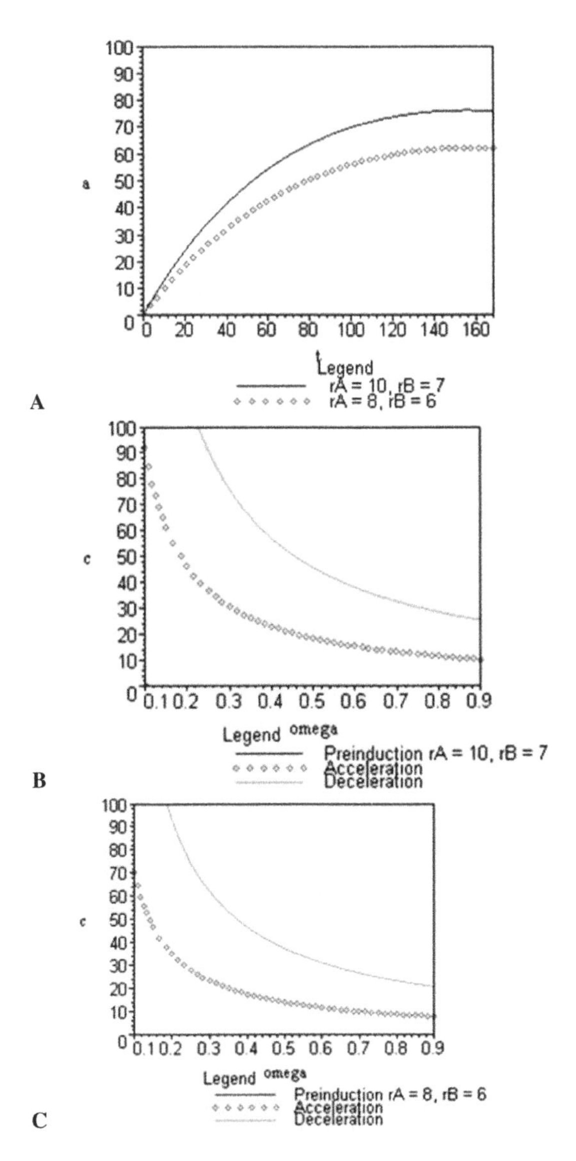

Fig. 2 Degree of hydration, B, C: Relative water consumption

curves [2]. In Table 2, the relative water consumption is shown as a function of the water/cement ratio, being this inversely proportional to the w/c ratio, ω.

Plotting the relative water consumption versus the water/cement ratio can be done directly from the functions provided by the algorithm, as it is shown in Fig. 3.1, B and C. After the preinduction period, the volume of lost water is so small that the corresponding curve it is not appreciated in the plot.

Table 1 Degree of hydration. Ratio between volume of reacted cement to initial cement volume

Hydration Stage	20 min Preinduction	28 hours Acceleration	7 days Deceleration
a = Degree of hydration r_A=10, r_B=7	0.39	30.62	75.93
a = Degree of hydration r_A=8, r_B=6	0.29	23.32	61.96

Table 2 Relative water consumption

Hydration stage	20 min Preinduction	28 hours Acceleration	7 days Deceleration
Relative consump. water, r_A=10, r_B=7	0.027/	9.188/	22.778/
Relative consump. water, r_A=8, r_B=6	0.019/	6.995/	18.587/

4 Conclusions

This work demonstrates that the agent-based molecular formation algorithm for simulation of cement Portland hydration process at the nanoscale, is compatible with the current understanding of the physical-chemical behaviour of the cement grains when they are in contact with water. A great advantage of this computational technique is that the emergent phenomena can be modelled through very simple rules governing the behaviour of each agent as well as the discrete evolution of the system configuration. This computation technique can be helpful for better understanding the correlation between the formation of micro and nano structure as well as the strength development and other mechanical properties of cementitious systems.

References

1. Cerro-Prada, E., Vázquez-Gallo, M.J., Alonso-Trigueros, J., Romera-Zarza, A.L.: Modelling hydration process of cement nanoparticles by using an agent-based molecular formation algorithm. In: Proceedings of IASTED International Conference on Nanotechnology and Applications (NANA 2008), Crete, Greece, pp. 66–71 (2008)
2. Van Breugel, K.: Simulation of hydration and formation of structure in hardening cement-based materials, Ph.D. Thesis, Delft University of Technology, Delft, The Netherlands (1991)
3. Van Breugel, K.: Numerical simulation of hydration and microstrutural development in hardening cement-based materials. Cement Concrete Res. 25, 319–331 (1995)
4. Kondo, et al.: 5th ICCC, Tokyo, vol. II, pp. 203–248 (1968)
5. Stroeven, M., Stroeven, P.: SPACE system for simulation of aggregated matter application to cement hydration. Cement Concrete Res. 29, 1299–1304 (1999)
6. Maple 7. 2001 edn. Copyright@Maplesoft. Waterloo Maple Inc. (2001)

CHH Cement Composite

A. Cwirzen, K. Habermehl-Cwirzen, L.I. Nasibulina, S.D. Shandakov, A.G. Nasibulin, E.I. Kauppinen, P.R. Mudimela, and V. Penttala

Abstract. The compressive strength and electrical resistivity for hardened pastes produced from nanomodified Portland SR cement (CHH- Carbon Hedge Hog cement) were studied. The nanomodification included growing of carbon nanotubes (CNTs) and carbon nanofibers (CNFs) on the cement particles. Pastes having water to binder ratio of 0.5 were produced. The obtained hardened material was characterized by increased compressive strength in comparison with the reference specimens made from pristine SR cement, which was attributed to reinforcing action of the CNTs and CNFs. The electrical resistivity of CHH composite was lower by one order of magnitude in comparison with reference Portland cement paste.

1 Introduction

Carbon nanotubes (CNTs) were first described in 1991 [1] and since that time an intensive research has been initiated in number of institutes around the world. CNTs are characterized by high tensile strength and elastic modulus as well as exceptional electrical and thermal conductivity. CNTs were incorporated into matrixes based on metals, ceramics and to a limited extend into Portland cement [2-5]. The main problems while incorporating CNTs into any matrix is to obtain their uniform dispersion and sufficient bond with the binder matrix. Incorporation of CNTs into Portland cement-based matrixes was done mainly by water dispersion, [6, 7]. The main drawback of this method is the limitation of the maximum amount of the CNTs to around 1-1.5% (according to the cement weight, with water to binder ratio of 0.5). Higher additions usually deteriorate severely workability, [6]. Some solution of this problem was to add surfactants, usually polycarboxylated-based superplasticizers, which facilitated dispersion processes and enhanced the workability. Another solution is to chemically functionalize CNTs surfaces with "polar impurities" e.g. OH, COOH end groups [8, 9, 10] or oxidation, [11]. One of the main drawbacks of functionalization in the case of Portland cement-based binders is strong loss of workability due to absorption of surfactants

A. Cwirzen, K. Habermehl-Cwirzen, and V. Penttala
Helsinki University of Technology, Faculty of Engineering and Architecture

L.I. Nasibulina, S.D. Shandakov, A.G. Nasibulin, E.I. Kauppinen, and P.R. Mudimela
Helsinki University of Technology, NanoMaterials Group, Department of Applied Physics and Center for New Materials

on the functionalized surfaces, [7, 12]. The present paper will describe new type of Portland cement incorporating in-situ grown CNTs and carbon nano fibers (CNF) named as CHH cement (Carbon Hedge Hog cement). Basic mechanical and electrical properties of the obtained hydrated matrix will be described.

2 Production of CHH and Experimental Setup

The CHH cement is produced from Portland cement by growing the CNTs/CNFs from Fe catalyst particles naturally occurring in cement. The processes included in production of the CHH cement composite are shown schematically in Figure 1. The CNTs/CNFs were grown using a modified Chemical Vapor Deposition (CVD) method, [11]. The modification included addition of a screw feeder, which allowed continuous productions of the CHH cement.

The obtained hybrid material was called Carbon Hedge Hog (CHH) cement. A SEM image of cement particles with grown CNTs/CNFs is shown in Figure 2.

The amount and morphology of the grown CNTs/CNFs depended on the applied temperatures, which varied from 500 to 700^{0}C, on the flow speed of pristine cement and on the type/amount of the introduced gas. In this paper only results obtained from CEM I 42.5 N (SR) produced by Finnsementti Oy will be presented. The main chemical phases of that cement were C_3S (68%), C_2S (13%), C_3A (1%) and of C_4AF (13%). The specific surface area was 360 m^2/kg and the bulk density was 3100 kg/m^3.

Seven mixes were studied in this part of the research. Constant water to binder ratio of 0.5 was used for all mixes. The amount of added superplasticizer equaled 1.5% as calculated by weight of the binder. Each mix contained CHH produced under different conditions, which resulted in different amounts of the grown

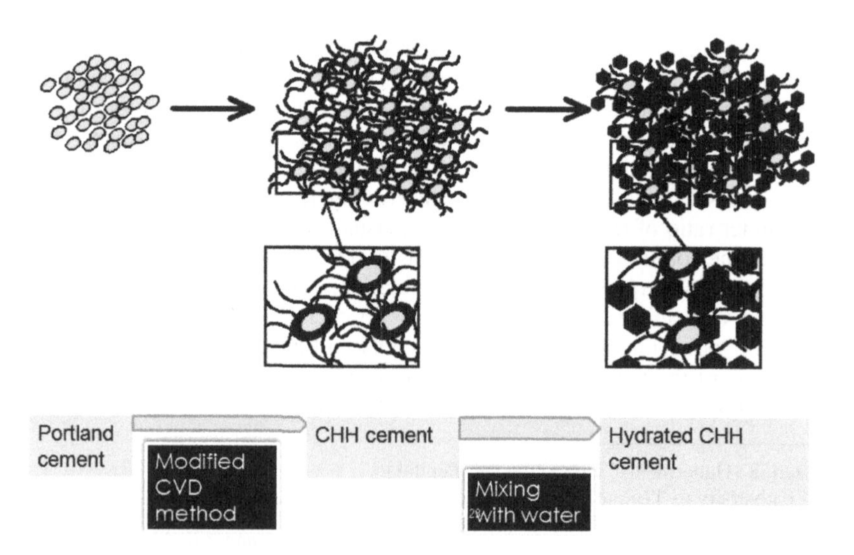

Fig. 1 Production process for CHH cement composite

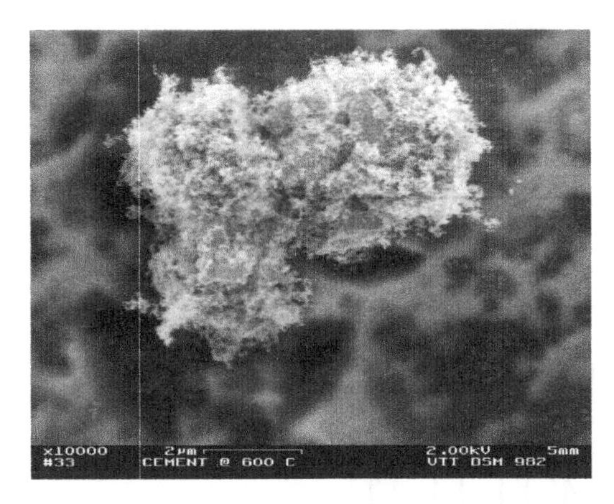

Fig. 2 SEM image of CHH cement particle as produced by the modified CVD method

CNTs/CNFs varying from 2 to 20% by weight of produced binder, [13]. The obtained workability varied from fluid to plastic and enabled to fill the formworks with just light vibration.

The compressive and flexural strengths were determined by using beams having dimensions of 10 x 10 x 60 mm^3. These small sizes of the specimens were dictated by the small amount of available materials. In order to increase the homogeneity of the mix, and especially to reduce the amount of entrapped air, a special small volume vacuum mixer was used.

3 Tests Results and Discussion

The studies of the mechanical properties revealed no significant change of the flexural strength of any of the studied mixes in comparison with the reference paste. On the other hand, the measured compressive strength, Figure 3, showed a remarkable increase in the case of mixes C3 and C6. Mixes C1, C2, C4, and C5 revealed lower strength in comparison with the reference pristine cement.

The main reason for these variations is the amount of the CNTs/CNFs present on the cement particles. The TG studies which results are published elsewhere [12] have confirmed that conclusion. Furthermore, the recorded electrical resistivity values showed significant differences. The highest conductivity was obtained for mixes C1 and C2. At the same time, these mixes showed the lowest compressive strength values. On the bases of SEM and TG results, it can be attributed mainly to the high amount of the CNTs/CNFs, which covered tightly the surface of the cement particles. As a result, the access of water to cement was significantly limited with resulted in a lower hydration degree. Furthermore, the morphology of the grown nanofibers could affect the obtained results. For instance, longer, more spiral-like shaped and with more surface defects, CNTs/CNFs could be characterized by better bond with the binder matrix. Additionally bond strength could be also increased by actual embedment of the catalyst particle into the hydration phases which adds a chemical bond possible otherwise by functionalization of the surfaces.

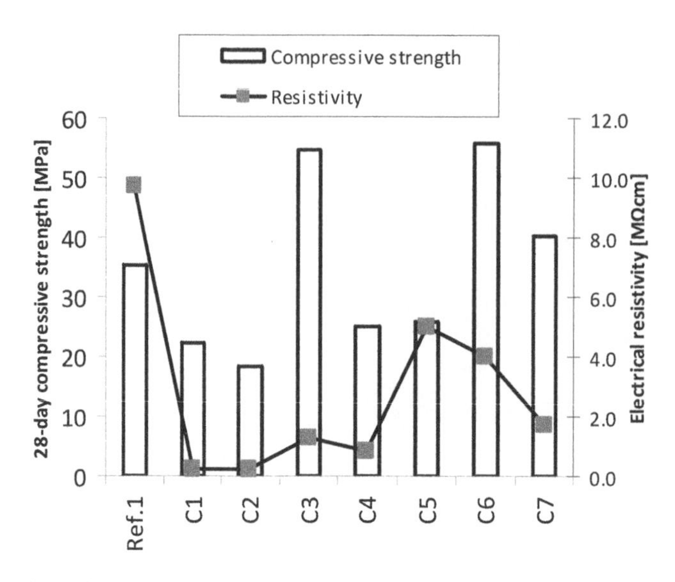

Fig. 3 28-day compressive strength and electrical resistivity [13]

The SEM studies showed that during hydration processes the CNTs/CNFs were imbedded into the CSH matrix. Furthermore, as seen in Figure 4, a bridging of voids occurred. Studies of the resin impregnated and polished specimens' revealed that the microstructure was regular for any binder matrix made of Portland cement at water to binder ratio of 0.5. Interestingly, there was significantly less Portlandite present in the hydrated CHH matrix in comparison with specimens produced from pristine SR cement as observed on polished section in the SEM using back scattered electron detector (BSE). These results were also confirmed by XRD measurements which results were published elsewhere, [11, 12].

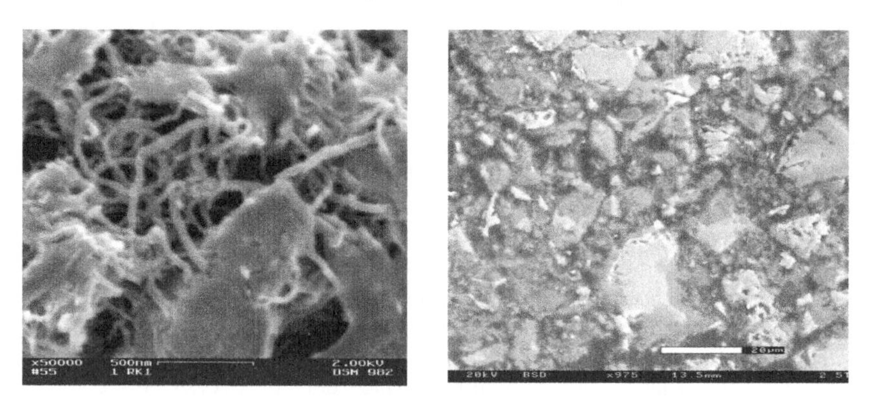

Fig. 4 Hydrated CHH binder: a) fractures surface, b) polished surface [13]

4 Conclusions

The CHH cement composite is a new material obtained after hydration of Portland cement modified with CNTs/CNFs by application of chemical vapor deposition method. The resulting material is characterized by high amount of incorporated CNTs/CNFs, which are uniformly distributed thought out the binder matrix and partly chemically bonded to it through a catalyst particle originating from Portland cement. It was possible to obtain a composite containing up to 20% of nanofibers. The mechanical properties of the composite were improved providing that the optimum amount of the nanofibers was grown. The electrical conductivity was increase by one order of magnitude allowing to classify this materials a semiconductor.

Acknowledgments. The results presented in this paper belong to a research project financed by Finnish Academy of Sciences, which is gratefully acknowledged.

References

1. Iijima, S.: Helical microtubules of graphitic carbon. Nature 354, 56–58 (1991)
2. Gao, L., Jiang, L., Sun, J.: Carbon nanotube-ceramic composites. J. Electroceram, 1751–1755 (2006)
3. Dubios, P.H., Alexandre, M.: Performant Clay/Carbon Nanotube Polymer Nanocomposites. Adv. Eng. Mater. 8, 147–154 (2006)
4. Girifalco, L.A., Hodak, M., Lee, R.S.: Carbon nanotubes, buckyballs, ropes, and a universal graphitic potential. Phys. Rev. B 62, 13104–13110 (2000)
5. Li, G.Y., Wang, P.M., Xiaohua, Z.: Pressure-sensitive properties and microstructure of carbon nanotube reinforced cement composites. Cement Concrete Comp. 29, 377–382 (2007)
6. Kowald, T.: Influence of surface-modified Carbon Nanotubes on Ultrahigh Performance Concrete. In: Proceedings of International Symposium on Ultra High Performance Concrete, September 2004, pp. 195–203 (2004)
7. Cwirzen, A., Habermehl-Cwirzen, K., Penttala, V.: Surface decoration of carbon nanotubes and mechanical properties of cement/CNT composites. Adv. Cem. Res. 20, 65–73 (2008)
8. Yao, N., Lordi, V., Ma, S.X.C., Dujardin, E., Krishnan, A., Treacy, M.M.J., Ebbesen, T.W.: Structure and oxidation patterns of carbon nanotubes. J. Mater. Res. 13, 2432–2437 (2000)
9. Ebbesen, T.W., Hiura, H., Bisher, M.E., Treacy, M.M.J., Sheeve Keyer, J.L., Haushalter, R.C.: Decoration of Carbon Nanotubes. Adv. Mater. 8, 155–157 (1996)
10. Moisala, A., Nasibulin, A.G., Kaupinen, E.I.: The role of metal nanoparticles in the catalytic production of single-walled carbon nanotubes – A review. J. Phys-Condens Mat. 15, 3011–3035 (2003)
11. Cwirzen, A., Habermehl-Cwirzen, K., Shandakov, D., Nasibulina, L.I., Nasibulin, A.G., Mudimela, P.R., Kauppinen, E.I., Penttala, V.: Properties of high yield synthesized carbon nano fibers/Portland cement composite. Adv. Cem. Res. (submitted) (2008)
12. Cwirzen, A., Habermehl-Cwirzen, K., Shandakov, D., Nasibulina, L.I., Nasibulin, A.G., Mudimela, P.R., Kauppinen, E.I., Penttala, V.: CHH cement composites - microstructure and hydration processes. Cement Concrete Res. (submitted) (2008)
13. Cwirzen, A., Habermehl-Cwirzen, K., Shandakov, D., Nasibulina, L.I., Nasibulin, A.G., Mudimela, P.R., Kauppinen, E.I., Penttala, V.: Mechanical and selected physical properties of cement pastes produced by using Portland cement modified with multi-walled carbon nanotubes. In: Proc. 8th International Symposium on Utilization of High-Strength and High-Performance Concrete, Tokyo, Japan (2008)

Modeling of Nanoindentation by a Visco-elastic Porous Model with Application to Cement Paste

D. Davydov and M. Jirásek

Abstract. In order to derive a meaningful constitutive relationship for the Calcium-Silicate-Hydrate (*CSH*) phase in hardened cement paste, a micromechanical approach is used. It is well known that *CSH* gel can be considered as a porous medium composed of certain particles, but the precise shape of these particles has not been well established yet, neither experimentally nor theoretically. Different authors consider spheres, platelets or fibers. In the present study we propose a viscoelastic constitutive relationship for *CSH* spherical particles and upscale it to the level at which nanoindentation takes place. The resulting contact problem can be solved semi-analytically using the Laplace-Carson transform. Material properties can be identified by solving an inverse problem based on the nanoindentation experiment. An example of application to white cement paste is presented.

1 Introduction

Nanoindentation is a widely used technique for measuring properties of materials at the micron and submicron levels. It consists of establishing contact between a substrate (sample) and an indenter with known properties and geometry. The force P acting on the indenter is applied as the control variable and the corresponding penetration depth h is recorded. Commonly, a trapezoidal loading program with loading, holding and unloading periods is used (Fig. 1).

Elastic properties of the substrate can be evaluated from the unloading part of the load-penetration curve by the standard Oliver-Pharr procedure [5], which is based on the analytical solution of the contact problem for an axisymmetric indenter and a linear elastic isotropic homogeneous infinite half-space [6]. Although this procedure is based on the assumption of material homogeneity, it has been applied to heterogeneous materials as well [3].

D. Davydov
Department of Mechanics, Faculty of Civil Engineering, Czech Technical University in Prague
e-mail: denisdavydov@fsv.cvut.cz

M. Jirásek
Department of Mechanics, Faculty of Civil Engineering, Czech Technical University in Prague
e-mail: milan.jirasek@fsv.cvut.cz

Another problem with the interpretation of indentation results is caused by deviations from an ideal shape of the indenter. The geometry of the actual indentation tip varies due to the production process and wear out during experiments. The asssociated error can be reduced by calibration based on series of indentations into a reference material with known properties [5]. The results can be used to determine an approximation of the tip shape for which the Oliver-Pharr procedure gives the correct elastic modulus. Usually, the cross-sectional area $A_{tip} = \rho^2\pi$ is approximated by several terms of the series

$$A_{tip} = C_0 f^2 + C_1 f + C_2 f^{1/2} + C_3 f^{1/4} \ldots \tag{1}$$

where ρ and f are respectively the radius of the section and its distance from the apex of the axisymmetric tip (Fig.1), and C_i are constants describing the tip shape.

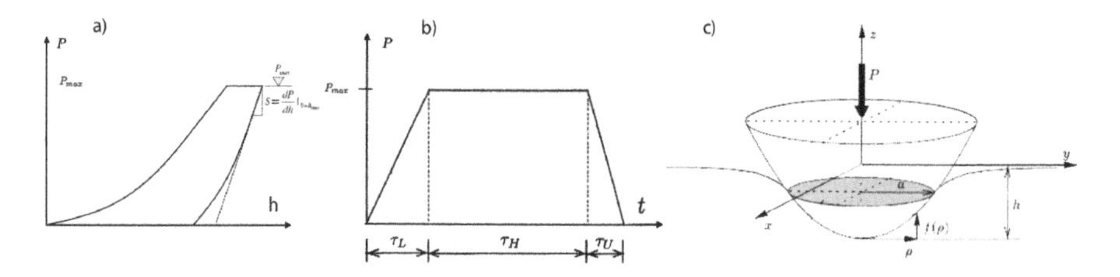

Fig. 1 a) Illustration of the load-penetration P-h curve; b) Trapezoidal loading program used in a nanoindentation experiment; c) Indentation of material by a rigid axysimmetric indenter

Vandamme [8] has developed a viscoelastic solution of the contact problem for a conical indenter using several simple creep models. In view of the current understanding of the structure of *CSH* – the main creeping phase in the cement paste – as a porous medium [3,4], it is necessary to extend this procedure to visco-elastic two-phased porous materials.

2 Viscoelastic Solution of Contact Problem

Analysis of the indentation problem with a rigid indenter of an arbitrary axisymmetric shape can be based on the so-called Galin-Sneddon solution [6]. The relation between the penetration depth h and the corresponding load P is parametrically described by

$$h = a\int_0^a \frac{f'(\rho)d\rho}{\sqrt{a^2-\rho^2}}, \quad P = 2\frac{E}{1-\nu^2}\int_0^a \frac{\rho^2 f'(\rho)d\rho}{\sqrt{a^2-\rho^2}} \tag{2}$$

where a is the radius of the projected contact area A_c (Fig. 1), ρ is the tip radius, and $f(\rho)$ is a smooth function describing the tip shape and implicitly defined by Eq. (1). An ideal conical indenter is described by

$$f(\rho) = \rho / tan(\alpha) \tag{3}$$

where α is the semi-apex angle. In Eq. (1), this case would correspond to $C_0 = \pi \tan^2 \alpha$ and $C_i = 0, i = 1, 2, \ldots$. For $f(\rho)$ given by Eq. (3), the improper integrals in Eq. (2) can be calculated analytically, and the following formula can be derived for evaluation of the indentation modulus from the initial slope of the unloading part of the indentation curve:

$$\frac{E}{1-\nu^2} = \frac{1}{2} \frac{dP}{dh} \frac{\sqrt{\pi}}{\sqrt{A_c}} \tag{4}$$

The foregoing elastic solution can be extended to the viscoelastic case by the method of functional equations [2]. It consists of replacing elastic constants in the contact problem solution by Laplace-Carson transform (*LCT*) of bulk and shear relaxation functions. The second part of Eq. (2) can be rewritten in the form

$$P(t) = M\ F(a(t)), \quad F(a(t)) = 2\int_0^a \frac{\rho^2 f'(\rho) d\rho}{\sqrt{a^2 - \rho^2}} \tag{5}$$

Here, function $F(a(t))$ depends only on the tip geometry, whereas the material properties are reflected by the indentation modulus $M \equiv E/(1-\nu^2)$. The visco-elastic solution of the contact problem can then be written as

$$\hat{P}(s) = \hat{M}(s)\hat{F}(a(s)) \tag{6}$$

where $\hat{\bullet}(s)$ denotes the *LCT* of $\bullet(t)$. The *LCT* of the indentation modulus can be expressed in terms of the *LCTs* $\hat{K}(s)$ and $\hat{G}(s)$ of the bulk and shear relaxation functions as

$$\hat{M}(s) = 4\hat{G}(s)(3\hat{K}(s) + \hat{G}(s))/(3\hat{K}(s) + 4\hat{G}(s)) \tag{7}$$

Using homogenization procedures in the Laplace-Carson domain, it is possible to upscale the visco-elastic properties from the microstructure level to the level of the indentation experiment. For the self-consistent homogenization scheme, the effective moduli K_{hom} and G_{hom} are obtained by solving the system of two nonlinear equations

$$\begin{aligned} K_{hom} &= \sum_r f_r K_r [1 + \frac{\alpha_{hom}}{K_{hom}}(K_r - K_{hom})]^{-1} / \sum_\omega f_\omega [1 + \frac{\alpha_{hom}}{K_{hom}}(K_\omega - K_{hom})]^{-1} \\ G_{hom} &= \sum_r f_r G_r [1 + \frac{\beta_{hom}}{G_{hom}}(G_r - G_{hom})]^{-1} / \sum_\omega f_\omega [1 + \frac{\beta_{hom}}{G_{hom}}(G_\omega - G_{hom})]^{-1} \end{aligned} \tag{8}$$

in which f_r are the volume fractions of individual phases, K_r and G_r are their bulk and shear moduli, and

$$\alpha_{hom} = \frac{3K_{hom}}{3K_{hom}+4G_{hom}}, \quad \beta_{hom} = \frac{6}{5}\frac{K_{hom}+2G_{hom}}{3K_{hom}+4G_{hom}}$$

A rearrangement of Eq. (6) yields an expression for the LCT of $F(a(t))$,

$$\hat{F}(a(s)) = \hat{P}(s)/\hat{M}(s) \equiv \hat{P}(s)\hat{Y}(s) \tag{9}$$

where we have denoted $\hat{Y}(s) = 1/\hat{M}(s) =$ the LCT of the indentation compliance.

Exploiting the properties of the LCT, Eq.(9) can be written in the time domain as

$$F(a(t)) = \int_0^t Y(t-\tau)\,\dot{P}(\tau)\,\mathrm{d}\tau \tag{10}$$

where the dot over P denotes the time derivative. To calculate the improper integrals in Eq. (2), $f'(\rho)$ is needed. Thus, Eq. (1) has to be solved for $f(\rho)$. For a simple shape of the indenter (two constants) these integrals can be calculated analytically [1], but for the case when the contact area of the tip is approximated using three or more terms, a closed-form solution is complicated or even not available. Numerical integration can then be used. Eq. (1) is solved for $f(\rho)$ at $\rho_i \in [0, \rho_{max}]$, $i = 1,2,...N$. Afterwards, $f(\rho_i)$ is interpolated by piecewise cubic Hermite polynomials. The Monte-Carlo integration method is used to compute the integrals. The obtained values $h_i = h(a_i)$ and $F_i = F(a_i)$ is then interpolated for later usage.

For very simple visco-elastic models, such as the Kelvin-Maxwell-Voight model, the exact form of $Y(t)$ can be derived analytically [8]. However, for a logarithmic creep model, even in its simplest form, numerical inversion has to be done. Since the basic creeping phase *CSH* is believed to be porous, we consider a two-phase polycrystal composite with one phase being viscoelastic and another corresponding to the pores. From given viscoelastic model parameters and volume fraction of pores, $\hat{K}_{hom}(s)$ and $\hat{G}_{hom}(s)$ for the homogenized medium can be calculated by solving (8) in the Laplace-Carson domain. Substituting the result into Eq. (7), the indentation modulus $\hat{M}(s)$ and thus $\hat{Y}(s) = 1/\hat{M}(s)$ can be calculated. The Stehfest algorithm [7] is used for the inverse transform and calculation of $Y(t)$.

Indentation experiments are usually conducted under trapezoidal load control (Fig. 1), described by

$$P(t) = \begin{cases} P_L(t) = t/\tau_L \, P_{max} & 0 \le t \le \tau_L \\ P_H(t) = P_{max} & \tau_L \le t \le \tau_L + \tau_H \\ P_U(t) = (\tau_L + \tau_H + \tau_U - t) \, P_{max} / \tau_U & \tau_L + \tau_H \le t \le \tau_L + \tau_H + \tau_U \end{cases} \tag{11}$$

where τ_L, τ_H and τ_U are the loading, holding and unloading durations, respectively. Considering this load history in Eq. (10) leads to

$$\begin{aligned} F(a(t)) &= \frac{P_{max}}{\tau_L} \int_0^t Y(t-\tau)\,\mathrm{d}\tau \quad 0 \le t \le \tau_L \\ F(a(t)) &= \frac{P_{max}}{\tau_L} \int_0^{\tau_L} Y(t-\tau)\,\mathrm{d}\tau \quad \tau_L \le t \le \tau_L + \tau_H \end{aligned} \tag{12}$$

Knowing $Y(t)$, the integrals in Eq. (12) are numerically calculated using an adaptive Simpson quadrature. With $F(t)$ at hand, the previously found interpolation of the values $F(a_i)$ and $h(a_i)$ allows obtaining $h(t)$. Thus, an optimization procedure to fit the experimentally observed penetration depth history can be invoked.

3 Example of Application and Conclusions

In order to check the proposed method, one and a half year old cement paste CEM I 42.5 prepared at $w/c = 0.4$ has been tested. Indentation has been performed by a Hysitron nanoindenter using a trapezoidal loading program with $\tau_L = 10\,\mathrm{s}$, $\tau_H = 100\,\mathrm{s}$ and $\tau_U = 10\,\mathrm{s}$ and the maximum applied force $P_{\max} = 1000\,\mu\mathrm{N}$. One of the obtained indentation curves has been considered. The indenter geometry parameters have been determined as $C_0 = 24.5$, $C_1 = 4793$, $C_2 = -82009$, $C_3 = 30934$, $C_4 = -236130$. The Poisson ratio has been set to 0.24. From the Oliver-Pharr procedure, elastic modulus of 31.64 GPa has been obtained. The porosity of *CSH* has been taken as 0.26 [3,4]. The deviatoric creep has been described by the logarithmic compliance function

$$J(t - t_0) = 1/G_0 + J \log(1 + (t - t_0)/\tau) \tag{13}$$

with $J = 0.0082\,\mathrm{GPa}^{-1}$ and $\tau = 0.0287\mathrm{s}$.

Fig. 2 shows the best fit of function $F(t)$ and the corresponding fit of the experimental penetration depth curve.

The considered example shows the applicability of the proposed method to extraction of viscoelastic properties of porous media from the nanoindentation experiment. This method could be useful in further studies of the *CSH* gel in cement paste as well as of other porous media.

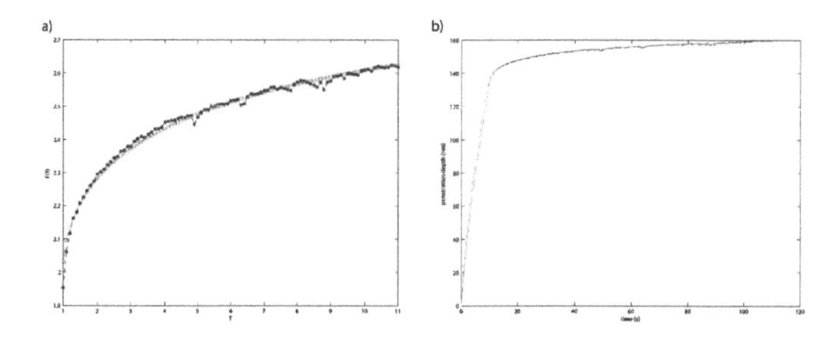

Fig. 2 a) Experimental F(t) curve and its fit in dimensionless form; b) Experimental penetration depth h(t) and the corresponding fit by a porous logarithmic creep model

Acknowledgments. The authors are grateful to the European Community for the full support of Denis Davydov under the Marie Curie Research Training Network MRTN-CT-2005-019283 "Fundamental understanding of cementitious materials for improved chemical, physical and aesthetic performance" and to the Ministry of Education, Youth and Sports of the Czech Republic for the full support of Milan Jirásek under the Research Plan MSM6840770003.

References

1. Jager, A., et al.: Identification of viscoelastic properties by means of nanoindentation taking the real tip geometry into account. Meccanica 42, 293–306 (2007)
2. Lee, E., et al.: The contact problem for viscoelastic bodies. J. Appl. Mech. 82, 438–444 (1960)
3. Constantinides, G.: The nanogranular nature of C-S-H. J. Mech. Phys. Solids. 55, 64–90 (2007)
4. Jennings, H., et al.: A multi-technique investigation of the nanoporosity of cement paste. Cem. Concr. Res. 37, 329–336 (2007)
5. Oliver, W., et al.: An improved technique for determining hardness and elastic modulus using load and displacement sensing indentation experiments. J. Mater. Res. 7, 1564–1582 (1992)
6. Sneddon, I.: The relation between load and penetration in the axisymmetric Boussinesq problem for a punch of arbitrary profile. Int. J. Eng. Sci. 3, 47–57 (1965)
7. Stehfest, H.: Numerical inversion of Laplace transforms. CACM 13(1), 438–444 (1970)
8. Vandamme, M., et al.: Viscoelastic solutions for conical indentation. Int. J. Sol. Struct. 43, 3142–3165 (2006)

Multi-scale Study of Calcium Leaching in Cement Pastes with Silica Nanoparticles

J.J. Gaitero, W. Zhu, and I. Campillo

Abstract. Calcium leaching is a degradation process consisting in the progressive dissolution of the cement paste as a consequence of the migration of the calcium ions to the aggressive solution. Although the most important changes take place at the nano- and micro-scale, their consequences are observed at every length scale. Within this work, a multi-scale approach combining a wide variety of experimental techniques was used to study such phenomenon in cement pastes with silica nanoparticles. The experimental results proved that the pozzolanic reaction induced by the nanoparticles resulted in a C-S-H gel more stable chemically and with longer silicate chains. In addition, the reduction of the amount of portlandite gave place to pastes with improved microstructure. As a consequence, the performances of such pastes were greatly enhanced both before and during the degradation process.

1 Introduction

Calcium leaching is a degradation process consisting in the progressive dissolution of the cement paste as a consequence of the migration of calcium ions to the aggressive solution. The kinetics of the process depend of several parameters being the most important ones the porosity (the natural path for chemical attack), the composition of the paste (each phase degrades at a different rate), and the nature of the aggressive solution (generally soft water). These are the reasons for using nanosilica in order to reduce calcium leaching. Silica, generally in the form of silica fume, has long been used to improve the performance of the cement paste in terms of strength and refinement/reduction of the porosity [1, 2, 6, 7, 15, 17]. Furthermore, this takes place by mean of a pozzolanic reaction that results in the reduction of the amount of portlandite, the hydrous phase most severely affected by

J.J. Gaitero
Labein-Tecnalia, Parque Tecnologico de Bizkaia, Derio, Spain
e-mail: jjgaitero@labein.es

W. Zhu
University of the West of Scotland, Paisley, Scotland

I. Campillo
CIC nanoGUNE Consolider, Donostia, Spain

calcium leaching. The use of silica nanoparticles has several advantages in comparison to other types of silica [3, 13]. As a consequence of their reduced size the nanoparticles act as nucleation site for the growth of hydration products, accelerating the hydration rate. Furthermore, their large surface area and purity (up to 99 %), altogether with their amorphous nature, provides them a great reactivity.

2 Materials and Methods

The samples were prepared at a water-cement ratio w/c=0.4 using an Ordinary Type I 52.5R Portland cement and a 6 wt.% of four different types of commercial silica nanoparticles. The mixing process varied depending on whether the nanoparticles were in the form of colloidal dispersion (CS1, CS2, CS3) or dry powder (ADS). While in the first case, the dispersion was mixed with the water and stirred for 5 minutes before adding the solution to the cement, in the other one, the powder was mixed with the cement and agitated for a minute before pouring the water on it. One set without silica nanoparticles (REF) was also prepared for comparison. The samples were cast into $1\times1\times6$ cm^3 prisms using steel moulds, where they stayed for 24 hours in a chamber at 20 ºC and 100 % humidity. After such time, they were unmoulded and introduced in a saturated lime solution at room temperature for another 27 days of curing. At the end of this period (t_0), they were moved into a bath containing a 6 M amonium nitrate solution where they stayed for 9, 21, 42 or 63 days for the accelerated calcium leaching (t_1, t_2, t_3 and t_4 respectively). In order to prevent the samples from carbonation, the aggressive bath was maintained all the time in contact with a nitrogen atmosphere. Furthermore, the solution was renewed whenever its pH reached a value of 9.2 to ensure its aggressiveness.

3 Results and Discussion

The multi-scale approach used in this work consisted in the combination of several experimental techniques that provided information about every length scale of the material. The discussion of the obtained results will be made in order attending to the length scale of the phenomena involved; i.e. it will begin with the macroscopic techniques and finish with the atomistic ones.

From a macroscopic point of view the addition of silica nanoparticles to the cement paste increased the overall strength of the paste and helped retaining it along the degradation process, see Fig. 1(a). The improvement in performance was about 30 % in the cured specimens (t_0) and 100-700 % in the degraded ones (t_4). Comparison between the different types of additions used revealed that the colloidal dispersions were much more effective reducing the effects of the degradation than the agglomerated dry silica (ADS).

There was a very good correlation between the macroscopic strength and the porosity evolution, see Fig. 1(a)-(b), which followed exactly the opposite trend. At t_0 the reference specimen (REF) was the one with the highest porosity, followed by ADS and then the pastes with colloidal silica. As soon as the degradation began

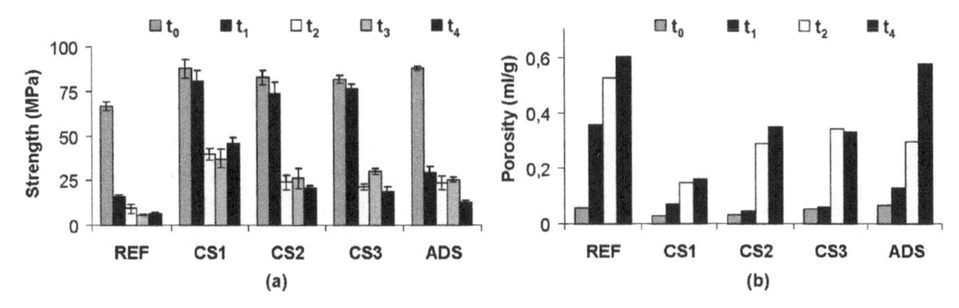

Fig. 1 (a) Compressive strength and (b) total pore volume, measured by mercury intrusion porosimetry, as a function of the degradation time

(t_1), the total pore volume increased dramatically in REF followed by ADS, being the rest of the specimens barely affected until t_2. This difference of behavior was attributed to the better dispersion of colloidal silica throughout the paste which resulted in a more homogeneous porosity and portlandite distribution compared to ADS. After such time, the increase in porosity was more progressive in all the pastes.

X-ray powder diffraction spectra proved that the great changes in porosity and strength undergone by the specimens during the early stages of the degradation process (t_1 for REF and ADS, and t_2 for CS1, CS2, and CS3) could be attributed almost entirely to the dissolution of portlandite, see Fig. 2. Therefore, the reduction in the amount of portlandite during the curing process because of the reaction with the silica nanoparticles contributed significantly to reduce the negative consequences of calcium leaching. Portlandite dissolution and the consequent porosity increase were also accompanied by a complete hydration of all the anhydrous cement present at the paste. The slower degradation observed afterwards was a consequence of the more progressive dissolution of other phases like ettringite or the C-S-H gel itself.

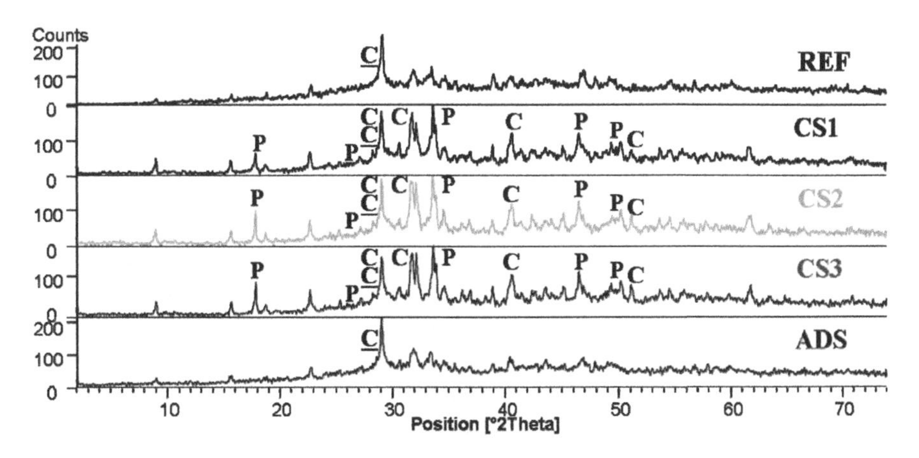

Fig. 2 X-ray powder diffraction spectra after 9 days of degradation (t_1). C: Clinker, C̲: Calcite, P: Portlandite

The elemental composition obtained by x-ray fluorescence confirmed these findings, see Fig. 3(a)-(b). The complete dissolution of portlandite at t_1 in REF and ADS and t_2 in CS1, CS2, and CS3 were clearly reflected in the evolution of C/S. The same happened with the dissolution of ettringite at t_3 in REF and t_4 in ADS and CS3. However, such dissolution was not complete in the latter because it was accompanied by the apparition of gypsum which, having a smaller calcium content than ettringite, was considered as an intermediate state in the degradation process.

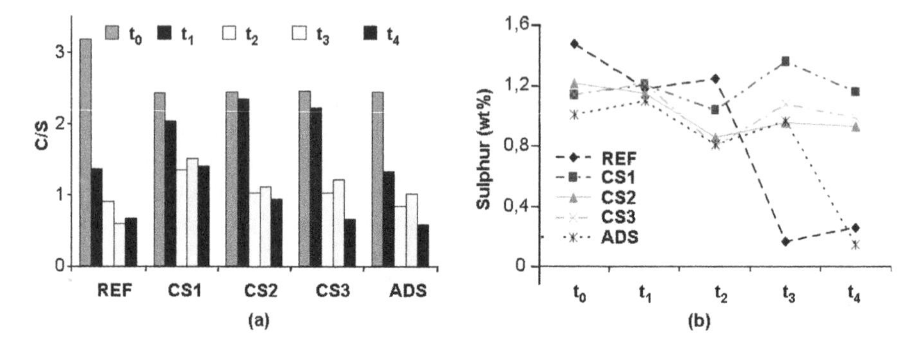

Fig. 3 (a) C/S and (b) sulphur content measured by x-ray fluorescence at different stages during the degradation process

Depth sensing nanoindentation [4, 5, 11, 12, 16, 18] was used to study the detrimental effects of the calcium loss in the mechanical properties of the C-S-H gel, see Table 1. Previous to the immersion of the specimens in the aggressive solution, no variation of the indentation modulus between the different pastes was observed. On the contrary, after 42 days of accelerated calcium leaching REF and ADS had undergone a severe and homogeneous degradation, while CS1 was almost intact at its centre with increasing loss of performance the closer to its outer surface.

Table 1 Indentation modulus

Sample	Degradation Time (Days)	Indentation Modulus (GPa)	
		LD-CSH	HD-CSH
REF	0	19±3	27±4
	42	2.3±0.5	3.6±0.6
ADS	0	23±2	27±3
	42	3.5±0.7	5.0±0.9
CS1	0	20±3	26±3
	42	4.6±1-14±2	7.1±0.7-24±3

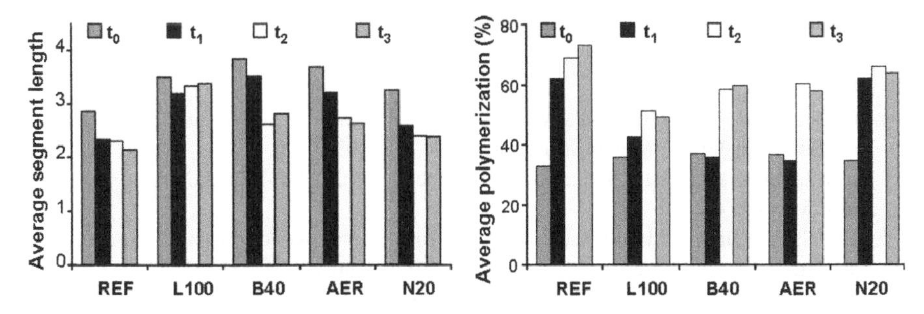

Fig. 4 Time evolution of the average segment length and the average polymeration calculated from the areas of the peaks of the ^{29}Si MAS-NMR spectra

^{29}Si MAS-NMR provided information about the changes taking place in the atomic structure of the C-S-H gel [8, 14]. Here the discussion will be made attending to two parameters calculated from the NMR spectra and defined elsewhere [9, 10]: the average polymerization, and the average segment length. According to Fig. 4, the addition of silica nanoparticles resulted in an increase in the average segment length. As soon as the degradation began, such average segment length was sharply reduced because of the hydration of the unreacted cement, which gave place to the apparition of an important number of new short silicate chains. Furthermore, as a consequence of the calcium loss, chains began to merge together resulting in an increase of the degree of polymerization. As calcium loss went on the average polymerization continued growing because of the progressive merging of the silicate chains. However, the fact that the average segment length barely varied could only be explained if such merging took place only at the chains ends. Therefore, it was concluded that longer chains improved calcium stability.

4 Conclusions

The addition of silica nanoparticles improved considerably the performance of the cement paste both before and during the degradation by calcium leaching. The multi-scale approach used for the characterization of the samples showed that the consequences of the addition of only 6 wt.% of silica nanoparticles affected to all aspects of the material. At the macroscopic level, they increased the macroscopic strength of the cured pastes and limited its reduction along the degradation process. This was a consequence of the improved microstructure (less porosity and portlandite) and the changes undergone by the C-S-H gel which made it more resistant to the decalcification.

References

1. Asgeorsson, H., Gudmundsson, G.: Pozolanic activity of silica dust. Cem. Concr. Res. 9, 249–252 (1979)
2. Bentur, A., Cohen, M.D.: Effect of condensed silica fume on the microstructure of the interfacial zone in Portland cement mortars. Am. Ceram. Soc. 70, 738–743 (1987)

3. Björnström, J., Martinelli, A., Matic, A., et al.: Accelerating effects of colloidal nano-silica for beneficial calcium-silicate-hydrate formation in cement. Chem. Phys. Let. 392, 242–248 (2004)
4. Constantinides, G., Ulm, F.-J., Van Vliet, K.J.: On the use of nanoindentation for cementitious materials. Mater. Struct. 36, 191–196 (2003)
5. Constantinides, G., Ulm, F.-J.: The nanogranular nature of C-S-H. J. Mech. Phys. Solids 55, 64–90 (2007)
6. Duval, R., Kandri, E.H.: Influence of silica fume on the workability and the compressive strength of high-performance concretes. Cem. Concr. Res. 28, 533–547 (1998)
7. FIP commission on concrete, Condensed silica fume in concrete. Thomas Telford, London (1988)
8. Gaitero, J.J., Sáez de Ibarra, Y., Erkizia, E., Campillo, I.: Silica nanoparticle addition to control the calcium-leaching in cement-based materials. Phys. Stat. Sol. (a) 203, 1313–1318 (2006)
9. Gaitero, J.J., Campillo, I., Guerrero, A.: Reduction of the calcium leaching rate of cement paste by addition of silica nanoparticles. Cem. Concr. Res. 38, 1112–1118 (2008)
10. Gaitero, J.J., Manzano, H., Campillo, I., Dolado, J.S.: Multi-scale approach for the study of cement pastes: calcium leaching, a case of study. Int. J. Mater. Prod. Tec. (accepted)
11. Hughes, J.J., Trtik, P.: Micro-mechanical properties of cement pate measured by depth-sensing nanoindentation: a preliminary correlation of physical properties with phase type. Mater. Charact. 53, 223–231 (2004)
12. Oliver, W.C., Pharr, G.M.: An improved technique for determining hardness and elastic modulus using load and displacement sensing experiments. J. Mater. Res. 7, 1564–1579 (1992)
13. Qing, Y., Zenan, Z., Deyu, K., Rongshen, C.: Influence of nano-SiO2 addition on properties of hardened cement paste as compared to silica fume. Cons. Build. Mat. 21, 539–545 (2005)
14. Richardson, I.G.: The nature of C-S-H in hardened cements. Cem. Concr. Res. 29, 1131–1147 (1999)
15. Taylor, H.F.W.: Cement chemistry. Academic Press, London (1990)
16. Velez, K., Maximilien, S., Damidot, D., et al.: Determination by nanoindentation of elastic modulus and hardness of pure constituents of Portland cement clinker. Cem. Concr. Res. 31, 555–561 (2001)
17. Wu, Z.-Q., Young, J.F.: The hydration of tricalcium silicate in the presence of colloidal silica. J. Mater. Sci. 19, 3477–3486 (1984)
18. Zhu, W., Bartos, P.J.M.: Assessment of interfacial microstructure and bond properties in aged GRC using a novel micro indentation method. Cem. Concr. Res. 27, 1701–1711 (1997)

Nanotechnologies for Climate Friendly Construction – Key Issues and Challenges

M.M. Andersen and M.R. Geiker

Abstract. Expectations as to the climate potentials of nanotechnology are high, none the least related to the construction sector. This paper seeks to highlight key aspects in the early development and application of eco-innovative nanotech solutions in the construction sector, "nanoconstruction". The paper provides a framework for addressing relevant issues of green nanoconstruction and takes stock of current challenges. Eco-innovative nanoconstruction has the potential to simultaneously enhance the competitiveness and climate potential of the construction sector and could become a key strategic factor for the sector ahead. However, the considerable lack of knowledge both on the eco-opportunities and risks of nanoconstruction and the industrial dynamics involved forms a serious barrier for pursuing nanoconstruction as a serious strategic target for business and policy.

1 Introduction

Despite the enormous and still rising research and development (R&D) investments in nanotechnology worldwide, nanotechnology is still at an early formative stage of development; much nanoscience is still pre-commercial [1, 9, 10, 29, 31, 33, 42, 43].

The hype (extensive focus, debate and phantazising) related to nanotechnology is considerable, with grand expectations of nanotechnology to restructure the world atom by atom. There are especially high expectations to nanotech's *eco-innovative* (climate friendly, 'green') potential. It is difficult to find a nanoreport or policy document where major environmental benefits are not a main or important claim (see e.g. [13, 22, 31, 35, 37, 39, 40]). As the climate agenda is becoming increasingly important for the competitiveness and development of the construction sector, the "nanoconstruction" eco-innovative potentials are increasingly interesting.

M.M. Andersen
Department of Management Engineering, Technical University of Denmark, Lyngby, Denmark
email: mmua@man.dtu.dk
www.man.dtu.dk

M.R. Geiker
Department of Civil Engineering, Technical University of Denmark, Lyngby, Denmark
email: mge@byg.dtu.dk
www.byg.dtu.dk

At the same time concerns regarding possible environmental and health risks related to nanotechnology are increasingly being addressed by policy makers, NGOs, and more lately also nanoscientists and companies and [1, 5, 7, 11, 20, 24, 28, 36, 39]. There are concerns, that regulation is lacking behind the rapid advances in nanotechnology and a precautionary principle is called for by the European Commission [17]. There is, as this paper will unfold, overall still much uncertainty both as to the environmental risks and opportunities related to nanotechnology [5].

The construction sector was among the first to be identified as a promising application area for nanotechnology back in the beginning of the 1990s; but today we see that the fragmented, generally low tech-tech and conservative construction sector is falling behind other sectors in applying nanotechnology [25]. When talking about eco-innovation in nanoconstruction we are still dealing more with potentialities than with actual developments. Data and analyses are lacking [6, 26, 27].

This paper seeks to highlight key aspects in the early development and application of eco-innovative nanotech solutions in the construction sector. The paper provides a framework for addressing relevant issues of green nanoconstruction and takes stock of current challenges.

2 Trends and Key Issues in Climate Friendly Nanoconstruction

The great diversity of nanotechnologies means that it is not easy to identify what green nanotechnology could mean for construction. The high environmental expectations to nanotechnology are related to some fundamental features of nanotechnology. Potentially the atom-by-atom construction of nanomaterials may lead to optimised tailoring of materials and products without dangerous and messy by-products. Self-assembly, i.e. the attempt to mimic nature's intrinsic way of building on the nanometre scale, molecule by molecule through self-organization, has eco-potentials because it is extremely resource efficient ([36] p.39). Also the large surface area of nanoparticles leads to a high reactivity which may lead to higher energy efficiency, e.g. increasing absorption rates for light and facilitating reaction processes at reduced temperatures and with less materials loss ([35] p.89). An important feature of relevance for nanoconstruction is that nanotechnology allows the design of materials with multifunctional properties. A single nanomaterial can replace several traditional ones potentially increasing the resource efficiency. E.g. nanocomposites can be made strong, light, thin, electrically conductive and fireproof. Nanocoatings can be self-cleaning, de-polluting and antimicrobial. See also [6, 23, 27, 35, 41] for early but not very thorough discussions on green nanotech opportunities in construction.

The goal of a recent Danish report [6] was to identify the potentials of nanotechnology to meet the needs and solve the problems of the construction sector including the environmental challenges. In this work *six nanopillars* emerged that systematize the potentials of nanotechnology in relation to the construction industry: 1) nanostructured materials, 2) nanostructured surfaces, 3) nanooptics, 4) nanosensors & electronics, 5) nanointegrated energy production & storage, 6) nanointegrated environmental remediation. Table 1 gives an overview of nanoresearch and technology areas and their construction relevance.

Table 1 Overview – nanorelated areas and their relevance for the construction sector

Nanorelated research and technology areas	Relevance for the construction sector (main topics)	
Topics	Application in	Important environmental properties
1. Nanostructured materials a) Nanoporous materials, incl. cement and wood based materials b) Polymers c) Composites d) Other materials	Construction materials in general Insulation materials Load carrying materials	Multifunctional, including: Strength – weight ratio Durability Fire resistance Self-cleaning, impact on indoor and outdoor climate Energy & resource efficiency Recyclability Degradability
2. Nanostructured surfaces as coatings and thin films a) Chemically modified surfaces b) Physically modified surfaces	Everywhere in buildings and civil works, none the least renovation	Multifunctional, including: Strength and toughness Durability incl. aesthetics Impact on indoor climate Hygiene Maintainability Self-cleaning, see Environmental remediation, 6.
3. Nanooptics a) Planar light wave circuits b) Photonic crystal fibers c) Light emitting diodes, LED & OLED d) Integrated optical sensors	Integrated functions in general Electrical and lighting systems Climate control	Energy efficiency Fire and other safety
4. Nanosensors & electronics For monitoring and transmission a) Biosensors, b) Optical sensors, c) Chemical sensors, d) Gas sensors, e) Microorganisms, f) Electro active materials	Monitoring and control every where in buildings and civil works	Embeddedness Durability Maintainability Resource efficiency

5. Nanointegrated energy production & storage a) Solar cells b) Fuel cells c) Other	Heating and cooling systems Building envelope Electricity supply	Energy self-sufficiency and – efficiency in buildings and utility systems
6. Nanointegrated environmental remediation a) Catalytic cleaning b) Other separation and purification processes	Air purification in buildings and infrastructures Water systems (supply and waste) Waste systems	Inbuilt air- and water cleaning Environmental remediation in general Indoor climate, incl. cleaning and hygiene Degradability Resource efficiency Substitution of hazardous materials

Source: Modified after Andersen and Molin 2007

The overview illustrates the great variety and scope of the many emerging nanotechnological areas, and the broad application opportunities which address almost all aspects of construction. They are interesting because they point to novel climate solutions for achieving resource efficient and intelligent buildings and cities and because many can be applied in existing buildings where the climate potential is considerable; e.g. via surface treatments, applications of thin panels and high efficient insulation.

The majority of the novel solutions are in an early stage of development, but some are fully commercial. Data are poor but two recent consultancy reports on green nanoconstruction [27, 41] identifies a wide range of commercially available products worldwide, illustrating that much is beginning to happen in this area.

However, so far, it is the major industrial multinational players who are pioneering the development and application of nanotechnologies in construction, while the majority of (predominantly small) construction companies, universities and other knowledge institutions have little insight and experience with nanotechnology [6, 8, 12, 23, 26, 32, 44]. A Danish innovation analysis shows a generally weak demand for nanotechnology in the construction sector:

'The overall picture of the demand for, knowledge of, and views on nanotechnology in the construction sector is that knowledge and expertise are currently too fragmented to allow for a substantial uptake, diffusion and development of nanotechnological solutions in the construction industry. At present, only very vague ideas of the possible benefits can be identified among key agents of change such as architects, consulting engineers and facility managers. Furthermore the demand side will be reluctant about introducing nanotechnological materials until convincing documentation about functionalities and long-term effects is produced. A need for documentation of the consequences for health and safety is evident'. ([6] p.32).

According to the two recent reports on green nanoconstruction [27, 41] barriers for the wider development of climate friendly nanoconstruction are considerable and lie mainly in four areas: a) the lack of knowledge of nanotech opportunities in the construction sector b) reluctance of the sector towards (radical) innovation, c) the high costs of some, but not all, nanotechnologies, and d) public concern about nanorisks.

But another, and often overlooked factor, is that there are also barriers on the nanoside. Today most nanotechnologies are targeted at other applications than construction, mainly more knowledge-intensive areas such as medico, food and military [5, 6, 31, 34]. It will take effort and time to shift the attention and capabilities among the nano scientific community towards the construction area. Also on the innovation dynamics of nanoconstruction are analysis and insights lacking.

3 Strategies for Climate Friendly Nanoconstruction

In the strategic priorities of the European Construction Technology Platform (ECTP) see Table 2, climate issues, here 'becoming sustainable', is given considerable attention. The ECTP defines sustainable development in construction quite broadly, encompassing resource efficiency, environmental impact, utility networks, and the cultural heritage and safety issues.

Table 2 List of strategic research priorities ECTP 2005

A: Meeting client / user requirements	
A1	Healthy, safe and accessible indoor environment for all
A2	A new image of the cities
A3	Efficient use of underground city space
A4	Mobility and supply through efficient networks
B: Becoming sustainable	
B1	Reduce resource consumption (energy , water, materials)
B2	Reduce environmental and man-made impacts
B3	Sustainable management of transport and utilities networks
B4	A living cultural heritage for an attractive Europe
B5	Improve safety and security
C: Transformation of the construction sector	
C1	A new client-driven, knowledge-based construction process
C2	ICT and automation
C3	High added-value construction materials
C4	Attractive workplaces

Source: http://www.ectp.org/documentation/ECTP-SRA-2005_12_23.pdf

Generally nanotechnology offers opportunities for meeting many of the challenges addressed by the ECTP strategy. These include A) meeting the user requirements both in terms of developing intelligent, fashionable and efficient buildings and cities and an improved indoor environment; B) achieving high resource efficiency and contributing to environmental remediation and energy production; as well as C) renewing the sector in making it more knowledge based and automated. Nanotechnology may particularly address the climate/sustainability agenda in simultaneously contributing to making the construction sector both more clever and clean and hence improve the innovation capacity and competitiveness of the sector [2, 5]. In doing so, the nanoconstruction area fits well with the rapidly rising policy and business interest into "eco-innovation"[1]. The climate agenda is increasingly moving away from the more general sustainable development agenda towards the market oriented "eco-innovation" agenda [3, 4, 30].

The eco-innovation policy perspective seeks to address the specific challenges different sectors and types of companies face when they are eco-innovative. "Sustainable construction" has recently been identified as one of the six 'lead markets' for innovation in the EU and one out of four eco-innovation policy priority areas [16, 18, 19]. Germany has recently launched a major programme within nanoconstruction, which includes climate issues [23]. This underlines the recent considerable interest in promoting eco-innovation in the construction sector which could form a window of opportunity for the development of nanoconstruction.

4 Conclusions

Eco-innovative nanoconstruction has the potential to simultaneously enhance the competitiveness and climate potential of the construction sector and could become a key strategic factor for the sector ahead. This paper has shortly discussed a wide range of potential nanotechnologies applicable for construction with promising climate impacts, most, however, in an early stage of development. They are none the least interesting because they point to novel climate solutions for achieving resource efficient and intelligent buildings and cities and because many can be applied in existing buildings where the climate potential is considerable. It is essential, however, that the current knowledge gap on risk issues, eco-opportunities and industrial dynamics are met if green nanoconstruction is to move from expectations to a serious strategic target for business and policy makers.

References

1. Aitken, R.J., Chaudhry, M.Q., Boxall, A.B.A., Hull, M.: Manufacture and use of nanomaterials: current status in the UK and global trends. Occupational Medicine 56, 300–306 (2006)
2. Andersen, M.M.: Embryonic innovation – path creation in nanotechnology. In: DRUID conference Copenhagen (2006a) (June 18-20, 2008) (accessed December 15, 2008), http://www2.druid.dk/conferences/viewpaper.php?id=703&cf=8

[1] Eco-innovations are innovations which create value on the market while pursuing reductions in net environmental impacts, i.e. eco-innovations are seen as a business opportunity, breaking with 35 years tradition of treating the environment as a burden to business.

3. Andersen, M.M.: Review: System transition processes for realising Sustainable Consumption and Production. In: Tucker, A., et al. (eds.) System Innovation for Sustainability, vol. 1, pp. 320–344. Green Leaf Publishing, Sheffield (2008a)
4. Andersen, M.M.: Eco-innovation – towards a taxonomy and a theory. DRUID conference Copenhagen (2008b) (June 18-20, 2008) (accessed December 15, 2008), http://www2.druid.dk/conferences/userfiles/file/June_13b.pdf
5. Andersen, M.M., Rasmussen, B.: Environmental opportunities and risks from nanotechnology. Risoe-R-1550-EN Risø National Laboratory, Roskilde (2006)
6. Andersen, M.M., Molin, M.: NanoByg: A survey of nanoinnovation in Danish construction. Risoe-R-1234(EN) Risø National laboratory, Roskilde (2007) (accessed December 15, 2008), http://www.risoe.dk/rispubl/reports/ris-r-1602.pdf
7. Arnall, A.H.: Future Technologies, Today's Choices: Nanotechnology, Artificial Intelligence and Robotics - A technical, political and institutional map of emerging technologies. Greenpeace Environmental Trust, London (2003), http://www.greenpeace.org.uk/MultimediaFiles/Live/FullReport/5886.pdf
8. Bartos, P.J.M., Hughes, J.J., Trtik, P., Zhu, W. (eds.): Nanotechnology in Construction XVI. Springer, Heidelberg (2004)
9. BMBF Nanotechnology conquers markets: German innovation initiative for nanotechnology. Federal Ministry of Education and Research (BMBF) (2004) (accessed December 15, 2008), http://www.bmbf.de/pub/nanotechnology_conquers_markets.pdf
10. Build-NOVA, Characteristics of the construction sector – technology and market tendencies. Europe INNOVA EC, Bruxelles (2006)
11. Colvin, V.: Nanotechnology: environmental impact. Presentation at National Center for Environmental Research (NCER). EPA (2002)
12. CRISP/SPRU, The Emperor's New Coating: New Dimensions for the Built Environment: the nanotechnology revolution. CRISP, London (2003), http://www.crispuk.org.uk/REPORTS/NanoReportFinal270103.pdf
13. EC, Towards a European Strategy for Nanotechnology. European Commission (2004) (accessed December 15, 2008), http://cordis.europa.eu/ nanotechnology/actionplan.htm
14. EC, Putting knowledge into practice: A broad-based innovation strategy for the EU. COM, 502 final, Brussels (2006) (accessed December 15, 2008), http://eur-lex.europa.eu/LexUriServ/site/en/com/2006/com2006_0502en01.pdf
15. EC (2008) (accessed December 15, 2008), http://www.cordis.europa.eu/nanotechnology
16. EC. Coordinated action to accelerate the development of innovative markets of high value for Europe – the Lead Markets Initiative. MEMO/08/5, Brussels (January 7, 2008) (accessed December 15, 2008) (2008a), http://ec.europa.eu/enterprise/leadmarket/leadmarket.htm
17. EC, Eco-innovation - When business meets the environment. Call for proposals 2008, CIP Eco-innovation and pilot and market replication projects (2008b) (accessed December 15, 2008), http://ec.europa.eu/environment/etap/ecoinnovation/index_en.htm

18. EC SANCO, Nanotechnologies: A Preliminary Risks Analysis, report on the basis of a workshop organized in Bruxelles on 1-2 March by the Health and Consumer Protection Directorate General of the European Commission (SANCO), European Communities, Bruxelles (2004)
19. ECTP, Strategic Research Agenda for the European Construction Sector: Achieving a sustainable and competitive construction sector by 2030, European Construction Technology Platform, Brussels (2005), http://www.ectp.org
20. European Parliament Scientific Technology Options Assessment Committee, The Role of Nanotechnology in Chemical Substitution (2007) (accessed December 15, 2008), http://www.nanowerk.com/spotlight/spotid=2212.php
21. Fellenberg, R., Hoffschulz, H.: Nanotechnologie und Bauwessen (Nanotecture). VDI Technologiezentrum, Düsseldorf (2006)
22. Friends of the Earth Germany - BUND, For the Responsible Management of Nanotechnology (2007) (accessed December 15, 2008), http://www.bund.net/lab/reddot2/pdf/bundposition_nano_03_07.pdf
23. Gann, D.: A Review of Nanotechnology and its Potential Applications for Construction. SPRU/CRISP (2003) (accessed December 15, 2008), http://www.crispuk.org.uk/REPORTS/LongNanotech240203.pdf
24. Geiker, M.R., Andersen, M.M.A.: Nanotechnologies for sustainable construction. In: Khatib, J. (ed.) Sustainability of Construction Materials. Woodhead Publishing Ltd, UK (forthcoming)
25. Green Technology Forum, Nanotechnology for Green Buildings, Indianapolis (2007), http://www.greentechforum.net
26. Hansen, S.F., Larsen, B.H., Olsen, S.B., Baun, A.: Categorization framework to aid hazard identification of nanomaterials. Nanotoxicology, 1–8 (2007)
27. Hullmann, A.: The economic development of nanotechnology - An indicator based analysis (2006) (accessed December 15, 2008), ftp://ftp.cordis.europa.eu/pub/nanotechnology/docs/nanoarticle_hullmann_nov2006.pdf
28. Kemp, R., Andersen, M.M.: Strategies for eco-efficiency innovation. In: Strategy paper for the Informal Environmental Council Meeting, July 16-18, Maastricht. VROM, Den Haag (2004)
29. Luther, W.: International Strategy and Foresight Report on Nanoscience and Nanotechnology. VDI Technologiezentrum for Risoe National Laboratory, Düsseldorf (2004)
30. Luther, W., Zweck, A.: Anwendungen der Nanotechnologie in Architektur und Bauwesen. VDI Technologiezentrum, Düsseldorf (2006)
31. Lux Research, The Nanotech Report. Lux Research, New York (2004)
32. Malinowski, N., Luther, W., Bachmann, G., Hoffknecht, A., Holtmannspötter, D., Zweck, A.: Nanotechnologie als wirtschaftlicher Wachstumsmarkt: Innovations- und Technikanalyse, VDI Technologiezentrum, Düsseldorf. 53 (2006)
33. Nanoforum, Nanotechnologies help solve the world's energy problems. Nanoforum (2003) (accessed December 15, 2008), http://www.nanoforum.org
34. Nanoforum, Benefits, risks, ethical, legal and social aspects of nanotechnology. Nanoforum (2004) (accessed December 15, 2008), http://www.nanoforum.org
35. NSET, The National Nanotechnology Initiative: Research and Development Leading to a Revolution in Technology and Industry: Supplement to the Presidents FY 2004, Budget. National Science and Technology Council, Washington D.C (2003)

36. Royal Society, Nanoscience and nanotechnologies: opportunities and uncertainties. The Royal Society & The Royal Academy of Engineering (2004) (accessed December 15, 2008), http://www.nanotec.org.uk/finalReport.htm
37. Schmidt, K.F.: Green Nanotechnology. Woodrow Wilson International Center for Scholars - Project on Emerging Nanotechnologies (2007) (accessed December 15, 2008), http://www.nanotechproject.org/process/assets/files/2701/187_greennano_pen8.pdf
38. Scientifica, Nanotech: Cleantech - Quantifying the Effect of Nanotechnologies on CO2 Emissions. Scientifica (2007) (accessed December 15, 2008), http://www.cientifica.eu/index.php?option=com_content&task=view&id=73&Itemid=118
39. Willems, van den Wildenberg: NRM nanoroadmap project - Work document on Nanomaterials. Willems and van den Wildenberg Espana s.l (2004)
40. Wood, S., Geldart, A., Jones, R.A.L.: The social and economic challenges of nanotechnology. Economic & Social Research Council, Swindon (2003)
41. Zhu, W., Bartos, P., Porro, A.: Application of nanotechnology in construction: Summary of a state-of-the-art report. Mater. Struct. 37, 649–658 (2004)

The Potential Benefits of Nanotechnology for Innovative Solutions in the Construction Sector

F.H. Halicioglu

Abstract. The world of the construction sector is being changed by new technologies, new materials, new building typologies, new concerns and opportunities. The construction sector has been slow to embrace nanotechnology, but nanotech innovations have an enormous impact on building design and construction. Nanotechnology represents a major opportunity for the construction sector to develop new products, substantially increase quality, and open new markets. The paper aims to describe and examine the potential benefits of nanotechnology for innovative solutions in the construction sector. It offers a possibility of a revised understanding of the relationship between nanotechnology and the building design and construction in the understanding of innovative approaches.

1 Introduction

Nanotechnology has the potential to transform the built environment in ways almost unimaginable today. Nanotechnology is already employed in the manufacture of everyday items from sunscreen to clothing, and its introduction to architecture is not far behind. On the near horizon, it may take building enclosure materials (coatings, panels and insulation) to dramatic new levels of performance in terms of energy, light, security and intelligence. Even these first steps into the world of nanotechnology could dramatically alter the nature of building enclosure and the way our buildings relate to environment and user. At mid-horizon, the development of carbon nanotubes and other breakthrough materials could radically alter building design and performance [3]. Novel construction materials could result from the application of nano-technology (e.g. through the use of nano-particles, nano-tubes and nano-fibres), offering new combinations of strength, durability and toughness. Examples are bio-mimetic materials based on structures and compounds found in nature, composites with self adjusting interfaces, shape-memory, self-repairing and strain hardening materials [13].

F.H. Halicioglu
Dokuz Eylul University Faculty of Architecture, Izmir, Turkey
e-mail: hilal.halicioglu@deu.edu.tr

If nanotechnology is to change how we design and how we live, then a study of nanotechnology's implications for architecture is clearly needed. Many nano-engineered materials are already available to architects and builders, and are beginning to transform our buildings. Looking further ahead, nanotechnologies now in research and development will likely have a significant impact on building within the next twenty years. For example, carbon nanotubes, fifty to one hundred times stronger than steel at one-sixth of the weight, could bring unprecedented strength and flexibility to our buildings. On the far horizon, the full impact of nanotechnology on our lives and our environment into the next century and beyond is impossible to predict but important to consider [4].

With a view to executing significant innovations in nanotechnology, specifically in the construction sector, it is necessary to do research in their development. The paper aims to describe and examine the potential benefits of nanotechnology for innovative solutions in the construction sector. It offers a possibility of a revised understanding of the relationship between nanotechnology and the building design and construction in the understanding of innovative approaches.

2 Nanotech Innovations in Building Construction and Potential Benefits of Nanotechnology for Innovative Solutions in Construction Sector

Innovations with nanotechnologies in construction sector depend on technological developments. Nanotechnology has the potential to create radical innovations in buildings. Since materials are construction's core business the sector is expected to be an important beneficiary of nanomaterials [1]. Already, dozens of nanomaterials are available in the architectural marketplace, yet their chemistry, performance capabilities, environmental and health effects, costs, risks and benefits remain a mystery to most designers. Some, for example, may be familiar with the self-cleaning windows marketed by PPG, Pilkington and others, or with the depolluting or "smog-eating" concrete used in Richard Meier's Jubilee Church (Fig. 1), but only a handful could cite titanium dioxide nanoparticles as the material that makes these marvels possible. The wide range of nanocoatings available today are also relatively unknown despite their promising potential to dramatically improve insulation, kill bacteria, prevent mildew, and reduce maintenance and environmental harm [4].

Nanotechnology will result in a unique next generation of bio-products that have hyper-performance and superior serviceability. These products will have strength properties now only seen with carbon-based composites materials. These new hyper-performance bioproducts will be capable of longer service lives in severe moisture environments. Enhancements to existing uses will include development of resin-free biocomposites or wood-plastic composites having enhanced strength and serviceability because of nanoenhanced and nanomanipulated fiber-to-fiber and

Fig. 1 Church Dio Padre Misericordioso (Jubilee Church), Rome, Richard Meier 2003 [2]

fiber-to-plastic bonding. Nanotechnology will allow the development of intelligent wood- and biocomposite products with an array of nanosensors to measure forces, loads, moisture levels, temperature, pressure, and chemical emissions [14].

Two nano-sized particles that stand out in their application to construction materials are titanium dioxide (TiO2) and carbon nanotubes (CNT's). The former is being used for its ability to break down dirt or pollution and then allow it to be washed off by rain water on everything from concrete to glass and the latter is being used to strengthen and monitor concrete. CNT's though, have many more properties, apart from exceptional strength, that are being researched in computing, aerospace and other areas and the construction industry will benefit directly or indirectly from those advancements as well [9].

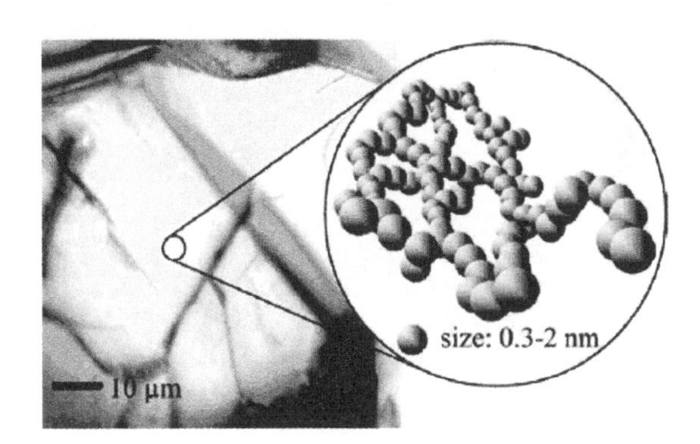

Fig. 2 Structure of the nanoporous SiO2 network of silica aerogel [12]

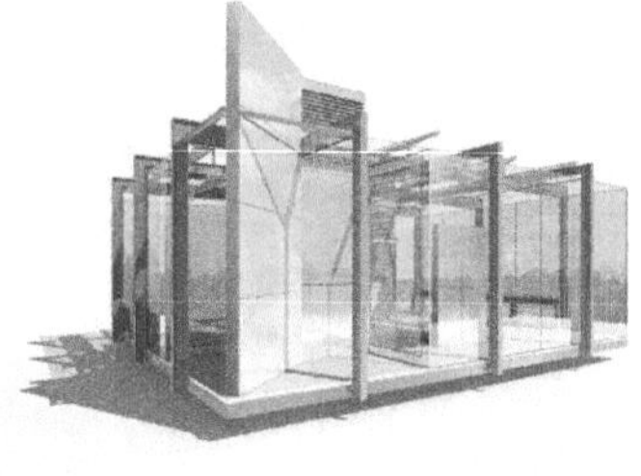

Fig. 3 The Nanohouse Initiative [5]

Silica aerogel is a translucent material consisting of a nanostructured SiO2 network (Fig. 2) with a porosity of up to 99%. Silica aerogel is a promising material for applications in building envelopes because of its high visual transmittance and its low thermal conductivity [7] Besides its low thermal conductivity the aerogel is load bearing which makes it attractive for evacuated transparent insulation applications. An interesting application for aerogels in buildings is in evacuated solar collectors [11, 12].

The Nanohouse Initiative (Fig. 3) is a collaboration between the best of Australia's scientists, engineers, architects, designers and builders - working together to design and build a new type of ultra-energy efficient house and exploiting the new materials being developed by nanotechnology [5]. The NanoHouse Initiative, conceived in 2002 by Dr Carl Masens at the Institute for Nanoscale Technology and visualised and implemented by architect James Muir, has proven a successful method of explaining what nanotechnologies are

Fig. 4 The Carbon Tower Prototype [8]

and how they work; for example, how the latest technology windows clean themselves, how tiles might resist build up of soap scum, or timber surfaces resist UV damage [10].

The Carbon Tower Prototype (Fig. 4) is a 40-story mixed-use high-rise that incorporates five innovative systems: pre-compressed double-helix primary structure, tensile-laminated composite floors, two external filament-bound ramps, breathable thin-film membrane, and vritual duct displacement ventilation. Studies conducted by Arup suggest that, if built, the tower would the lightest and strongest building of its type [6].

3 Conclusions

Nanotech innovations represent the application of nanotechnologies in the field of materials science and engineering and have a great impact on construction sector. All of these potential uses and benefits of nanotechnology in construction, as described above, will play a key role in innovative approaches used in new building design and construction.

References

1. Andersen, M.M., Molin, M.: NanoByg – A survey of nanoinnovation in Danish construction – Executive summary English, Denmark (2007)
2. Church Dio Padre Misericordioso (Jubilee Church), Rome (accessed December 24, 2008), http://www.galinsky.com/buildings/jubilee/index.htm
3. Elvin, G.: Nanotechnology + Architecture (accessed December 24, 2008), http://www2.arch.uiuc.edu/elvin/nanotechnologyindex.htm
4. Elvin, G.: NanoBioBuilding: Nanotechnology, biotechnology and the future of building. Green Technology Forum (2006) (accessed December 24, 2008), http://www.greentechforum.net/wpcontent/uploads/2006/12/nanobiobuilding.pdf
5. Elvin, G.: The Nanohouse Initiative, Nanohouse - Architectural applications (accessed December 24, 2008), http://www2.arch.uiuc.edu/elvin/nanohouse.htm
6. Elvin, G.: Carbon Tower (accessed December 24, 2008), http://www2.arch.uiuc.edu/elvin/carbontower.htm
7. Fricke, J.: Preface—first Int. Symp. on Aerogels. Springer, Würzburg (1985)
8. Knecht, B.: Brave new solid-state, carbon-fiber world Architects Peter Testa and Sheila Kennedy are reinventing the design process through collaboration with industry (accessed December 24, 2008), http://archrecord.construction.com/innovation/2_Features/0310carbonfiber.asp
9. Mann, S.: Report on Nanotechnology and Construction. Institute of Nanotechnology European Nanotechnology Gateway (2006), http://www.nanoforum.org
10. Nanotechnology in Australia (accessed) http://www.nano.uts.edu.au/about/australia.html

11. Ortjohann, J.: Granular Aerogel for the Use in Solar Thermal Collector. ISES 2001, Solar World Congress (2001)
12. Reim, M., Körner, W., Manara, J., et al.: Silica aerogel granulate material for thermal insulation and day lighting. J. Solar Energy 79, 131–139 (2005)
13. Strategy for construction RTD. E- Core European Construction Research Network (accessed December 24, 2008), http://www.eccredi.org/pages/E-CORE_NL.pdf
14. Wegner, T.H., Winandy, J.E., Ritter, M.A.: Nanotechnology Opportunities in Residential and Non-Residential Construction. In: 2nd International Symposium on Nanotechnology in Construction, Bilbao, Spain (CD-ROM). Bagneux, France: RILEM: 9, November 13-16 (2005) (accessed December 24, 2008), http://www.fpl.fs.fed.us/documnts/pdf2005/fpl_2005_wegner003.pdf

Use of Nano-SiO_2 to Improve Microstructure and Compressive Strength of Recycled Aggregate Concretes

P. Hosseini, A. Booshehrian, M. Delkash, S. Ghavami, and M.K. Zanjani

Abstract. The purpose of this paper is to provide new type of recycled aggregate concrete (RAC) incorporating nano-SiO_2. In particular, we investigate the effects of colloidal nano-silica solution on the properties of fresh and hardened concrete. The main variables included the dosage of nano-silica (including 0%, 1.5%, and 3% of cement content) and the cement content of the concrete (including 400 and 450 kg/m3). Results were compared with plain concretes. Tests were conducted to determine the mechanical properties (compressive strength) and microstructure (SEM test) of the concretes.

1 Introduction

Concrete is the premier construction material across the world and the most widely used in all types of civil engineering works, including infrastructure, low and high-rise buildings, defence installations, environment protection and local/domestic developments. Concrete is a manufactured product, essentially consisting of cement, aggregates, water and admixture(s). Among these, aggregates, i.e. inert granular materials such as sand, crushed stone or gravel form the major part [1]. Nowadays, because of extreme use of these mineral aggregates and decreasing the amount of sources and mines all over the world, engineers are trying to find a way to reduce using these valuable materials given by nature. On the other hand, the amount of construction and demolition waste has been increased considerably over the last few years [2,3]. One of the construction sector's major contributions to the preservation of the Environment and sustainable development is the reuse and recycling of the waste materials it generates (reducing, reusing, recycling and regenerating the residues that originate the constructive activity) [4].

P. Hosseini, A. Booshehrian, M. Delkash, and S. Ghavami
Department of Civil Engineering, Sharif University of Technology, Tehran, Iran
e-mail: p.hosseini@civil.sharif.edu

M.K. Zanjani
Ati Saz Construction Co. Tehran, Iran

The investigation of the mechanical properties of recycled aggregate concrete is necessary to determine the feasibility of use as well as the impact on durability of structures. There have been an increasing number of studies on the influence of recycled concrete aggregate as partial or total replacement of natural aggregates and its effect on the mechanical properties and durability of the recycled aggregate concrete [4-7], but they always resulted in a lower level of concrete strengths and durability. This was due to the residual impurities on the surface of the recycled aggregates, which blocked the strong bond between cement paste and aggregate [7-10].

The lower mechanical properties of these recycled materials ought to be neutralized by some other substance or procedure to improve the insufficient strength and durability caused by replacing conventional aggregate.

Recently, nano-technology has attracted considerable scientific interest due to the new potential uses of particles in nanometer (10^9 m) scale [11]. Due to this ultrafine size, nano-particles show unique physical and chemical properties different from those of the conventional materials. Because of their unique properties, nano-particles have been gained increasing attention and been applied in many fields to fabricate new materials with novelty functions. If nano-particles are integrated with traditional building materials, the new materials might possess outstanding or smart properties for the construction of different parts and uses in civil structures [12].

The present study used nano-silica particles in binder to enhance the mechanical properties of recycled aggregate concrete in terms of the compressive strength. The influence of the nano-silica on the concrete strength was assessed by measuring the compressive strength at 3, 7, 14, 28 days.

2 Experimental Procedures

2.1 Materials and Mixture Proportions

The cement used is Portland cement type I-425, according to the Iranian national standards (389). Fine aggregate is natural river sand with a fineness modulus of 3.2 and used in all mixtures. The coarse aggregate used is crushed limestone with a nominal maximum size of 20 mm for production of basic mixtures (original mixtures) which used for demolition process. The fine and coarse aggregates had specific gravities of 2.74 and 2.65, and water absorption of 0.1% and 0.8%, respectively. Also, the coarse recycled aggregates had specific gravity of 2.42 and water absorption of 4.8% with a nominal maximum size of 20 mm similar to natural coarse aggregate. The grading of coarse and fine aggregates was in accordance with the requirements of ASTM C143-99.

The water reducer superplasticizer (naphthalene-type with a solid content of 40%) is employed to aid the dispersion of nano-particles in concrete and achieve good workability of concrete. The colloidal nano-silica was purchased from Asan ceram Co (Tehran, Iran) with solid content of 30%. Application of colloidal nano-particles aids better dispersion of nano-particles in the concrete matrix and decreases agglomeration of nano-particles which improves nano-particles performance in concrete. Physical and chemical properties of cement and nano-silica particles are provided in Table 1, respectively.

Table 1 Chemical composition and physical properties of binder materials

	Cementitious materials(%)	
Item	Cement	Nano-silica
SiO_2	21.4	99.8
Al_2O_3	6	-
Fe_2O_3	3.4	-
CaO	64	-
MgO	1.8	-
Cl	-	-
SO_3	1.4	-
KO+Na_2O	1	-
L.O.I	3	2.8
Particle Size (nm)	-	40
Specific Gravity (m^2/kg)	3110	200000

The mix proportions of control concretes (NC-400 and NC-450), concretes containing just coarse recycled concrete aggregates (RA1-400 and RA1-450) and concretes containing coarse recycled concrete aggregates and different dosages of nano-silica particles (1.5% and 3%), are summarized in Table 2. The compressive strength of basic concretes, placed in the range of 30-40 MPa. For producing different mixtures of this study, mix proportions (except the amount of binders) assumed similar to the basic mixtures demolished for producing recycled aggregates.

2.2 Test Procedure

To fabricate the recycled aggregate concrete containing nano-particles, superplasticizer was firstly mixed into water in a mortar mixer, and then colloidal nano-particles are added and stirred at a high speed for 1 min. The cement, sand and coarse aggregate were mixed at a low speed for 2 min in a concrete rotary mixer, and then the mixture of water, water-reducing agent, nano-particles was slowly poured in and stirred at a low speed for another 2 min to achieve good workability.

Table 2 Concrete mixture proportions used in the study

Mix No.	Description	w/b	Water (kg)	Cement (kg)	Gravel (kg)	Sand (kg)	Superplasticizer (kg)	Nano-silica (solid) (kg)
NC-400	Control mix	0.4	177.2	400	646.4	1200.4	1.2	0.0
RA1-400	CRA	0.4	201.6	400	625.1	1160.9	1.4	0.0
RA2-400	CRA + 1.5% NS	0.4	201.6	394	611.6	1135.8	1.4	6
RA3-400	CRA + 3% NS	0.4	201.6	388	598.1	1110.7	1.4	12
NC-450	Control mix	0.4	196.3	450	612.3	1137.0	1.6	0.0
RA1-450	CRA	0.4	219.4	450	592.1	1099.7	1.6	0.0
RA2-450	CRA + 1.5% NS	0.4	219.4	443.25	576.9	1071.4	1.6	6.75
RA3-450	CRA + 3% NS	0.4	219.4	436.5	561.7	1043.1	1.6	13.5

To fabricate control and recycled aggregate concrete, superplasticizer is firstly dissolved into water. Cement, sand and coarse aggregate were mixed uniformly in a concrete rotary mixer, then the mixture of water and superplasticizer is poured in and stirred for several minutes. Finally, the fresh concrete is poured into oiled molds to form the cubes of the size 10×10×10 cm for the compressive strength testing. After pouring, an external vibrator is used to facilitate compaction and reduce the amount of air bubbles. The specimens are de-moulded at 24 hours and then cured in a curing room (relative humidity in excess of 95%, temperature 20±2) for 3, 7, 14 and 28 days. For each mix, 8 specimens of 100 mm cubes for compressive strength were made. Compressive strength was determined according to BS 1881: Part 117: 1993 at various ages (3,7,14 and 28 days). The workability tests were carried out in accordance with the requirements of ASTM C143-98.

3 Results and Discussion

The results of compressive and workability tests are shown in Table 3. According to these results, the strength of specimens produced by coarse recycled aggregates (RA1-400, RA1-450) is lower than those for control mixes (respectively NC-400, NC-450). The main factor of strength reduction of concrete with recycled aggregates is the existence of old mortar adhered to recycled aggregates, since the strength of mortar is much lower than the strength of natural aggregates [13]. The workability of concrete with recycled aggregates is generally lower, mainly due to increased specific surface and porosity, leading to high absorption capacity of recycled aggregates. Note that the reduction of workability of recycled concrete is a helpful factor in the formation of micro-cracks which have negative effect on the compressive strength [14].

The presented results, on the hand, clearly confirm that by adding nano-silica particles to the concrete matrix can significantly increase the overall strength. This observation is directly linked to supper-pozzolanic behavior of nanoparticles

Table 3 Workability and average compressive strength (MPa) at different test ages (3, 7, 14 and 28 days)

Mix No.	Curing days				Workability (mm)
	3	7	14	28	
NC-400	15.2	24.3	29.6	34.2	120-130
RA1-400	12.6	19.6	24.3	28.1	90-100
RA2-400	13.9	19.6	24.3	28.1	60-70
RA3-400	15.9	24.6	30.1	35.3	30-40
NC-450	19.4	31.4	37.3	41.8	130-140
RA1-450	16.3	25.1	31.2	35.3	90-100
RA2-450	17.6	29.3	35.4	40.1	60-70
RA3-450	20.9	32.1	39.1	43.7	30-40

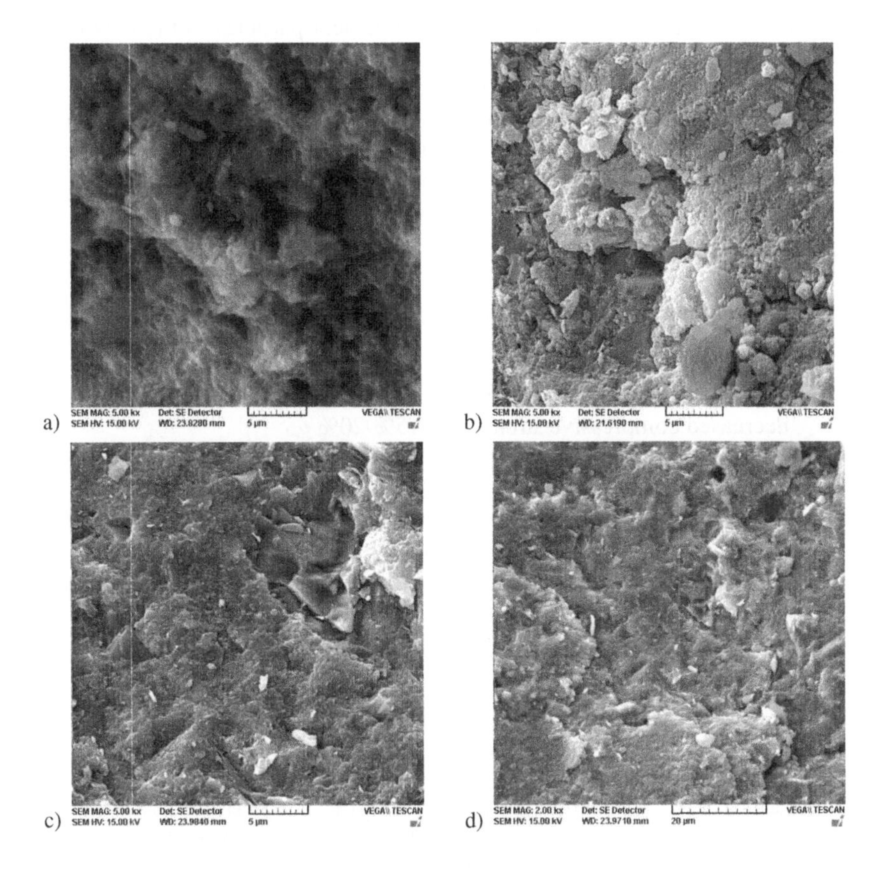

Fig. 1 Results of SEM studies; (a) Control mix (NC-400), (b) Recycled concrete & 0% Nano (RA1-400), (c) RC & 1.5% Nano (RA2-400), (d) RC & 3% Nano (RA3-400)

[16, 17]. Finally, data shown in Table 3 illustrate a considerable reduction of slump in fresh concrete [14] resulting from extremely large specific area of nanoparticles.

The results of SEM studies appear summarized in Figure 1. As shown in Figure 1a) and 1b), the application of recycled aggregates leads to formation of voids with small dimensions due to absorbing cement particles on the surface of aggregates. Therefore, the porosity of ITZ of these types of concrete is slightly higher than that of conventional concretes. On the other hand, when comparing Figures 1b) and 1c), it can be seen that by adding silica nano-particles, the transition zone of recycled concrete has become denser and more uniform and even extremely small voids have been omitted, especially because of exclusive performance of nano-particles. These particles combine with calcium hydroxide crystals ($Ca(OH)_2$) and produce dense calcium-silicate-hydrate gelatin by making pozzolanic reaction and gelatin is the main reason of more uniform and denser transition zone [15, 18,19]. Nano-particles are scattered in cement paste and placed on the surface of recycled aggregates to disperse cement particles in the new transition zone (between cement and recycled aggregate) in a better way and produce a more uniform and denser zone. Regarding to the individual and exclusive performance of nano-particles in the transition zone and in the mortar between aggregates, nano-particles condense these two principal sections of concrete and improve mechanical properties of concrete produced by recycled aggregates.

4 Conclusions

Under the conditions laid down in this research work, the following conclusions can be drawn:

- Replacement of coarse natural aggregates by 100% recycled aggregates decreased compressive strength by 15%-20%
- Increasing the amount of cement leads to an increase in compressive strength.
- It was observed that by increasing the amount of nano-silica as replacement of cement, the compressive strength of specimens increases as well.
- Adding 3% of nano-silica leads to values of compressive strengths higher than for conventional concrete.
- Increasing nano-silica quantity, as well as the amount of recycled aggregates, resulted in lower workability of concrete.

References

1. Limbachiya, M.C., Koulouris, A., Roberts, J.J., Fried, A.N.: Performance of recycled aggregate concrete. In: RILEM International Symposium on Environment-Conscious Materials and Systems for Sustainable Development, pp. 127–136 (2004)
2. Symonds: European Commission. Construction and demolition waste management practices, and their economics impacts. Report to DGXI, European Commission (1999)

3. Hendriks, C.F., Pietersen, H.S., Fraay, A.A.: Recycling of building and demolition waste. An Integrated approach. In: Proceedings of the International Symposium on Sustainable Construction: Use of Recycled Concrete Aggregate, London, UK, pp. 419–431 (2000)
4. Fonteboa, B.G., Abella, F.M.: Concretes with aggregates from demolition waste and silica fume. Materials and mechanical properties. Build Envir. 43, 429–437 (2008)
5. Rahal, K.: Mechanical properties of concrete with recycled coarse aggregate. Buil. Envir. 42(1), 407–415 (2007)
6. Buck, A.D.: Recycled concrete as a source of aggregate. ACI J. 74(5), 212–219 (1977)
7. Hansen, T.C., Hedegard, S.E.: Properties of recycled aggregate concrete as affected by admixtures in original concretes. ACI J. 81(1), 21–26 (1984)
8. Poon, C.S., Shui, Z.H., Shui, L., et al.: Effect of microstructure of ITZ on compressive strength of concrete prepared with recycled aggregates. Cons. and Build. Mat. 18, 461–468 (2004)
9. de Oliveira, M.B., Vazquez, E.: The influence of retained moisture in aggregates from recycling on the properties of new hardened concrete. Waste Manage 16, 113–117 (1996)
10. Tu, T.Y., Chen, Y.Y., Hwang, C.L.: Properties of HPC with recycled aggregates. CemConcr. Res. 36, 943–950 (2006)
11. Jo, B.W., Kim, C.H., Tae, G.H., Park, J.B.: Characteristics of cement mortar with nano-SiO_2 particles. Cons. and Build. Mat. 21, 1351–1355 (2007)
12. Li, H., Xiao, H.G., Yuan, J., et al.: Microstructure of cement mortar with nano-particles. Comp: Part B 35, 185–189 (2004)
13. Larranaga, M.E.: Experimental study on microstructure and structural behaviour of recycled aggregate concrete, Doctoral thesis, Catalunya polytechnic university, Barcelona (March 2004)
14. Li, H., Zhang, M.H., Ou, J.P.: Abrasion resistance of concrete containing nano-particles for pavement. Wear 260, 1262–1266 (2006)
15. Li, Z., Wang, H., He, S., et al.: Investigations on the preparation and mechanical properties of nano-alumina reinforced cement composites. Mat. Lett. 60, 356–359 (2006)
16. Khaloo, A.R., Hosseini, P.: Investigation on relationship between compressive strength and microstructure of cement mortars containing different pozzolanic materials and nano particles. J. Iranian Concr. Ins. 30, 17–22 (2008) (in Persian)
17. Khaloo, A.R., Bahadori, H., Hosseini, P., et al.: Investigation on effect of cement replacement with nano particles on mechanical properties and microstructure of concretes. Submitted to 8th Iranian International Congress on Civil Engineering. Shiraz, Iran (2009) in Persian
18. Tao, J.: Preliminary study on the water permeability and microstructure of concrete incorporating ano-SiO_2. Cem. Concr. Res. 35, 1943–1947 (2005)
19. Li, H., Xiao, H.G., Yuan, J., et al.: Microstructure of cement mortar with nano-particles. Comp: Part B 35, 185–189 (2005)

The Effect of Various Process Conditions on the Photocatalytic Degradation of NO

G. Hüsken, M. Hunger, M.M. Ballari, and H.J.H. Brouwers

Abstract. This paper presents the research conducted on photocatalytic concrete products with respect to the evaluation of the effect of varying process conditions on the degradation of nitric oxide (NO). The degradation process under laboratory conditions is modeled using the Langmuir-Hinshelwood kinetic model as basic reaction model. The suitability of the model is validated by experimental data as well as data obtained from literature. Furthermore, the effect of variations of process conditions like irradiance and relative humidity on the reaction rate constant k and adsorption equilibrium constant K_d are considered in the model.

1 Introduction

Air-quality in inner city areas is a topic which already receives a lot of attention nowadays. But in the coming years, the overall interest in this special topic will be bigger as there will be a number of conflicts with the limiting values given by the European Council [1]. These conflicts are caused on the one hand by a reduction of already existing limiting values till 2010 and increasing traffic rates, especially for diesel powered passenger cars and freight vehicles, on the other hand.

A promising approach for solving the problem of nitrogen oxides (NO_x) is the photochemical conversion of nitrogen oxides to low-dosed nitrates due to heterogeneous photocatalytic oxidation (PCO). The reaction products in form of nitrate compounds are water soluble and will be flushed from the active concrete surface by rain. The nitrate compounds can subsequently be extracted from the rain water by a standard sewage plant. For the photocatalytic oxidation, titanium dioxide (TiO_2) in low concentration is applied as photocatalyst. This photocatalyst uses light in the UV-A range of the sun light for the chemical conversion of nitrogen oxides.

G. Hüsken, M. Hunger, M.M. Ballari, and H.J.H. Brouwers
Department of Construction Management & Engineering, Faculty of Engineering Technology, University of Twente, Enschede, The Netherlands
e-mail: `g.husken@ctw.utwente.nl`

The basic working principle of the photocatalytic process is well know and discussed manifold in literature [2, 3]. First efforts in large scale applications of the photocatalytic reaction for air-purifying purposes have been made for about 10 years in Japan and have been also adopted by the European market [4]. But these investigations, as well as the fundamental research, are limited to single products only. A comparative study on the NO_x degradation of active concrete surfaces is discussed in [5]. The present paper is a continuation of this research using the results derived with one of the evaluated paving blocks as basis for further analysis. The basic principle of the test setup used as well as the conduction of the measurements are discussed in [6] and oriented on [7].

A reaction model is derived by using varying volumetric flow scenarios as well as different pollutant concentrations while the remaining test conditions have been kept constant. The reaction model describes the sample in terms of a reaction rate constant k and an adsorption equilibrium constant K_d. As changes of the relative humidity and UV-A irradiance can considerably change the efficiency of the PCO, their influence on the reaction is investigated and described by the model.

2 Reaction Model

The PCO at the surface of the paving block can be considered as heterogeneous catalysis [2] and is therefore characterized by adsorption of the pollutant molecules and desorption of the reaction products. It is demonstrated in [8] that not the diffusion but the conversion is the rate limiting step. For the prevailing photocatalytic gas-solid reaction, only adsorbed NO can be oxidized. Therefore, the Langmuir-Hinshelwood rate model is used for the modeling as suggested by [3, 9] and will also be applied here. According to the model, the disappearance rate of reactants reads:

$$r_{NO} = \frac{kK_d C_g}{1 + K_d C_g} \tag{1}$$

with k as reaction rate constant (mg/m^{3}s), K_d as the adsorption equilibrium constant (m^3/mg) and C_g as the NO concentration (mg NO per m^3 air) in the inlet gas. The NO balance equation now reads:

$$v_{air} \frac{dC_g}{dx} = -r_{NO} = -\frac{kK_d C_g}{1 + K_d C_g} \tag{2}$$

Supposing that $C_g = C_{g,in}$ and considering the reactor geometry, integration of Eq. (2) yields:

$$\frac{1}{k} + \frac{1}{kK_d} \frac{Ln \frac{C_{g,in}}{C_{g,out}}}{(C_{g,in} - C_{g,out})} = \frac{L}{v_{air}(C_{g,in} - C_{g,out})} = \frac{V_{Reactor}}{Q(C_{g,in} - C_{g,out})} \tag{3}$$

as $V_{Reactor} = LBh$ and $Q = v_{air}Bh$, again, $C_{g,out} = C_g(x = L)$. In Table 1 the values for $C_{g,out}$ for the experiments with one paving block are summarized. The inlet concentration $C_{g,in}$ was adjusted to 0.1, 0.3, 0.5 and 1 ppm NO (equivalent to 0.135, 0.404, 0.674 and 1.347 mg/m^3) and the flow rate Q was 1, 3 and 5 l/min.

Table 1 NO outlet concentrations of the reactor considering varying inlet concentrations and flow rates for the photocatalysis of the paving block example

$C_{g,in}$ [ppm]	$C_{g,out}$ [ppm] Volumetric flow rate Q [l/min]			NO_x removal rate [%] Volumetric flow rate Q [l/min]		
	1	3	5	1	3	5
0.1	0.011	0.032	0.041	89.0	68.4	59.4
0.3	0.039	0.157	0.197	87.1	47.6	34.3
0.5	0.210	0.309	0.356	58.0	38.3	28.9
1.0	0.334	0.729	0.779	66.6	27.1	22.1

In Figure 1a, $y = V_{Reactor}/Q(C_{g,in}-C_{g,out})$ is set out versus $x = Ln(C_{g,in}/C_{g,out})/(C_{g,in}-C_{g,out})$ and the data fit with the line $y = 1.19x+2.37$ obtained by the regression analysis. The intersection with the ordinate corresponds to $1/k$ so that k = 0.42 mg/m^{3}s and the slope of the regression line is corresponding to $1/kK_d$ so that K_d = 2.00 m^3/mg. By means of the obtained values of k and K_d the conversion rate and diffusion rate can be compared (cp. [8]).

Further relevant data on photocatalytic reactions could be found in [10]. Here, the paving block type NOXER was exposed to varying NO concentrations. With the help of background information regarding the conduction of the measurement, the Langmuir-Hinshelwood model could be applied. Using the data given in [10], a linear fit was derived as given in Figure 1b. The given data result in k = 3.54 mg/m^{3}s and K_d = 0.538 m^3/mg. Compared with own experiments, the conversion

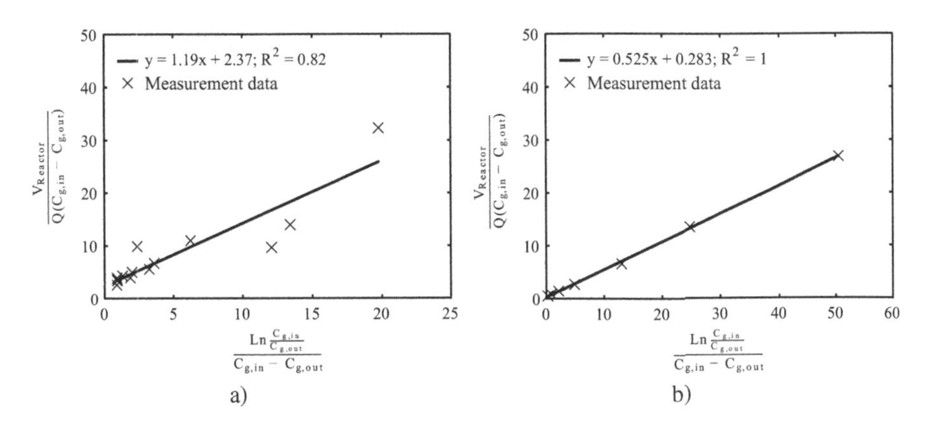

Fig. 1 Regression results of experimental data. a) own measurements presented in Table 1. b) data taken from [10]

rate is notedly higher and relatively less smaller than the diffusion transfer rate. This can be explained for the conversion by varying amounts and types of TiO_2 whereas the diffusion could be influenced by different surface morphology of the paving blocks. However, the data given in [10] are in good agreement with the model.

3 Modeling of External Influences

The degradation of NO and therewith the performance of the photocatalytic reaction is governed by physicochemical as well as product related parameters. In the previous section, a general model for the reaction kinetics has been derived considering reaction kinetics and flow related parameters. However, for a comprehensive modeling two further external influencing factors of the degradation process have to be considered. These are the UV-A irradiance and the concentration of water, expressed by the relative humidity.

To start the photocatalytic degradation process, UV-A light of the appropriate wavelength λ and irradiance E is necessary. According to [2], the increasing photocatalytic activity caused by increased irradiance can be divided into two regimes: i) for $E \leq 250$ W/m^2 the degradation increases proportional to E and ii) for $E > 250$ W/m^2 the photocatalytic activity grows as the square root of E. This linear behavior in the range of low irradiance ($E < 15$ W/m^2) could not be confirmed be own experiments [11]. In order to incorporate the dependency of the reaction constant k on the UV-A irradiance, a suitable mathematical expression can be found in [12]. Therewith, the reaction rate constant k would read:

$$k = \alpha_1\left(-1 + \sqrt{1 + \alpha_2 E}\right) \tag{4}$$

With α_1 and α_2 being factors to be fitted from the experiment. The expression considers the linear and nonlinear behavior of the degradation process for varying UV-A irradiance. The experimental data as well as the fit of Eq. (4) are depicted in Figure 2a and show good agreement. It is assumed that the adsorption equilibrium constant K_d is not influenced by the UV-A irradiance. This assumption is also confirmed by the experimental data in Figure 2b. The grey marked values in Figure 2b are considered as outliers due to remarkable scattering in the measurement caused by low flow values combined with low inlet pollution.

The influence of the relative humidity RH is caused by the hydrophilic effect of TiO_2 under exposure to UV-A light. According to [4], the hydrophilic effect at the surface is gaining over the oxidizing effect when high relative humidity values are applied. The water molecules adsorbed at the surface prevent the pollutants to react with the TiO_2. Therefore, it is assumed that both the conversion of NO and the adsorption of NO at the surface is affected. Considering the experimental data

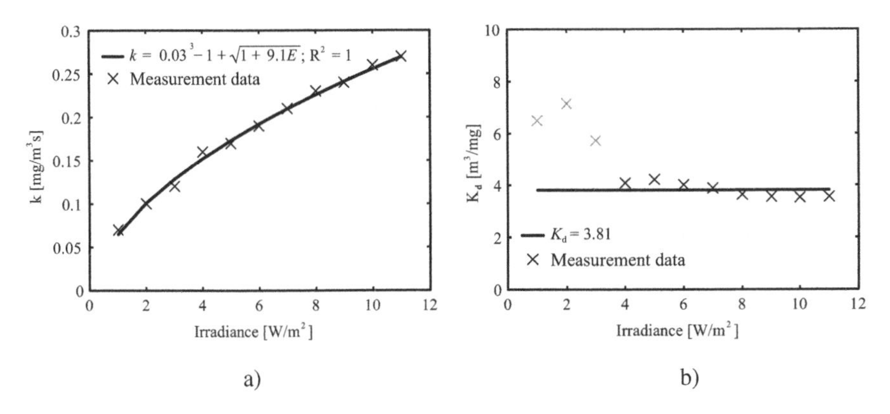

Fig. 2 Influence of UV-A irradiance. a) reaction rate constant k. b) adsorption equilibrium constant K_d.

given in Figure 3a and 3b, the influence of the relative humidity on the reaction rate constant k can be explained by:

$$k = \alpha_3 \alpha_4^{RH} RH^{\alpha_5} \tag{5}$$

while the dependency of the adsorption equilibrium constant K_d is expressed by a quadratic function:

$$K_d = \alpha_6 RH^2 + \alpha_7 RH + \alpha_8 \tag{6}$$

The fitting of the parameters α_3 to α_8 showed a good agreement with the experimental data (cp. Figure 3a and 3b).

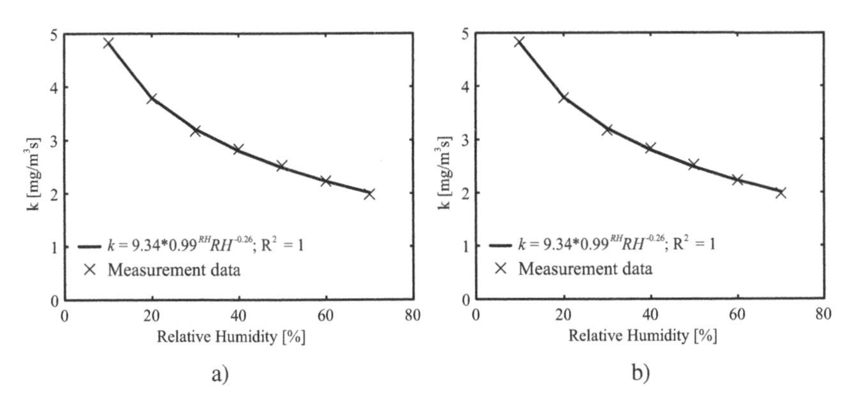

Fig. 3 Influence of relative humidity. a) reaction rate constant k. b) adsorption equilibrium constant K_d.

4 Conclusions

The heterogeneous photocatalytic oxidation seems to be a promising technique for reducing air pollution in inner city areas with high emissions of nitrogen oxides caused by increasing traffic loads. Numerous measurements of this research project carried out within the last two years showed that the concentration of nitrogen oxides in the ambient air can be effectively reduced by the photocatalytic oxidation using TiO_2. The experimental data provide a basis for the modeling of the degradation process using the Langmuir-Hinshelwood kinetics. The prediction of the performance of certain air-purifying concrete products can now be predicted by the derived model. Furthermore, mathematical expressions are proposed describing both the kinetic boundary conditions as well as the process conditions. The latter influences and the transformation of the results to practical applications is part of ongoing research.

Acknowledgments. The authors wish to express their thanks to the following sponsors of the research group: Bouwdienst Rijkswaterstaat, Rokramix, Betoncentrale Twenthe, Graniet-Import Benelux, Kijlstra Beton, Struyk Verwo Groep, Hülskens, Insulinde, Dusseldorp Groep, Eerland Recycling, ENCI, Provincie Overijssel, Rijkswaterstaat Directie Zeeland, A&G maasvlakte, BTE, Alvon Bouwsystemen, and v. d. Bosch Beton (chronological order of joining).

References

1. The Council of the European Union, Council Directive 1999/30/EC - Relating to limit values for sulphur dioxide, nitrogen dioxide and oxides of nitrogen, particulate matter and lead in ambient air (1999)
2. Herrmann, J.M., Péruchon, L., Puzenat, E., Guillard, C.: Photocatalysis: From fundamentals to self-cleaning glass application. In: Baglioni, P., Cassar, L. (eds.) Proceedings International RILEM Symposium on Photocatalysis, Environment and Construction Materials, Florence, Italy, October 8-9, 2007. RILEM Publications, Bagneux (2007)
3. Zhao, J., Yang, X.: Photocatalytic oxidation for indoor air purification: a literature review. Build Environ. (2003) doi:10.1016/S0360-1323(02)00212-3
4. Beeldens, A.: Air purification by road materials: results of the test project in Antwerp. In: Baglioni, P., Cassar, L. (eds.) Proceedings International RILEM Symposium on Photocatalysis, Environment and Construction Materials, Florence, Italy, October 8-9, 2007. RILEM Publications, Bagneux (2007)
5. Hüsken, G., Hunger, M., Brouwers, H.J.H.: Comparative study on cementitious products containing titanium dioxide as photo-catalyst. In: Baglioni, P., Cassar, L. (eds.) Proceedings International RILEM Symposium on Photocatalysis, Environment and Construction Materials, Florence, Italy, October 8-9, 2007. RILEM Publications, Bagneux (2007)
6. Hunger, M., Hüsken, G., Brouwers, H.J.H.: Photocatalysis applied to concrete products – Part 1: Principles and test procedure. ZKG International 61(8), 77–85 (2008)

7. ISO 22197-1, Fine ceramics (advanced ceramics, advanced technical ceramics) – Test method for air-purification performance of semiconducting photocatalytic materials – Part 1: Removal of nitric oxide (2007)
8. Hunger, M., Brouwers, H.J.H., Ballari, M.M.: Photocatalytic degradation ability of cementitious materials: a modeling approach. In: Sun, W., Breugel, K., van Miao, C., Ye, G., Chen, H. (eds.) Proceedings of 1st International Conference on Microstructure related Durability of Cementitious Composites, Nanjing, China, October 13-15 (2008)
9. Dong, Y., Bai, Z., Liu, R., Zhu, T.: Decomposition of indoor ammonia with TiO_2-loaded cotton woven fabrics prepared by different textile finishing methods. Atmos Environ. (2007) doi:10.1016/j.atmosenv.2006.08.056
10. Mitsubishi Materials Corporation: NOx removing paving block utilizing photocatalytic reaction. Brochure Noxer – NOx removing paving block (2005)
11. Hunger, M., Hüsken, G., Brouwers, H.J.H.: Photocatalysis applied to concrete products – Part 2: Influencing factors and product performance. ZKG International 61(10), 76–84 (2008)
12. Imoberdorf, G., Irazoqui, H.A., Cassano, A.E., Alfano, O.M.: Photocatalytic Degradation of Tetrachloroethylene in Gas Phase on TiO_2-Films: A Kinetc Study. Ind. Eng. Chem. Res. 44, 6075–6085 (2005)

Molecular Dynamics Approach for the Effect of Metal Coating on Single-Walled Carbon Nanotube

S. Inoue and Y. Matsumura

Abstract. The functionalized single-walled carbon nanotube (SWCNT) is focused lately, but there is no guarantee to keep its outstanding properties. In this paper the physical strength of a SWCNT is derived in terms of a stress-strain curve by molecular dynamics simulation. The breaking stress of a metal-coated SWCNT was lower than that of an uncoated SWCNT; however, the force constant increased by 17%, which can be attributed to the effect of the metal coating on the SWNCT. With regard to the rupture phenomena, it was observed that the uncoated SWCNT ruptured more easily than the metal-coated SWCNT at the rupture point. The rupture phenomenon was initiated by a local distortion of the metal atoms of the SWCNT.

1 Introduction

Among various types of carbon nanotubes [1, 2], single-walled carbon nanotube (SWCNT) [3] has been attracting considerable attention since their discovery in 1993. An SWCNT exhibits several useful properties such as high thermal conductivity due to its unique quasi-one-dimensional structure. It can be synthesized using different techniques such as a laser furnace technique [4], arc discharge technique [5, 6], and various chemical vapor deposition (CVD) techniques [7-13]. A large number of SWCNT can be grown inexpensively using a super growth technique [14]; however, their growth mechanism has not yet been elucidated completely. Each of the abovementioned techniques involves a different growth mechanism of SWNCT; however, a well-established and acceptable growth model does not exist. Thus far, several theoretical and practical growth models have been introduced [15-22]. Modifying the properties of SWCNT can enhance their applicability in various engineering fields. Recently, Ishikawa et al. [23] deposited metal species onto a vertically aligned SWCNT (VA-SWCNT) film, and Zhang et al. [24] coated an isolated SWCNT with several metal species. Such experiments

---

S. Inoue and Y. Matsumura
Department of Mechanical System Engineering, Hiroshima University

are very interesting because it is known that the SWCNT exhibits high thermal conductivity; thus, the VA-SWCNT is also expected to exhibit high thermal conductivity; however, it should be conjugated with a metal species before use. Inappropriate conjugation would decrease the efficiency of the SWCNT; however, a suitable metal species coated on the SWCNT would enhance its efficiency. The electrical properties of SWCNT are strongly dependent on its chirality, which we cannot control at present. By coating the SWCNT with metal species, we can prepare a metallic SWCNT whose electrical properties are independent of chirality. However, this may affect the original properties of SWCNT, such as high thermal conductivity and high physical strength.

In this study, we determined the stress–strain curve of SWCNT and observed their rupture phenomena by molecular dynamics simulation. It was observed that the breaking stress of the metal-coated SWCNT was lower than that of the uncoated SWCNT; however, there was an increase of 17% in the force constant, which was caused by the incorporation of the coating metal. With regard to the rupture phenomena, it was observed that the uncoated SWCNT ruptured more easily than the metal-coated SWCNT due to rupture stress. The rupturing phenomena are initiated by amplifying the local distortion for the uncoated SWCNT and by the metal atoms tear the C-C bond at the local distortion for the metal-coated SWCNT.

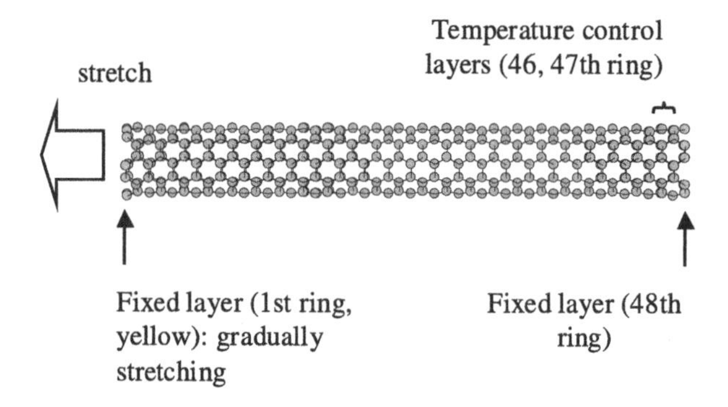

Fig. 1 Calculated system. SWCNT with (5, 5) chirality consists of 480 carbon atoms. The right end ring (48th ring) is fixed and the next two layers (46, 47th rings) work to control the temperature. The left end ring (1st ring) is also fixed by their y and z direction, but gradually expands with a certain displacement in each step

2 Methods

An isolated SWCNT with (5, 5) chirality is used as an object whose carbon-carbon interaction parameters are determined using a simplified Brenner-Tersoff potential [25, 26]. The carbon-metal and metal-metal interaction parameters of the isolated SWCNT are also determined using the Brenner type potential, and its potential parameters are employed from the results of Shibuta and Maruyama [27]. The metal-coated SWCNT is prepared by metal cluster deposition onto the isolated SWCNT, as described in our previous study [28]. The isolated SWCNT consists of 480 carbon atoms, and its one side (48th ring in Fig. 1) is fixed. The SWCNT is stretched by gradually pulling the other side (1st ring). The temperature is controlled by only the next two layers (46, 47th rings) on the fixed side as shown in Fig. 1 to avoid from unintentional sudden relaxation, if we also control the other end. Each stretch step should take an enough relaxation time; otherwise, there may arise an unrealistic distortion and that results in unreliable rupture.

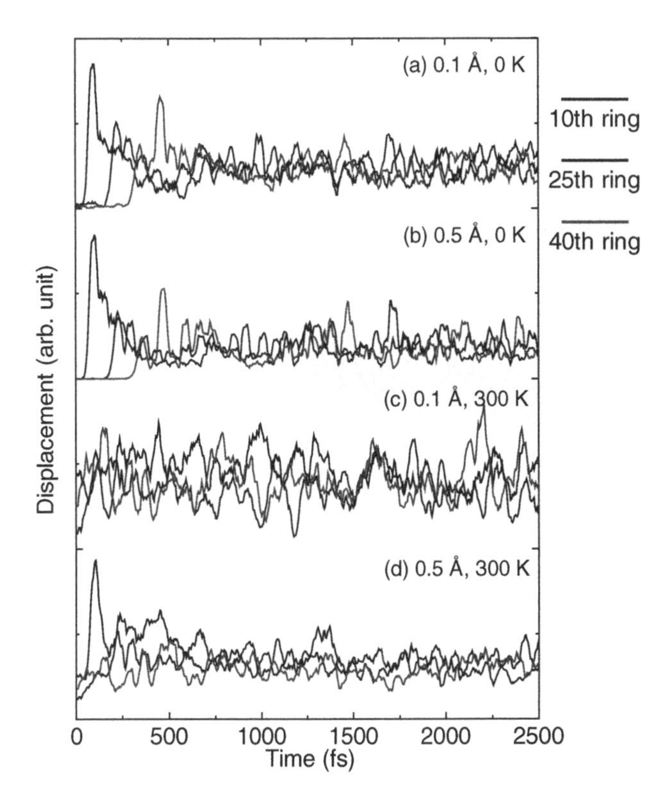

Fig. 2 The propagation of displacement. At time = 0 the first ring named in Fig. 1 is pulled from its equilibrium position by a certain length shown in this figure. The displacement propagates with some delay depends on the distance from the first ring and gradually converges to the new equilibrium position. The propagation delay does not depend on the displacement length in these range but depend on the temperature

Figure 2 shows the propagation of displacements of the 10th, 25th, and 40th rings at 0 K and 300 K. After the first ring is displaced by 0.1 Å and 0.5 Å at t = 0, the displacement propagates toward the opposite direction. The velocity of propagation does not depend on the initial displacement length in this study. Each displacement value is expressed as an average of 10 atoms in each ring. At 300 K, the displacement length is too small to be observed, due to the thermal vibration of each atom. If there is sufficient relaxation time, each atom or ring will be restored to its original position. The time profile of the 10th ring is shown in Fig. 3. At 300 K, an obvious convergence of the rings is not observed because the displacement length of the rings is smaller than that obtained due to thermal fluctuation; in contrast, at 0 K, an obvious convergence of the rings is observed. The atoms constantly vibrate around the equilibrium position, where vibrations are roughly equal to 785 fs at 0 K and 835 fs at 300 K. Even though 20 ps appears to be a sufficient relaxation time at 0 K, in this study, we consider 10 ps as the relaxation time after each instance of stretching by 0.1 Å.

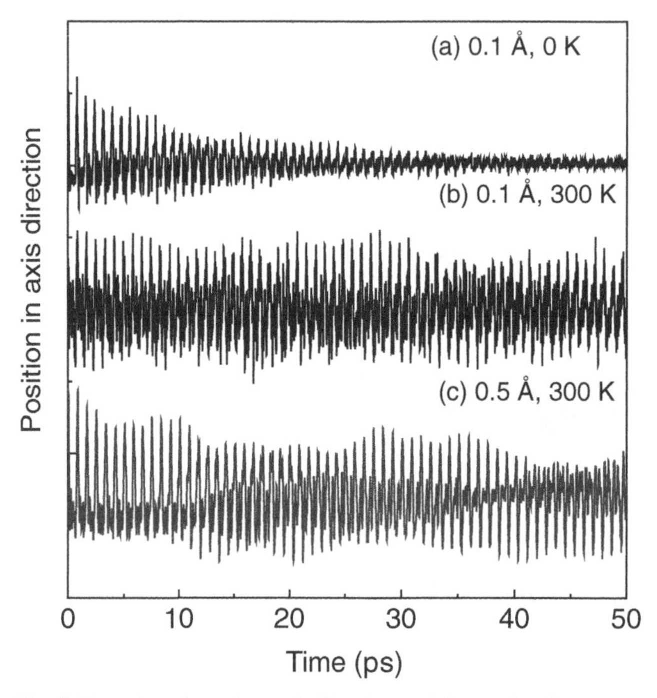

Fig. 3 The relaxation of stretch. The thermal fluctuation is comparable to the displacement that results in difficulty in discriminating the fluctuation and vibration in 300 K, but in 0K the relaxation can be seen. These vibration period is approximately 785 fs in 0 K and 835 fs in 300K

3 Results and Discussion

Figure 4(a) shows the stress–strain curve obtained by the molecular dynamics simulation. A suitable stress value is defined by assuming the SWCNT to be cylindrical, with the effective cross section as the diameter. In this study, diameters of the uncoated and metal-coated SWCNT are 6.93 Å and 8.32 Å, respectively. The value of the rupture stress of the metal-coated SWCNT was approximately half that of the uncoated SWCNT as shown in Fig. 4(b). This can be attributed partly to an increase in the cross-sectional diameter of the SWCNT by coating metals; however, because the rupture force also decreases, the physical strength certainly becomes weak by coating. It has been further observed that metal atoms tend to break carbon bonds. When the SWCNT exhibits local distortion due to stretching, metal atoms break the carbon bonds and stick to the defect. This breaking may be due to a difference in bond lengths. Assuming that the distortion of a hexagonal carbon network in the horizontal direction is negligible, the bond length of a C-C bond becomes 1.8 Å just prior to rupture. In reality, the C-C bond length could be reduced from 1.8 Å to approximately 1.75 Å by the distortion of the carbon hexagonal network. According to the Brenner potential, the influence on binding energy suddenly reduces as expressed by an attenuation function f in Eq. 1, where R1 is 1.7 Å and R2 Å is 2.0. On the other hand, the binding energy of a Ni-C bond, shows strongest between 1.76 Å and 2.0 Å that depends on the coordination number; therefore, nickel atoms tend to combine with carbon atoms firmly that results in the rupture of the SWCNT.

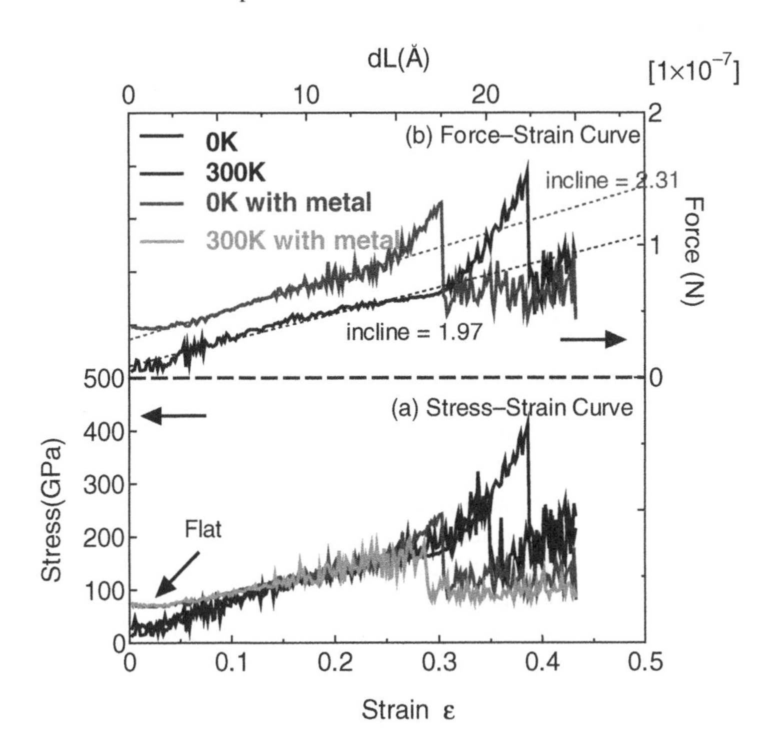

Fig. 4 The stress-strain curve (a) and force-strain curve (b). The stress is defined by assuming their cross section with 6.93 Å for the uncoated SWCNT and 8.32 Å for the metal-coated SWCNT in diameter. The metal-coated SWCNT meets earlier rupture point but has a larger force constant. The metal-coated SWCNT has a residual stress that makes the inclination flat in a small strain (displacement) range

$$f(r)=\begin{cases}1 & (r<R_1)\\ 0.5\cdot\left(1+\cos\dfrac{r-R_1}{R_2-R_1}\pi\right) & (R_1<r<R_2)\\ 0 & (R_2<r)\end{cases} \tag{1}$$

As to the force constant metal-coated SWCNT shows larger by 17% than that of uncoated one in Fig.4 (b) unlike the rupture stress and/or force. It was speculated that the force constant of the metal-coated SWCNT should be decreased because the binding energy of the C-C bond became weaker by increasing the coordination number. On the contrary, the force constant increases owing to the metal contribution. When e = 0, stress (s) is a nonzero value in the case of the metal-coated SWCNT. This is attributed to a residual stress, which is usually present on the conjugating surface of different species in a macroscopic model. According to our previous work the reason of realizing smooth coating on SWCNT was the coincidence of the bond length (Ni-Ni, with a infinite coordinating number) and the distance of the center of a hexagonal carbon network, which is the most stable position for the nickel atoms absorbed on SWCNT. However, strictly speaking, this distance is longer a little (approximately 0.05 Å) for the nickel atoms with practical coordination number; thus, the residual stress works toward the direction of shorten the SWCNT length. This residual stress is clearly seen at the beginning of stretch. The stress does not increase at the beginning of stretch until around e = 0.05 owing to the cancel of pulling stress and shorten stress. This stress remains constant until approximately e = 0.05 due to the absence of pulling stress and shortening stress.

In the uncoated SWCNT, the ruptured strain is approximately e = 0.38; however, this value is not important for the molecular dynamics simulation. As mentioned by Agrawal et al. [29], the ruptured strain value is affected by the cutoff length at the Brenner potential In reality, at e = 0.38, the length of the C-C bond becomes 2.0 Å; this is equivalent to the cutoff length at the Brenner potential. The rupture strain at 300 K is less than that at 0 K. This difference in strains is attributed to the thermal fluctuations and not the decrease of binding energy due to the increase of the temperature. At high temperature, the velocity of the atoms increases, which results in a large fluctuation, as shown in Fig. 5. The upper part of this figure shows of the results obtained at 300 K, and the lower part shows those results obtained at 0 K. This figure shows that the thermal fluctuation is considerably larger in the case of 300 K. After each instance of stretching, there is a sufficient relaxation period during which the fluctuation decreases and the center of vibration shifts to the new equilibrium position. However, when the SWCNT ruptures, the fluctuation diverges and does not decrease during the relaxation period. At 300 K and approximately e = 0.35, the C-C bond length can exceed 2.0 Å due to the thermal fluctuation. It is speculated that the force constant should become smaller in 300 K than that in 0 K and the incline of the stress-strain curve should

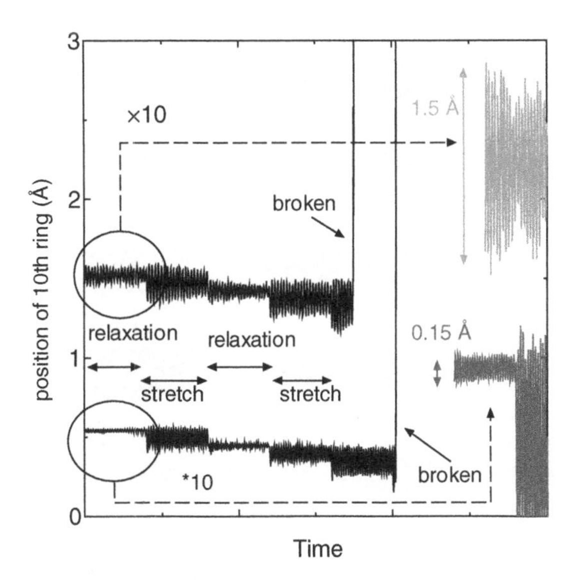

Fig. 5 The thermal fluctuation of the 10th ring in the axis direction in 0K (bottom) and in 300 K (top). Usually, after each stretch the fluctuation converges to the new equilibrium position during the following relaxation time, but just before the amputation the fluctuation diverges

become smaller. This is not shown in Fig. 4. This reason is not clear but in this study the procedure is continuous stretch that may include any fluctuation that conceals the temperature effect; however, this effect is clearly confirmed in the simple harmonic oscillation shown in Fig. 3. As we mentioned above, the force constant estimated by the simple formulation shown in Eq. 2 becomes smaller by 13%. (The harmonic period is 785 fs at 0 K, and 835 fs at 300 K).

$$T = 2\pi\sqrt{\frac{m}{k}} \tag{2}$$

Figure 6 shows images of the stretched SWCNT just before and after rupture. In the case of the uncoated SWCNT, the nanotube easily ruptures after acquiring a string-like shape; however, in the case of the metal-coated SWCNT, the nanotube does not rupture completely. These phenomena can be explained by the fact that the carbon atom prefers to exhibit an sp2 structure or at least tends to maintain this structure at the Brenner potential; thus, once a particular bond is broken, all the carbon atoms saturate the dangling bonds by forming a spherical structure. This causes the SWNCT to rupture easily. On the other hand, in the case of the metal-coated SWCNT, numerous metal atoms, which are not strictly defined the coordination number like the carbon atoms are, can terminate the dangling bond of the carbon atoms and form bonds with each other. As a result, the SWNCT does not

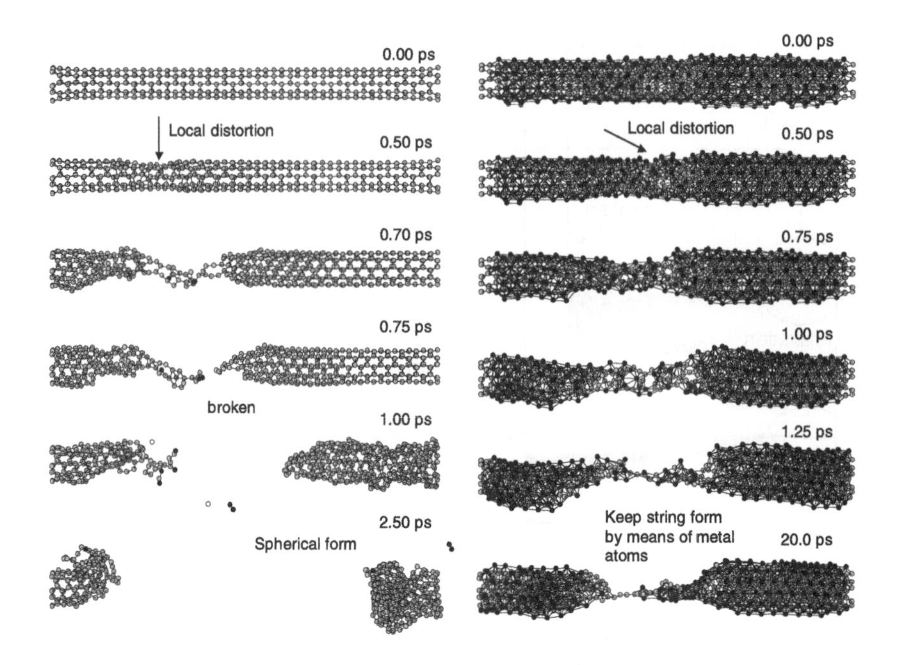

Fig. 6 The snap shots of stretches. The time shown in each figure denotes the time progress from the last stretch. The green, orange and red ball stands carbon atoms with 0, one, two dangling bonds respectively. The blue ball stands metal atoms. The uncoated SWCNT is broken lightly broken but the metal-coated SWCNT is not completely broken, because the carbon atoms tend to take a spherical form after arising a local distortion and/or local defect to keep sp2 structure; however, in case of the metal-coated SWCNT the metal atoms can terminate the dangling bonds of carbon atom that results in avoid or delay the complete amputation

rupture easily because of the linking of the metal atoms; a subsequent annealing process could result in the formation of hexagonal or pentagonal carbon rings.

4 Conclusion

Molecular dynamics simulation of metal-coated and uncoated SWCNT was performed, and the stress-strain curves were derived. With regard to the rupture point, the metal-coated SWCNT meets earlier rupture point than uncoated SWCNT. This is not due to the binding energy of C-C but rather than interferences of the coating metal. The binding energy of C-C must be weaken by increasing the coordination number but the earlier rupture is caused by the coating metal atoms. When the strain becomes approximately 0.3, the binding energy of the C-C bond becomes extremely low; on the other hand, the binding energy of the Ni-C bond becomes nearly maximum. The metal atoms tend to combine with the carbon atoms by breaking C-C bond that results in the rupture of SWCNT. Because the carbon atom prefers to keep sp2 structure, the carbon atom whose pair is robbed by the metal atom forms a spherical structure to saturate its dangling bond.

The force constant of the metal-coated SWNCT increases by 17% due to the effect of the coating metal. The disadvantage of metal coating is that the rupture stress becomes approximately half that of uncoated SWCNT and reduces by 25% in comparison with the rupture force. This implies that the metal-coated SWCNT keeps still a higher tensile strength than conventional materials. The advantages of coating the SWNCT with a metal outweigh its disadvantages because a metal-coated SWCNT can exhibit novel properties, and its electrical properties could be controlled. With regard to the rupture phenomena, the uncoated SWCNT ruptures faster in order to maintain the carbon sp2 structure by forming a spherical structure; on the other hand, the metal-coated SWCNT does not rupture completely because metal atoms that saturate the dangling bond of the carbon atoms and form bonds with each other are not clearly defined in terms of their coordination number in the potential function.

References

1. Iijima, S.: Helical microtubules of graphitic carbon. Nature 354, 56–58 (1991)
2. Dresselhaus, M.S., Dresselhaus, G., Saito, R.: Physics of carbon nanotubes. Carbon 33, 883–891 (1995)
3. Iijima, S., Ichihashi, T.: Single-shell carbon nanotubes of 1-nm diameter. Nature 363, 603–605 (1993)
4. Thess, A., Lee, R., et al.: Crystalline ropes of metallic carbon nanotubes. Science 273, 483–487 (1996)
5. Ajayan, P.M., Lambert, J.M., et al.: Growth morphologies during cobalt-catalyzed single-shell carbon nanotube synthesis. Chem. Phys. Lett. 215, 509–517 (1993)
6. Journet, C., Maser, W.K., et al.: Large-scale production of single-walled carbon nanotubes by the electric-arc technique. Nature 388, 756–758 (1997)
7. Dal, H.J., Rinzler, A.G., et al.: Single-wall nanotubes produced by metal-catalyzed disproportionation of carbon monoxide. Chem. Phys. Lett. 260, 471–475 (1996)
8. Cheng, H.M., Li, F., et al.: Bulk morphology and diameter distribution of single-walled carbon nanotubes synthesized by catalytic decomposition of hydrocarbons. Chem. Phys. Lett. 289, 602–610 (1998)
9. Kong, J., Cassel, A.M., Dai, H.J.: Synthesis of individual single-walled carbon nanotubes on patterned silicon wafers. Nature 395, 878–881 (1998)
10. Hafner, J.H., Cheung, C.L., et al.: High-yield assembly of individual single-walled carbon nanotube tips for scanning probe microscopies. J. Phys. Chem. B 105, 743–746 (2001)
11. Li, Y.M., Kim, W., et al.: Growth of single-walled carbon nanotubes from discrete catalytic nanoparticles of various sizes. J. Phys. Chem. B 105, 11424–11431 (2001)
12. Zhang, Y.G., Chang, A., et al.: Electric-field-directed growth of aligned single-walled carbon nanotubes. Appl. Phys. Lett. 79, 3155–3157 (2001)
13. Maruyama, S., Kojima, R., et al.: Electric-field-directed growth of aligned single-walled carbon nanotubes. Chem. Phys. Lett. 360, 229–234 (2002)

14. Hata, K., Futaba, D.N., et al.: Water-assisted highly efficient synthesis of impurity-free single-waited carbon nanotubes. Science 306, 1362–1364 (2004)
15. Yudasaka, M., Yamada, R., Iijima, S.: Mechanism of the effect of NiCo, Ni and Co catalysts on the yield of single-wall carbon nanotubes formed by pulsed Nd: YAG laser ablation. J. Phys. Chem. B 103, 6224–6229 (1999)
16. Kataura, H., Kumazawa, Y., et al.: Diameter control of single-walled carbon nanotubes. Carbon 38, 1691–1697 (2000)
17. Dai, H.J., Rinzler, A.G., et al.: Single-wall nanotubes produced by metal-catalyzed disproportionation of carbon monoxide. Chem. Phys. Lett. 260, 471–475 (1996)
18. Shibuta, Y., Maruyama, S.: Molecular dynamics simulation of formation process of single-walled carbon nanotubes by CCVD method. Chem. Phys. Lett. 382, 381–386 (2003)
19. Ding, F., Rosen, A., Bolton, K.: Molecular dynamics study of the catalyst particle size dependence on carbon nanotube growth. J. Chem. Phys. 121, 2775–2779 (2004)
20. Fan, X., Buczko, R., et al.: Nucleation of single-walled carbon nanotubes. Phys. Rev. Lett. 91, 145501 (2003)
21. Yudasaka, M., Kasuya, Y., et al.: Causes of different catalytic activities of metals in formation of single-wall carbon nanotubes. Appl. Phys. A 74, 377–385 (2002)
22. Inoue, S., Kikuchi, Y.: Diameter control and growth mechanism of single-walled carbon nanotubes. Chem. Phys. Lett. 410, 209–212 (2005)
23. Ishikawa, K., Duong, H.M., et al.: Extended abstracts ASME-JSME Thermal Eng. HT2007-32783 (2007)
24. Zhang, Y., Franklin, N.W., et al.: Metal coating on suspended carbon nanotubes and its implication to metal-tube interaction. Chem. Phys. Lett. 331, 35–41 (2000)
25. Brenner, D.W.: Empirical potential for hydrocarbons for use in simulating the chemical vapor-deposition of diamond films. Phys. Rev. B 42, 9458 (1990)
26. Yamaguchi, Y., Maruyama, S.: A molecular dynamics simulation of the fullerene formation process. Chem. Phys. Lett. 286, 336–342 (1998)
27. Shibuta, Y., Maruyama, S.: Bond-order potential for transition metal carbide cluster for the growth simulation of a single-walled carbon nanotube. Comput. Mat. Sci. 39, 842–848 (2007)
28. Inoue, S., Matsumura, Y.: Molecular dynamics simulation of physical vapor deposition of metals onto a vertically aligned single-walled carbon nanotube surface. Carbon 46, 2046–2052 (2008)
29. Agrawal, P.M., Sudalayandi, B.S., et al.: Molecular dynamics (MD) simulations of the dependence of C-C bond lengths and bond angles on the tensile strain in single-wall carbon nanotubes (SWCNT). Comput. Mat. Sci. 41, 450–456 (2008)

Polymer Nanocomposites for Infrastructure Rehabilitation

M.R. Kessler and W.K. Goertzen

Abstract. Polymer matrix composites (PMCs) are becoming increasingly important in the structural repair and rehabilitation of damaged infrastructure – from pipelines to buildings to bridges. For example, composite overwraps are used to repair corroded steel pipelines because the repair can be completed in a relatively short amount of time and the fluid transmission in the piping system can remain undisrupted while the repair is being made. Often in these applications, a primer and filler adhesive is used to fill defects in the substrate so that load can be adequately transferred to the continuous fiber composite. In this work we discuss various nano-scale reinforcements such as fumed silica, alumina, nanoclay, and carbon nanotubes as additives to this filler adhesive in order to improve mechanical properties and to tailor the thermal expansion of the composite to match the underlying substrate being repaired. The thermal expansion mismatch is especially important in applications where temperature fluctuations are present. We highlight our results from rheology, thermal expansion, and dynamic mechanical analysis testing of nanosilica/cyanate ester composites and show that the incorporation of the nano-scale fillers can result in improvement of the thermo-mechanical behavior of the composites.

1 Introduction

Corrosion has a costly and deleterious effect on aging infrastructure throughout the world. As such, considerable attention has been focused on innovative techniques to arrest corrosion in the carbon steel found in bridges, pipelines and pipework, water and wastewater systems, and electric power generation facilities and

M.R. Kessler
Department of Materials Science and Engineering, Iowa State University,
Ames, Iowa, USA
e-mail: mkessler@iastate.edu
http://mse.iastate.edu/polycomp/

W.K. Goertzen
Department of Materials Science and Engineering, Iowa State University,
Ames, Iowa, USA

to restore the structural integrity of these systems, especially pipelines and bridges [1-4]. Many of these repair technologies utilize fiber-reinforced polymer matrix composites.

In damaged pipelines, composite overwraps can be used for timely, cost-effective repair of external corrosion (as shown in Figure 1) without the need to disrupt fluid transmission in the piping system while the repair is being made. In order for the composite overwrap to be effective, a putty (filler adhesive) is used to fill the defect region to allow a uniform surface for the outer composite wrap to be applied. The filler adhesive is the medium by which the pipe pressure is transferred to the outer fiber-reinforced composite wrap. Because of the processing and performance requirements on the filler adhesive, nanoparticles may be added to the thermosetting resin to increase the thixotropy of the prepolymer resin and the thermomechanical properties of the cured adhesive. In our work we have investigated systems with fumed nanosilica, nanoalumina, multiwalled carbon nanotubes, and carbon nanofibers.

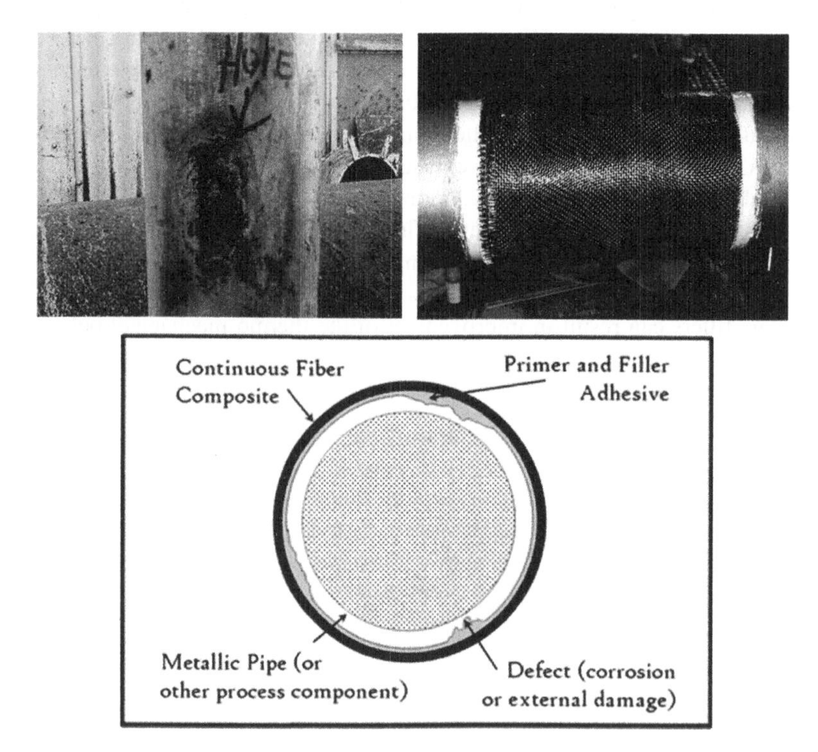

Fig. 1 Typical composite repair system for damaged pipeline showing (top left) damaged pipe with external corrosion damage, (top right) repaired pipe with composite overwrap installed, (bottom) cross-sectional schematic showing metallic pipe, filler adhesive, and external composite wrap

The steps involved in making the repair are shown in Figure 2. First the prepolymer resin is mixed with appropriate curing agents and nanoscale fillers (using a combination of high shear mixing and ultrasonication) and applied to the steel substrate to restore the original dimension of the pipe. Next the primer is applied to the remaining substrate surface. The outer composite overwrap is applied after first impregnating the reinforcement (in this case a carbon fiber fabric) by hoop wrapping the reinforcement around the defect region followed by curing of the thermosetting polymer matrix.

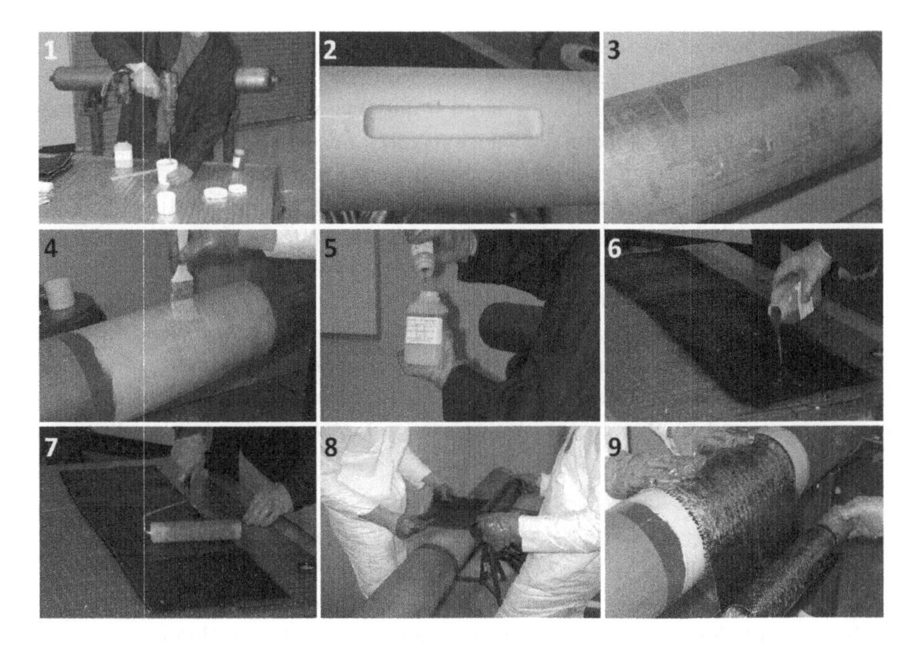

Fig. 2 Steps in repairing a damaged pipe using a composite overwrap. The epoxy putty used in step 3 to fill the defect is rheologically engineered with nanosilica (Photo courtesy of Jeff Wilson)

2 Nano-fillers

There are several nanoscale fillers that may be added to the thermosetting prepolymer prior to application on the repair substrate. The purpose of adding these fillers is to (1) increase the thixotropic behavior of the prepolymer to prevent sag of the putty before the outer composite is applied and cured; (2) increase the stiffness and strength of the cured network, thereby increasing the load that is transferred to the outer structural composite; and (3) reduce the thermal expansion mismatch between the polymer filler putty and the underlying substrate. This last purpose is especially important for materials which operate at elevated temperatures such as

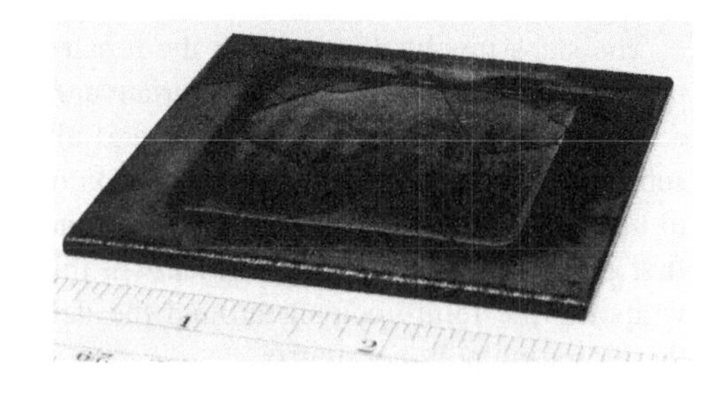

Fig. 3 Thermosetting filler adhesive (cyanate ester resin) cured in a simulated steel defect. The cracks in the polymer are a result of the large thermal expansion mismatch between the polymer and the steel substrate

systems that utilize pressurized steam and other process piping, pressure vessels and storage tanks, heat exchangers, burners, furnaces, and industrial exhaust systems. Figure 3 below illustrates the large strains that can develop in the polymer filler due to coefficient of thermal expansion (CTE) mismatch.

The incorporation of nanoscale filler dispersed at a molecular level results in an ultra-large interfacial area per unit volume between the nano-filler and the matrix polymer. It is this large internal interfacial area, coupled with the nanoscale dimensions constraint on the polymer matrix that is largely responsible for the unique features in polymer nanocomposites compared to polymers filled with conventional microscale filler. There are numerous nanoscale fillers that can be considered in composites for infrastructure rehabilitation applications; however, many of them share common features with regard to processing, morphology, and reinforcement effect. Several of these nanoscale fillers are discussed next.

2.1 Metallic Oxides—Nanosilica, Nanoalumina, Nanotitania

Much of the work we have performed to date with modifying the filler adhesive with nanoparticles has been with fumed nanosilica and nanoalumina. Fumed nanosilica is made by a vapor phase flame hydrolysis process of silicon tetrachloride. In this process, SiO_2 molecules condense and form spherical primary nanoparticles from 5 to 40 nm. These primary particles form mostly aggregates (primary particles sintered together) that are about 0.2 to 0.3 μm in diameter [5]. Fumed silica is used extensively as an agent to reinforce and modify the rheological properties of liquids, adhesives, and elastomers. Thermosetting polymers such as epoxies [6-10], polyurethanes [11-12], and polyesters [13] have used fumed silica to modify rheological (thixotropy, sag resistence, and anti-settling agent) and end-use mechanical properties. Figure 3 below shows TEM micrographs of silica aggregates used in the present work.

Nanoscale aluminum oxide particles and nanoscale titanium dioxide particles are processed by a similar flame hydrolysis process as the fumed silica, but with other metallic chloride precursors. TEM micrographs of nano-alumina particles used in the present study are shown in Figure 5.

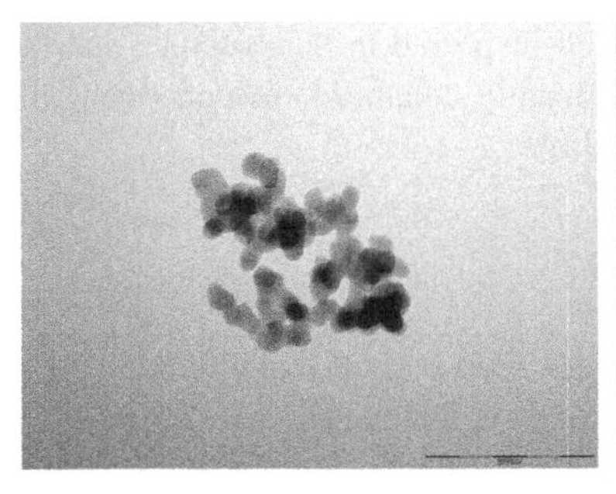

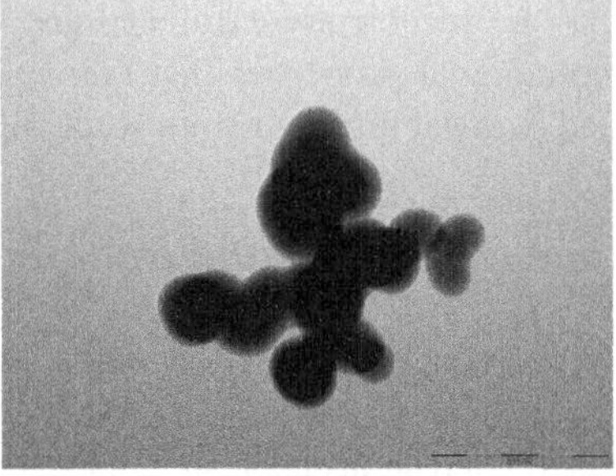

Fig. 4 TEM of fumed silica aggregates. The image on the left has a primary particle size of 12 nm (AEROSIL 200). The image on the right has a primary particle size of 40 nm (AEROSIL OX 50). The scale bar is 200 nm

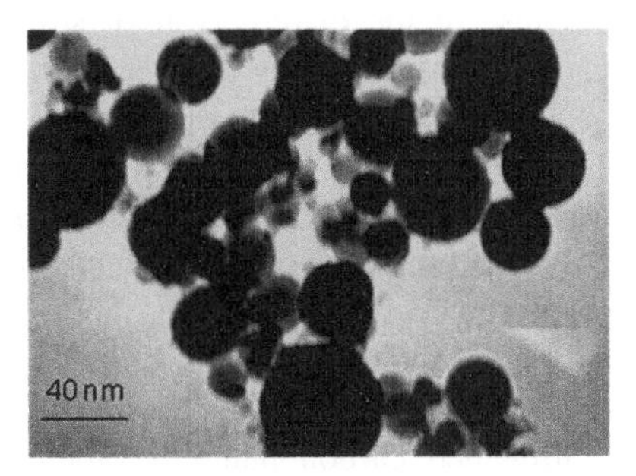

Fig. 5 TEM of nanoalumina particles used in this work (image courtesy of Mufit Akinc)

2.2 Nanoclay

Perhaps the most widely investigated nanoparticles in polymer composites are montmorillonite nanoclays. The key to obtaining well dispersed, effective nanocomposites with clays is to obtain exfoliation of the particles, which is complicated by the coupling of the particles due to surface charges and self-attraction. Nanoclays are often referred to as crystals or tactoids, but they are actually composed of thousands of silicate layers (platelets), geometrically stacked like a "deck of cards."

The surface of a platelet has a relative positive charge (cations). These charges can be shared between adjacent platelets and promote adhesion of the platelets. In addition, bonding of the platelets can also occur by weak van der Waals bonds, further promoting the "deck of cards" cubic structure and preventing mixing in organic solutions. In order to promote compatibility of the particles in organic materials, such as thermosetting prepolymers, the surfaces of the particles are typically made hydrophobic. This is usually accomplished by "modifying the surface with an organic surfactant," such as ammonium cations that have an alkyl chain [14]. Basically, these chains act as a tie layer—one end of the molecule has an affinity for the cation surface of the platelet and the other end has an affinity for organic molecules. This helps the particles to be mixed into an organic solution. If the clay particles are sufficiently exfoliated and well dispersed in the polymer matrix some of the material properties that are significantly enhanced include strength, stiffness, and permeability (moisture susceptibility).

2.3 Carbon Nanotubes and Nanofibers

Multi-walled carbon nanotubes [15] and single-walled carbon nanotubes [16,17] are now nearly 15 years old. Since their discovery and synthesis a decade and a half ago, much interest has been shown by researchers and business leaders within the polymer and composites community. The high strength and elastic modulus, and low CTE ($\alpha_{axial} \sim -1.5\times10^{-6}$ K^{-1}, $\alpha_{transverse} \sim -0.15\times10^{-6}$ K^{-1}) [18] combined with the high aspect ratio of the nanotubes make them ideal candidates for nano-reinforcement in polymer matrix composites for infrastructure rehabilitation. Because of the extremely high strength of the individual nanotubes, failure of nanocomposites nearly always occurs at the interface between the matrix and the nanotubes, and adequate dispersion in the host matrix can be an issue during processing. However, when these obstacles are overcome (such as by chemical functionalization of the nanotube surface), the benefits of adding carbon nanotubes to polymers include increased dimensional stability, conductivity, improved thermal properties (T_g and flame resistance), improved mechanical properties (strength and stiffness), and significantly reduced thermal expansion coefficients [19] even at relatively low loading levels. Figure 6 below shows TEM micrographs of a single multiwalled carbon nanotube and the nanotubes dispersed in a thermosetting polymer matrix used by our research group.

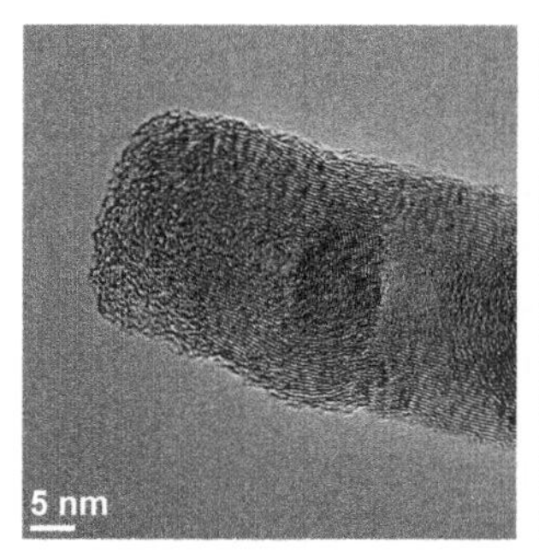

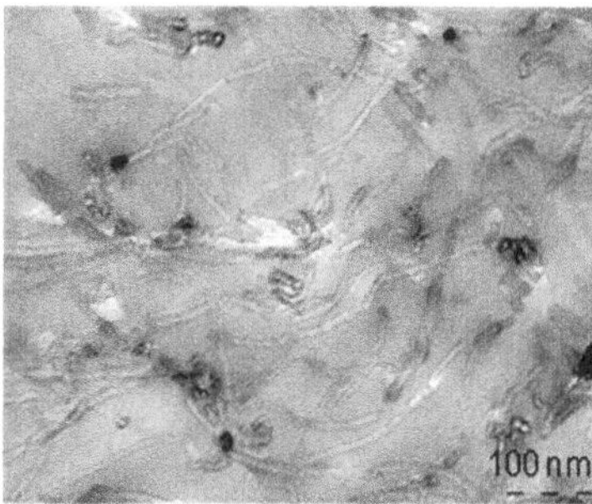

Fig. 6 TEM images of (left) an individual multiwalled carbon nanotubes and (right) carbon nanotubes dispersed in a thermosetting polymer matrix (images courtesy of Wonje Jeong)

Carbon nanofibers (CNFs) are similar to carbon nanotubes (CNT), but can be produced at a lower cost. CNFs with diameters ranging from 50 to 200 nm are larger than CNTs but smaller than continuous carbon fiber.

3 Experimental

The remainder of the paper will focus on our results in modifying a filler adhesive for pipeline repair applications with nanoscale fumed silica. The adhesive was a bisphenol E cyanate ester (BECy), EX-1510 from Bryte Technologies (Morgan Hill, CA) cured with a polymerization catalyst (EX-1510-B) at the manufacturer's suggested loading of 3 phr (parts per hundred resin). Hydrophilic fumed silica, Aerosil 200 (referred to as 12 nm) and Aerosil OX 50 (referred to as 40 nm), was supplied by Degussa (Frankfurt, Germany) shown previously in Figure 4.

Cyanate ester/silica suspensions were prepared by slowly adding the fumed silica during mixing of the filler putty pre-polymer with a high-shear mixer and further mixed briefly with a sonic dismembrator (3.2 mm diameter probe tip, frequency of 23 kHz, power ranged between 16 and 18 W during sonication). Figure 7 shows the difference in dispersion in cured composites with and without ultrasonic mixing.

Portions of the mixed resins were analyzed using parallel plate oscillatory rheology. After resin was mixed, it was degassed at 60 °C for 1 h under vacuum and then placed in a convection oven for the final curing process (180 °C for 2 h, 250 for 2 h, ramp of 1 °C/min between each step). Samples were machined from the solid block of material for dynamic mechanical analysis and thermomechanical analysis. Characterization equipment included a Q400 thermomechanical analyzer, a Q800 dynamic mechanical analyzer with gas cooling accessory, and a AR 2000ex controlled stress rheometer with environmental test chamber all from TA Instruments (New Castle, Deleware).

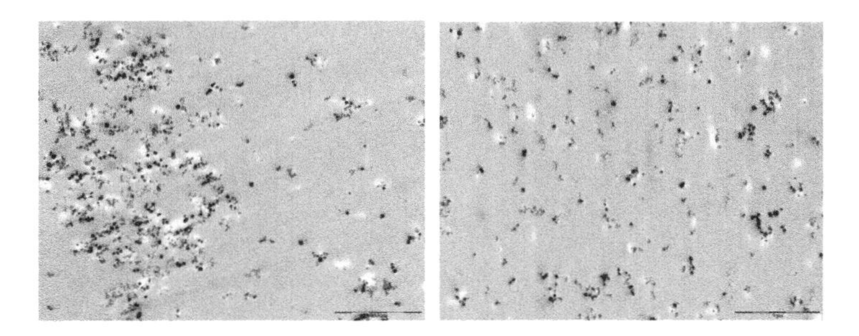

Fig. 7 TEM images showing the effect of sonication on dispersion for 5 phr 40 nm silica in cured cyanate ester resin (left) no sonication performed before curing the composite and (right) sonication at approximately 17 Watts for 75 s before cure. Scale bar is 2 microns

4 Results and Discussion

Figure 8 shows the intense shear thickening and pseudoplasticity in the suspension with increasing filler content. It also shows that the system with the smaller primary particles (12 nm) has a greater level of thixotropy, likely due to increased surface-surface interactions from hydrogen bonding and subsequent agglomeration.

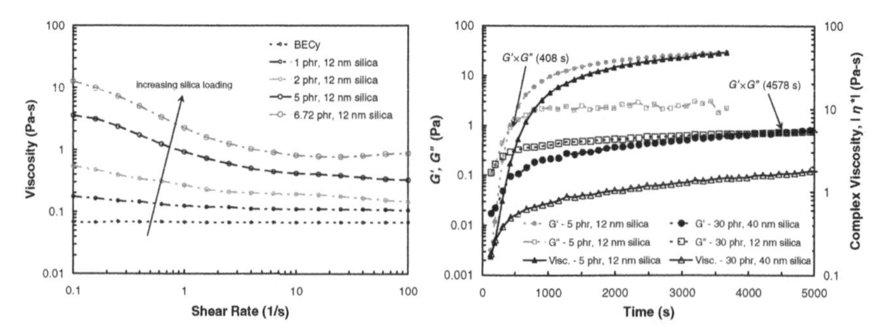

Fig. 8 Rheological behavior of the nanosilica/BECy suspension prior to curing showing, (left) the viscosity versus shear rate for the 12 nm silica suspension, (right) the time-dependent viscosity (thixotropy) in both the 12 nm and 40 nm suspension (results adapted from ref. [20])

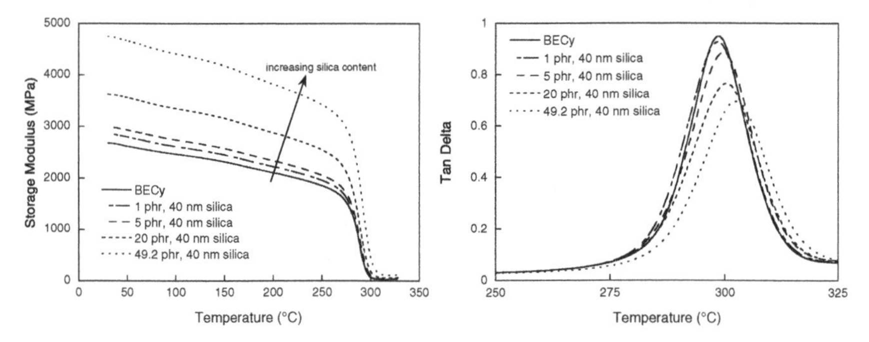

Fig. 9 Dynamic mechanical behavior of nanocomposites with 40 nm fumed silica (results adapted from ref. [21])

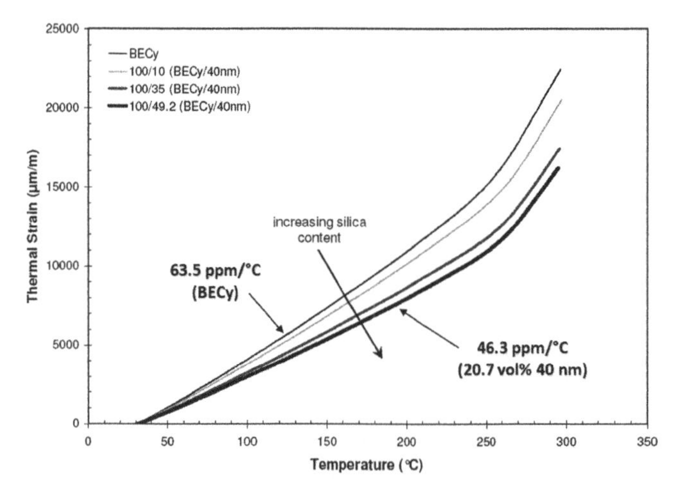

Fig. 10 Thermomechanical analysis (TMA) data showing the reduction in thermal expansion with increasing silica loading (results adapted from ref. [22]).

Figure 9 shows the increase in the glassy and rubbery moduli with increased loading of the 40 nm nano-silica and the decrease in damping with silica loading. The results indicate that the incorporation of fumed silica has a pronounced effect on the modulus of the nanocomposites. The decrease in damping was used in Ref. [21] to estimate the interphase thickness in the polymer nanocomposite. The coefficient of thermal expansion for the nanocomposites decreased from 63.5 ppm/K for the neat resin to 46.3 ppm/K when 20.7 vol% nanosilica (40 nm) was incorporated into the resin (see Figure 10).

5 Conclusions

Polymer nanocomposites are beginning to be used in civil infrastructure rehabilitation, specifically in the dimensional restoration filler (filler adhesive) used in the repair of damaged pipelines and pipework. The addition of nano-scale fumed silica increased the thixotropic behavior of the prepolymer (reducing unwanted sag in the resin), decreased the compliance (inverse of the stiffnes) and thermal expansion of the nanocomposites.

References

1. Mableson, R., Patrick, C., Dodds, N., Gibson, G.: Refurbishment of steel tubulars using composite materials. Plastics, Rubbers, and Composites 29(10) (2000)
2. Cuthill, J.: Advances in materials, methods, help gain new users. Pipeline & Gas Journal 229(11), 64–66 (2002)
3. Meier, U.: Composite Materials in bridge repair. Appl. Compos. Mater. 7, 75–94 (2000)
4. Radford, D.W., et al.: Composite repair of timber structures. Constr. Build. Mater. 16(7), 417–425 (2002)
5. Product Technical Data, CAB-O-SIL® M-5, Cabot Corporation, Billerica, MA, USA (2000)
6. Miller, D.G.: Improving rheology control of epoxy hardeners. Adhes Age 29, 37–40 (1986)
7. Kang, S., et al.: Preparation and characterization of epoxy composites filled with functionalized nanosilica particles obtained via sol-gel process. Polymer 42(3), 879–887 (2001)
8. Preghenella, M., Pegoretti, A., Migliaresi, C.: Thermo-mechanical characterization of fumed silica-epoxy nanocomposites. Polymer 46(26), 12065–12072 (2005)
9. Jana, S.C., Jain, S.: Dispersion of nanofillers in high performance polymers using reactive solvents as processing aids. Polymer 42(16), 6897–6905 (2001)
10. Wichmann, M.H.G., Cascione, M., Fiedler, B., Quaresimin, M., Schulte, K.: Influence of particle surface treatment on mechanical behavior of fumed silica/epoxy resin nanocomposites. Compos. Interface 13(8-9), 699–715 (2006)
11. Torro-Palau, A.M., Fernandez-Garcia, J.C., Orgiles-Barcelo, A.C., Martin-Martinez, J.M.: Characterization of polyurethanes containing different silicas. Int. J. Adhes Adhes 21, 1–9 (2001)

12. Zhou, S., Wu, L., Shen, W., Gu, G.: Study on the morphology and tribological properties of acrylic based polyurethane/fumed silica composite coatings. J. Mater. Sci. 39, 1593–1600 (2004)
13. Lippe, R.J.: Thixotropy recovery as a measure of dag in Polyester/Silica systems. Mod. Plast. 54, 62–65 (1977)
14. Zeng, C., et al.: Structure of nanocomposite foams. In: ANTEC 2002 Conference Proceedings. Society of Plastics Engineers, Brookfield CT (2002)
15. Iijima, S.: Helical microtubules of graphitic carbon. Nature 354, 56 (1991)
16. Iijima, S., Ichihashi, T.: Single-shell carbon nanotubes of 1-nm diameter. Nature 363, 603 (1993)
17. Bethune, D.S., et al.: Cobalt-catalyzed growth of carbon nanotubes with single-atomic-layerwalls. Nature 363, 605 (1993)
18. Lusti, H.R., Gusev, A.A.: Finite element predictions for the thermoelastic properties of nanotube reinforced polymers. Model Simul. Mater. Sc. 12, S107–S119 (2004)
19. Breuer, O., Sundararaj, U.: Big returns from small fibers: A review of poly-mer/carbon nanotube composites. Polym. Composite 25(6), 630–645 (2004)
20. Goertzen, W.K., Sheng, X., Akinc, M., Kessler, M.R.: Rheology and curing kinetics of fumed silica/cyanate ester nanocomposites. Polym. Eng. Sc. 48, 875–883 (2008)
21. Goertzen, W.K., Kessler, M.R.: Dynamic mechanical analysis of fumed sil-ica/cyanate ester nanocomposites. Compos. Part A-Appl. S 39, 761–768 (2008)
22. Goertzen, W.K., Kessler, M.R.: Thermal expansion of fumed Silica/Cyanate Ester nanocomposites. J. Appl. Polym. Sci. 109, 647–653 (2008)

Nanotechnology Divides: Development Indicators and Thai Construction Industry

T. Kitisriworaphan and Y. Sawangdee

Abstract. Nanotechnology and disparity between developed and developing nations could increase the gap of global development while it also affects to construction industry where workers have potentially exposed to nanomaterials application. This research examined the influence of development indicators as demographic, social and economic factors on nanotechnology policy among 250 nations. Results revealed that 68.2% of developed countries have policy on nanotechnology while only 18% of developing countries have such a policy. Fertility and mortality declining with the increasing of literacy, urbanization and energy consumption provide significant positive effect on nanotechnology divides. Furthermore, results pointed out the existing gap of development between developed and developing worlds.

1 Introduction

Majority of world population is still in developing countries where are considered as low quality areas due to the people are facing basic needs scarcity like improper infrastructure and unhealthy condition. The Millennium Development Goals (MDGs), with the agreement of world leaders, would like to promote environmental sustainability (as MDGs 7) while a phenomenon of urbanization booming as well as increasing of urban poverty in many dimensions including the living place becomes the most considerable issue among developed and developing nations. Urbanization also causes the urban poverty increasing due to the poor who can not afford the basic infrastructures and utilities in the city. Not only urban poverty becomes more serious but also high energy consumption is concentrated

T. Kitisriworaphan
Institute for Population and Social Research, Mahidol University
e-mail: nanosoctk@gmail.com
http://www.ipsr.mahidol.ac.th

Y. Sawangdee
Institute for Population and Social Research, Mahidol University
e-mail: prysw@mahidol.ac.th
http://www.ipsr.mahidol.ac.th

in the place where urbanization spread through and this also creates disparity of energy consumption between the rich and poor in the urban. This gap forces unsustainable city growth and push difficulty for the development goal (MDGs 7) to succeed especially in developing countries where this gap clearly emerged. Urbanization and construction industry clearly relates to each other due to construction industry strongly supports urban process through infrastructure development. According to Salamanca-Buentello et al. (2005) mentions that nanotechnology could help MDGs achievement especially for construction development. However, there is a doubt about the difference of demographic and socio-economic backgrounds of each nation on how nanotechnologies can contribute the development equity especially in construction sector where a gap between technology-based and labor-based intensities is strongly appeared in both developed and developing worlds [1].

A debate between different perspectives of potential risk and benefit from nanotechnology application is seriously discussed, for instance, the Joint Center for Bioethic at University of Toronto mentioned the benefits for socio-economic development while Erosion, Technology and Concentration (ETC group) in Winnipeg, pointed out that it will increase the divide between rich and poor countries [2, 3]. The United Nations Industrial Development Organization (UNIDO) and UNESCO also launch the international conference for this emerging dialogue. For construction industry, an expectation that nanotechnology will help to reduce CO_2 emission from cement composing process due to it is considered as a source of GHGs emission as about 3 percent of global generators of GHG (13,500 million ton) comes from the cement industry [4]. Besides new materials are expected to be more durable against coming severe natural disaster such as earthquake, flooding, landslide, or even promote environmental quality through air and water purification in the future.

However, unknown potential risk of small particles could generate health problems if unprotected policy and practice is ignored especially in the place where unskilled worker concentration like Thai construction industry. Thailand has also tried to reduce the cement products in order to combat global warming and it seems like nanomaterials are outstanding materials for this purpose. The European Commission launched a survey on Nanotechnology and Construction Industry 2006 which mentioned that nanomaterials such as Carbon nanotube, TiO_2 and Aerogel will arrive in the European construction industry within 10 years and their application will be mostly on building, bridge and road construction. However, for developing nation like Thailand, only new imported construction materials are possibly expected due to there is little R&D support for nanotechnological research beneficial for construction market. While East Asian country like UAE, a major target of Thai exported construction workers, is interested in nanotechnological application for their many construction projects. This changing could bring about new obstacle for construction workers who need to compete at international level if the low awareness of nanotechnology among them has still been.

According to the Asia Pacific Nano Forum 2004 on the societal impact of nanotechnology, most Asia Pacific countries have launched policy to support but still lack public awareness [5]. Many studies have been conducted around the world about the social concerns on nanotechnology development and there is still a lack of explanation on social perception of nanotechnology. Furthermore, most countries have low public and industry awareness on benefits and risk of nanomaterials [6, 7]. Normally, most scholars who have published their researches on the social dimensions of nanotechnology perception are scientists so they need social science knowledge to understand the influential factors influencing on public opinion and decision making on nanotechnology [8]. Some scholars also mention about possible risks of nanoparticles in ecosystems and human health [9, 10, 11]. In construction process, worker can directly contact with particles through skin and respiratory and the particles also release to nature by unwanted construction waste dumping that lets nanoparticles accumulating in food chain as later will cause poor human health.

2 Methodology

Studying the nanodivide situation at macro level requires the international indicators as it could help countries to compare each other about their concerns through nanopolicy application. Having nanopolicy in this study means a country having its nanotechnology policy or at least providing some evidences pointed that the country aims to generate nanotechnology policy soon, not only scientific but also social dimension. The study also employs development indicators such as demographic and socio-economic indicators which based on available Population Reference Bureau 2007 to explain the divide of nanotechnology among countries through their policy making with an assumption that the demographic and socio-economic indicators have significantly affected on having nanopolicy among nations. From content analysis, there are about 250 nations but only 195 countries are accepted as United Nations' members. According to Maclurcan (2005), he mentions that among global nanotechnology activities not classified by number of patents, but policy is also concerned [12].

Demographic indicators can indicate the national development on human capital and economic growth [13, 14]. Fertility is tied to labor supply especially for labor based intensive industry like construction. The high fertility means excess labor supply which was once considered as an advantage for economic development. For technology based industry like IT industry or even nanotechnology, it is considered to be more appropriate in low fertility areas due to huge number of labor is inessential. High mortality rate is normally existed in the place where poor public health system such as African countries due to epidemic and malnutrition problems. For this lack of basic needs, nanotechnology seems to be new accessible gap for developing world. Percent of population growth indicates the high population growth area relateing with slow economic development due to government needs to distribute basic equity in everywhere instead of focusing only economic development [15]. Higher dependency ratio means burden for socio-economic development due to labors have to take care of their dependent persons like children and elderly. Urbanization indicators like urban population and density can explain the development process especially for construction market. Higher life expectancy at birth explains the better public health technology and distribution. The high percent of urban population and density point out the high construction concentration activities such as material consumption for building, bridge and road.

The high in social indicators such as literacy rate and percent of contraceptive use indicates the high social development also. While the higher rate of under weight of less than five years children means poor public health system and low technology society may also exist in such area. For the developed countries where literacy rate is high, the technology absorbability can be faster than the developing countries where literacy rate is low. This phenomenon can create a new gap of development between two worlds due to high technology often requires high educated workers. This will become a burden for high technology development countries due to scientists have to produce high technology materials which can be applied comfortably by low skilled worker. Majority Thai construction workers are male laborers, and that is to say, about 68.5 percent of construction workers completed less than certificate level [16]. This low skilled worker can promote their position by experience, not education attainment. However if the construction sector employs high materials technology, it can affect the majority workers in Thai construction industry certainly.

Economic indicators such as Gross National Product (GNI, PPP) and carbon dioxide emission can indicate the better economic development while low amount of people living under a US dollar per day and percent of natural remained means the better economic growth. Construction industry can be monitored through some indicators like percent of urban population, fertility rate and natural remained due to construction process closely relates with urbanization.

3 Results and Discussion

To examine the nanopolicy activities among 250 nations, the study employed the data from many resources i.e. European Commission on Nanotechnology, National Nanotechnology Initiative, online articles on nanotechnology regulation and policy worldwide, etc. Results showed that most developed and developing countries already recognized the benefits of nanotechnology and establish their policy for working with this tiny technology as shown in table 1 about distribution of nanopolicy among countries.

From findings of nanopolicy distribution, they were rearranged into dummy variable as 0 = having no nanopolicy and 1 = having nanopolicy while other variables were controlled.

Table 1 Nanotechnology policy divides between developed and developing countries

Country	Have nanotechnology policy		Total
	Have	Do not have	
Developed country	68.2 (15)	31.8 (7)	100.0 (22)
Less Developed country	18.9 (43)	81.1 (185)	100.0 (228)
Total	76.8 (192)	23.2 (58)	100.0 (250)

After employed all demographic and socio-economic development indicators from 250 nations, study showed that some countries have no indicators provided on PRB database then the missing case was treated and finally the analysis was conducted as follows.

Step 1 to explore the relationship of demographic and socio-economic development indicators on nanopolicy dividing between developed and developing nations, the t-test statistic was employed and the mean difference between each demographic and socio-economic development indicators on nanopolicy were conducted.

Step 2 to examine the causal relationship of demographic and socio-economic indicators on nanopolicy dividing, the simple dummy dependent variable on regression (Linear Probability Model) was employed. Results indicated that almost all development indicators have significantly associated with nanopolicy variable, except death rate and population density as shown in table 2.

To provide better clear picture of relationship among each demographic and socio-economic development indicators on nanopolicy, the study also analyzed through the dummy variable on regression analysis as shown in table 3.

Table 2 Relationship between demographics, socio-economic indicators on nanotechnology policy divides

Domain	Have nanopolicy		Mean difference	t-test	Sig.	Total
	Have	Don't have				
Demographic development factors						
-Birth rate	14.21 (58)	26.73 (151)	12.52	7.85	.000	100(209)
-Death rate	8.62 (58)	9.42 (151)	0.80	1.06	.291	100(209)
-Growth rate (percent)	0.55 (58)	1.73 (151)	1.18	8.52	.000	100(209)
-Infant Mortality rate	13.32 (58)	44.92 (149)	31.60	6.06	.000	100(207)
-Maternal Mortality Ratio	66.78(55)	448.58 (113)	381.80	5.88	.000	100(168)
-Total Fertility Rate	1.86(58)	3.51(150)	1.65	7.28	.000	100(208)

Table 2 (*continued*)

Domain	Have nanopolicy		Mean difference	t-test	Sig.	Total
	Have	Don't have				
-Child dependency ratio	0.32 (58)	0.58 (150)	0.26	8.24	.000	100(208)
-Elderly dependency ratio	0.17 (58)	0.09 (150)	-0.08	-8.20	.000	100(208)
-Life Expectancy at birth (all)	74.38 (58)	64.20(149)	-10.118	-5.724	.000	100(207)
-Urban population (percent)	67.28 (58)	50.58(151)	-16.69	-4.56	.000	100(209)
-Population density (sq.m.)	351.53(58)	469.88(188)	118.34	0.31	.758	100(246)
Social development factors						
-Literacy rate of population age 15-24, female	90.46 (37)	58.63 (141)	-31.83	-4.52	.000	100(178)
-Literacy rate of population age 15-24, male	91.11 (37)	63.33 (141)	-27.78	-3.97	.000	100(178)
-Percent of Contraceptive use among married women (modern)	55.23 (48)	30.15 (112)	-25.08	-7.46	.000	100(160)
-Under weight of under 5 yrs child (percent)	8.96 (27)	17.35(106)	8.39	3.27	.001	100(133)
Economic development factors						
-Gross National Index PPP	20,585.82 (55)	8,347.62 (127)	-12,238.2	-6.41	.000	100(182)
-Population live under $US 1 a day	2.51 (58)	9.39 (151)	6.875	2.91	.050	100(209)
-Carbon dioxide emission (metric ton per capita)	7.25 (55)	3.27 (128)	-3.98	-4.64	.000	100(183)
-Natural remain (percent)	57.17 (54)	69.29 (125)	12.12	2.77	.050	100(179)

Table 3 Influential of demographics, socio-economic indicators on nanotechnology policy divides

Domain	Constant		Wald	Sig.	Model X^2	Sig. (Model X^2)	N
Demographic development factors							
-Birth rate	2.09	-0.16	34.31	0.000	63.52	0.000	209
-Death rate	-0.64	-0.04	1.11	0.291	1.167	0.280	209
-Growth rate (percent)	0.65	-1.47	40.86	0.000	62.33	0.000	209
-Infant Mortality Rate	0.23	-0.05	21.46	0.000	45.61	0.000	207
-Maternal Mortality Ratio	0.39	-0.01	15.78	0.000	51.42	0.000	168
-Total Fertility Rate	2.06	-1.23	29.58	0.000	60.26	0.000	208
-Child dependency ratio	2.31	-7.61	36.55	0.000	64.49	0.000	208
-Elderly dependency ratio	-3.20	16.40	39.69	0.000	50.28	0.000	208
-Life Expectancy at birth (all)	-10.16	0.13	23.62	0.000	41.43	0.000	207
-Urban population (percent)	-2.71	0.03	17.57	0.000	19.89	0.000	209
-Population density (sq.m.)	-1.17	0.00	0.09	0.760	0.11	0.744	246
Social development indicator							
-Literacy rate of population age 15-24, female	-3.94	0.03	12.64	0.000	23.99	0.000	178
-Literacy rate of population age 15-24, male	-3.95	0.03	9.58	0.002	19.59	0.000	178
-Percent of contraceptive use (modern)	-3.52	0.06	32.05	0.000	46.03	0.000	160
-Under weight of under 5 yrs child (percent)	-0.42	-0.08	8.82	0.003	11.96	0.001	133
Economic development factors							
-GNI_PPP	-1.84	0.00	26.47	0.000	33.19	0.000	182
-Population live under $US 1 per day	-0.71	-0.05	6.60	0.010	11.01	0.001	209
-Carbon dioxide emission (metric ton per capita)	-1.48	0.13	15.29	0.000	18.87	0.000	183
-Natural remain (percent)	0.18	-0.02	7.19	0.007	7.32	0.007	179

Findings revealed that most development indicators have significantly affected to nanopolicy. When the demographic development indicators were considered, finding revealed that most fertility indicators have strong affecting. Birth rate and growth rate indicated that countries having low fertility are more likely to have nanopolicy at statistical significant level .001. Aging societies are more likely to have nanopolicy at statistical significant level .001 as well as percent of urban indicated that countries with more urbanized are more likely to have nanopolicy at statistical significant level .001. However, the death rate and population density do not have casual relation with nanopolicy among nations. Overall, it can be said that low fertility nations are more likely to focus on nanotechnology regulation and policy while nanotechnology has been promoted in most developed countries at present.

Consideration of social aspect also revealed that countries having more social development are more likely to have nanopolicy. Literacy indicator for both sex and percent of contraceptive use among married women (modern methods) pointed out the modern society which positive literacy rate countries (better education) are more likely to have nanopolicy at statistical significant level .001 while nations where better infant health are more likely to have nanopolicy at statistical significant level .010 also. These findings confirm the influence of social indicators on national development [17].

For economic indicators also confirmed the same direction that more economic development areas are more likely to have nanopolicy. Furthermore, these findings strongly confirmed the existing gap of nanotechnology regulation and policy application between developed and developing countries. The finding is in accordance with a study of Schummer (2005) which showed that nanotechnologies can simultaneously and unavoidably generate the disparity gap between the rich and the poor.

Among finding indicators, the study found that increasing of urban population could lead to technology consideration like nanotechnology as well as lower fertility and percent of natural remained. For labor intensive sector like Thai construction industry where depends on the labor supply. Majority of construction workers are labors who also low education (primary school) [19]. The adoption of nanotechnology could not be easier to the firm but it could be possible by comparing with the Computer Aid Drafting (CAD) technology boom in 2 decades ago. Besides, literacy rate can increase the technology application though policy formation. To promote nanotechnology in construction sector, the awareness among construction workers on nanotechnology application must not be focused only on skilled labor but also unskilled one.

4 Conclusion

The study strongly states that "Nanotechnology could increase the technology importing among developing countries, not equal opportunity due to high technology certainly needs high skill workers" In order to apply high technology, only imported technology is possible for developing countries [18]. Finally, the new technology can be a new burden for workers, especially the construction workers in developing world like Thailand.

References

1. Salamanca-Buentello, F., Persad, D.L., Court, E.B., Martin, D.K., Daar, A.S., Singer, P.A.: Nanotechnology and the Developing World. PLoS Medicine (2005)
2. Compañó, R., Hullmann, A.: Forecasting the development of nanotechnology with the help of science and technology indicators. Nanotechnology 13, 243–247 (2002)
3. Schummer, J.: The impact of nanotechnologies on developing countries. In: Allhoff, F., Lin, P., Weckert, J. (eds.) Nanoethics: The ethical and social implications of nanotechnology, Hoboken, NJ, pp. 291–307. Wiley, Chichester (2007)
4. Mann, S.: Nanotechnology and Construction. Institute of Nanotechnology. Nanoforum Consortium (2006)
5. Asia Pacific Nano Forum, Societal Impact of Nanotechnology in the Asia Pacific Region. Asia Pacific Nanotech Weekly 2,47. Nanotechnology Research Institute (2004)
6. Sheetz, T., Vidal, J., Pearson, T.D., Lozano, K.: Nanotechnology: Awareness and societal concerns. Technology in Society, 329–354 (2005)
7. Scheufele, D.A.: Scientists worry about nanotechnology's health and environmental impacts. The Center for Nanotechnology in Society, Arizona State University (2008)
8. Fujita, Y.: Perception of nanotechnology among general public in Japan. Asian Pacific Nanotech Weekly 4, 6. Nanotechnology Research Institute (2006)
9. Boccuni, F., Rondinone, B., Petyx, C., Iavicoli, S.: Potential occupational exposure to manufactured nanoparticles in Italy. J. Clean Prod. 16, 949–956 (2008)
10. Cheng, M.D.: Effects of nanophase materials ($\leq$ 20 nm) on biological responses. Journal of Environmental Science and Health A39, 2691–2705 (2004)
11. Monteiro-Riviere, N.A., Orsière, T.: Toxicological Impacts of Nanomaterials. In: Wiesner, M.R., Bottero, J. (eds.) Environmental Nanotechnology: Applications and Impacts of Nanomaterials, pp. 395–434. The McGrow-Hill Companies (2008)
12. Maclurcan, D.C.: Nanotechnology and developing countries part 2: what realities. AZojono Journal of Nanotechnology Online 1, 1–19 (2005)
13. Bloom, D.E., Williamson, J.G.: Demographic transition and economic miracles in emerging Asia. The World Bank Economic Review 12(3), 419–455 (1998)
14. Drèze, J., Murthi, M.: Fertility, education, and development: Evidence from India population and development. Review 27(1), 33–63 (2004)
15. Preston, S.H.: The changing relation between mortality and level of economic development. Int. J. Epidemiol. 36, 484–490 (2007)
16. NSO, Labor force survey round 1-3. National Statistical Office. Ministry of Interior (2006)
17. Andrews, F.M.: Population issues and social indicators of well-being. Population and Environment 6, 210–230 (1983)
18. Mayer, J.: Technology diffusion, human capital and economic growth in developing countries. In: United Nations Conference on Trade and Development (UNCTAD) (2001)
19. NSO, Labor force survey round 1-3. National Statistical Office. Ministry of Interior (2007)

Improvement of Cementitious Binders by Multi-Walled Carbon Nanotubes

T. Kowald and R. Trettin

Abstract. To improve the mechanical properties of building materials carbon nanotubes (CNTs) were incorporated into normal concretes, high and ultra-high performance concretes as well as into model systems. Besides their outstanding mechanical properties CNTs are a very promising reinforcement material for composites because of their high aspect ratio, high resistance to corrosion and low specific weight. Multi-walled carbon nanotubes (MWCNTs) showed an influence on the hydration of binders and their macro- and microscopic mechanical properties. The effects of the nanostructures on the hydration, the composition and the micro- and nanostructure of the composite were investigated by the use of the in situ x-ray powder diffraction, isothermal calorimetric measurements, an ultrasonic method, imaging methods and the grid nanoindentation technique. The results show that the MWCNTs influence the hydration, the microstructure and are leading to improved mechanical properties of the composite. This presentation shows thermal analysis and porosimetry data of samples prepared with MWCNTs.

1 Introduction

The brittle nature of high performance concrete (HPC) and ultra-high performance concrete (UHPC) is a weakness of modern building materials. To overcome this problem and to improve certain characteristics of the materials fibres are very often used and are mainly utilized to improve the flexural strength, the ductility or the fire resistance.

T. Kowald
Institute for Building & Materials Chemistry, University of Siegen
e-mail: kowald@chemie.uni-siegen.de
http://www.uni-siegen.de/fb8/bwc/

R. Trettin
Institute for Building & Materials Chemistry, University of Siegen
e-mail: trettin@chemie.uni-siegen.de
http://www.uni-siegen.de/fb8/bwc/

Since carbon nanotubes (CNTs) were discovered in 1991 [1] the nanostructures are well known for their outstanding mechanical properties. Additionally certain other characteristics of the CNTs make them a very promising reinforcement material for composites. These are their high aspect ratio, high resistance to corrosion and low specific weight which is only 1/6 that of steel. When CNTs were incorporated into different binder systems like normal concretes, high and ultra-high performance concretes as well as into model systems they led to improved mechanical properties. The known effects of the CNTs on the building materials are summarized below. Prior to a successful application of the CNTs some key issues have to be addressed.

- The first point is the strong agglomeration of the tubes due to van der Waals forces between the nanoparticles. So the CNT-bundles have to be dispersed prior to their application for a good separation and a homogeneous distribution of the individual tubes within the composite.
- The second point is the linkage between the CNTs and the binder matrix. This is a key factor for the improvement of the resulting material.

The dispersion of the CNTs and their linkage to the binder matrix has to be optimized to advance the mechanical properties of the nanocomposites. Further a deeper understanding of how the CNTs interact with the binder is needed to improve the performance of the resulting composite. In plain cement pastes multi-walled carbon nanotubes (MWCNTs) as well as single-walled carbon nanotubes (SWCNTs) were used and led to an increase in the compressive strength by 30% and 6%, respectively [2]. Considering the correct handling of the CNTs in cement-based binders their application may lead to an improvement of the mechanical properties [3-5]. Using tricalcium silicate (C_3S) as a model system for cementitious binders and 0.5 ma.% of different carbon nanostructures an improvement of up to 45% in the flexural strength of a sample could be achieved [5].

Additionally to the improved mechanical characteristics systems with CNTs also showed a different hydration behavior and a changed microstructure. The observations on the hydration were done by a complementary usage of x-ray powder diffraction (XRPD) and isothermal calorimetric experiments. The results showed a clear influence on the reaction kinetics particularly regarding the crystallization of the portlandite ($Ca(OH)_2$) and point to an accelerating effect with a crystallization of fewer and smaller portlandite crystals [5]. The formation of finer reaction products when using CNTs could also be determined by investigating the microstructure by scanning electron microscopy. A chemical interaction of the COOH groups present on the surface of the CNTs with the hydrating binder could be illustrated by spectroscopic investigations by Li et al. [6]. Considering the experimental data it seems that the functional groups act as crystallization seeds resulting in a chemical link between the CNTs and the reaction products resulting in an increased number of finer reaction products.

The analysis of the distribution of the elastic modulus and the indentation hardness from grid nanoindentation experiments showed that the MWCNTs incorporated to a C_3S model system seem to lead to an increased quantity of C-S-H with higher E-modulus and hardness [7].

2 Experimental Part

For the thermal analysis prisms of pure tricalcium silicate (C_3S) and with MWCNTs were prepared to minimize the complexity of the binder system. The mercury intrusion porosimetry was done on samples consisting of very fine cement and precipitated silica.

For each experiment two types of samples were prepared. The reference samples were made of C_3S respectively cement, superplasticizer and water. The MWCNTs samples also contained 1.0 ma.% MWCNTs referred to the binder.

In case of the prisms prepared for the thermal analysis pure triclinic C_3S had been synthesized from a stoichiometrical mixture of calcium carbonate and silicon dioxide at a temperature of 1450°C. The purity of the C_3S was controlled by XRPD and when the free lime content determined by the Rietveld method was below 0.5 ma.% it was controlled by the Franke method, additionally. A final free lime content of 0.18 ma.% was reached, the specific density was 3.12 g/cm³ and the d_{50}-value of the ground C_3S was 4 µm as determined by lasergranulometry.

MWCNTs purchased from Sun Nanotech Co. Ltd. were used for the experiments and had the following specifications: purity > 80 %, free amorphous content < 10 %, diameter 10 – 30 nm and length 1 – 10 µm. The dispersion of the MWCNTs was done by sonification within the mixing water for 28 minutes. Afterwards a polycarboxylate-based superplasticizer (1 ma.% by binder content) was added and the dispersion sonificated for 2 minutes, additionally. A water to binder ratio of 0.22 had been used.

The pastes were moulded into prism-shaped forms and compacted by applying a pressure of 125 N/mm² for 30 minutes. The dimension of the prisms were 8 mm · 8 mm · 30 mm. The samples were demoulded after 2 days of storage at a relative humidity of >90 % and 20°C. Then the prisms were cured 5 d storing them under water at 20 °C.

The samples prepared for the mercury intrusion porosimetry consisted of a CEM II / B-S 52.5 R (d_{50} = 4.4 µm), 1 ma.% precipitated silica, 1 ma.% superplasticizer and a water to cement ratio of 0.22. The dispersions of the MWCNTs were prepared like it was done in the case of the C_3S samples. The pastes were cast into prism forms with a dimension of 15 mm · 15 mm · 60 mm. The prisms were taken out of the molds after one day and put back into the climate-chamber for 27 days.

Thermogravimetry analysis (TGA) were conducted to quantitatively estimate the $Ca(OH)_2$ using a STA 449 C Jupiter by Netzsch. The TG and the DSC data were stored simultaneously. As purge gas nitrogen and as reference for the DSC Al_2O_3 were used. The measurement conditions were as follows: starting temperature 38°C, heating rate 10 K/min, max. temperature 1000°C, purge gas N_2, reference material Al_2O_3 and crucible material Pt.

The mercury intrusion porosimetry (MIP) was used for assessment of porosity and pore size distribution of the specimens. Prior to the measurement with an Autopore 9220 (Micrometrics Instruments Corp.) the samples were broken to dimensions between 2 mm and 4 mm and dried for one week at 35°C.

3 Results

In this chapter the results of the TGA and the MIP are shown. By x-ray powder diffraction a lower amount of the crystalline $Ca(OH)_2$ was found when MWCNTs were used in a C_3S model system. The TGA should give information if there is also an influence on the total amount of $Ca(OH)_2$ produced when MWCNTs are incorporated into the samples. Additionally the effect of the MWCNTs on the pore size distribution of samples consisting of finely ground cement, precipitated silica and superplasticizer were analyzed by MIP.

In Fig. 1 the TG and DSC data for the C_3S samples with 1 mass.% MWCNTs and without MCNTs are shown. The straight lines are showing the TG results and the dashed ones are showing the DSC results. The data for the C_3S sample is drawn in red and the data for the C_3S sample with the MWCNTs in black. The decomposition of the $Ca(OH)_2$ is typically found between 450 and 550 °C. Looking at the TG and DSC graphs in Fig. 1 the influence of the decomposition ($Ca(OH)_2 \rightarrow CaO + H_2O$) can be seen.

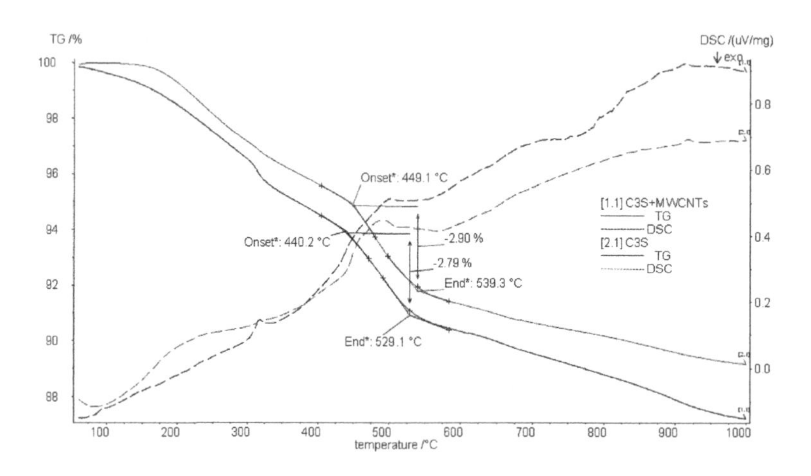

Fig. 1 Combined TG and DSC results for the C_3S samples with 1 mass.% MWCNTs and without MCNTs

Comparing the TG of the two samples it can be seen that the total amount of $Ca(OH)_2$ is nearly the same. In the case of C_3S the loss of water from the decomposition was 2.9% and in the case of C_3S+MWCNTs 2.79%. Having in mind that the x-ray powder diffraction showed a lower amount of crystalline $Ca(OH)_2$ when the MWCNTs were used but the TG showed nearly the same total amount of $Ca(OH)_2$ the conclusion is that there is an amount of x-ray amorphous $Ca(OH)_2$ in the samples with MWCNTs. Another indicator for the disordered or smaller $Ca(OH)_2$ crystals is the shift of the onset of the C_3S+MWCNTs to lower temperatures which is 440°C instead of 450°C for the C_3S. Comparing the TG at lower temperatures it can also been seen that there is higher amount of physically bound water in the C_3S+MWCNTs sample.

In Fig. 2 the pore size distributions for the cement samples without (CEM, red line) and with MWCNTs (CEM+MWCNTs, black line) are shown. In this work the pore size classification by Smolcyk (Table 1) was chosen for the interpretation of the MIP data.

Comparing the MIP data for the two samples it can be seen that in the region of the micro- or gel pores there was an increase when the MWCNTs were used and in the region of the meso- or capillary pores there was a slight decrease.

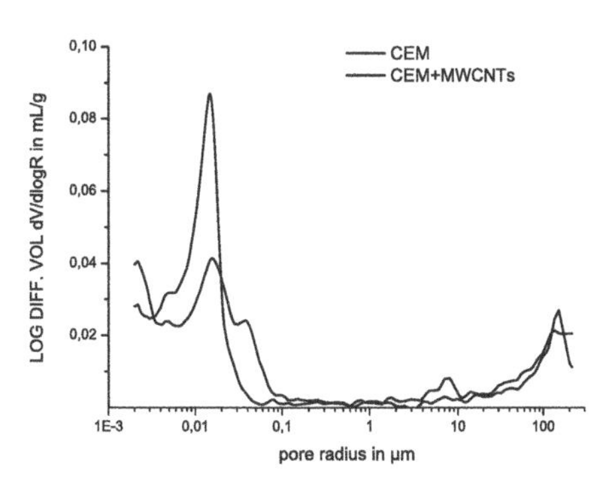

Fig. 2 Pore size distribution for the CEM samples with 1 mass.% MWCNTs and without MCNTs

Table 1 Pore sizes classifications

	Smolcyk	Setzer	IUPAC
Macropores or air pores	> 10 µm	50 µm – 2 mm	50 nm – 2 µm
Mesopores or capillary pores	0.03 – 10 µm	2 – 50 µm	2 nm – 50 nm
Micropores or gel pores	< 0.03 µm	< 2 µm	< 2 nm

4 Summary and Conclusions

CNTs improve the mechanical properties of building materials [2-5] but additionally the CNTs seem to influence the hydration [5, 6]. XRPD, the isothermal calorimetry and the nanoindentation method were used to study the influence of MWCNTs on the hydration of C_3S and the micromechanical properties as well as the surface fractions of the hydration products. Like already shown in [5] and [6] the MWNTs have an influence on the hydration and the resulting microstructure of the binder. The analysis of the distribution of the elastic modulus and the indentation hardness showed that the MWCNTs seem to lead to a higher quantity of C-S-H with higher E-modulus and hardness [7].

The results in this work show that the MWNTs are not changing the total amount of $Ca(OH)_2$ built during hydration but seem to have an influence on their crystallinity. Also a higher amount of micro- and gel pores have been found when MWCNTs were used.

Acknowledgments. The authors like to thank the DFG for financially supporting this work.

References

1. Iijima, S.: Helical microtubules of graphitic carbon. Nature 354, 56–58 (1991)
2. Campillo, I., Dolado, J.S., Porro, A.: High-performance nanostructured materials for construction. In: Proceedings of the 1st International Symposium on Nanotechnology in Construction, Paisley, pp. 110–121 (2003)
3. Kowald, T., Trettin, R.: Kohlenstoffbasierte nanostrukturen in modernen anorganischen Bindemitteln. In: Tagung Bauchemie, GDCh-Fachgruppe Bauchemie, pp. 162–165 (2004)
4. Kowald, T., Trettin, R.H.F.: Influence of surface-modified carbon nanotubes on Ultra-High Performance Concrete. In: Proceedings of the International Symposium on Ultra High Performance Concrete. Kassel university press GmbH, pp. 195–202 (2004)
5. Jiang, X., Kowald, T.L., Staedler, T., Trettin, R.H.F.: Carbon nanotubes as a new reinforcement material for modern cement-based materials. In: Proceedings of the 2nd International Symposium on Nanotechnology in Construction. RILEM Publications s.a.r.l. 26 (2005)
6. Li, G.Y., Wang, P.M., Zhao, X.: Mechanical behaviour and microstructure of cement composites incorporating surface-treated multi-walled carbon nanotubes. Carbon 43, 1239–1245 (2005)
7. Kowald, T., Trettin, R.: Improvement of modern building materials by carbon nanotubes. In: Proceedings of 8th International Symposium on Utilization of High-Strength and High-Performance Concrete (2008)

Effect of Nano-sized Titanium Dioxide on Early Age Hydration of Portland Cement

A.R. Jayapalan, B.Y. Lee, and K.E. Kurtis

Abstract. The effect of nano-scale non-reactive anatase titanium dioxide (TiO_2) on early age hydration of cement was experimentally studied. Isothermal calorimetry was performed on cement pastes with two different particle sizes of TiO_2 at replacement levels of 5, 7.5 and 10%. The addition of TiO_2 to cement increased the heat of hydration and accelerated the rate of reaction at early stages of hydration. This increase was found to be proportional to the percentage addition and the fineness of TiO_2. These results demonstrate that the addition of non-reactive nano-scale fillers could affect the rate of cement hydration by heterogeneous nucleation.

1 Introduction

The early age hydration of Portland cement remains of interest to researchers and to industry because of potential implications relating to setting time, dimensional stability and strength development. Researchers have noted an acceleration of cement hydration when fine fillers are added to cement [1, 2]. It has been observed that fine (0.5μm to 4μm) powders of limestone [3], quartz [4], silica fume [5] and pulverized fly ash [6], when added up to 15% cement replacement levels, can increase the rate of the cement hydration.

Addition of a fine non-reactive filler to cement modifies the hydration rate primarily due to dilution, modification of particle size distribution and heterogeneous nucleation [4]. For increasing dosage rates of inert filler (when used as a partial

A.R. Jayapalan
Georgia Institute of Technology
e-mail: amalrajpj@gatech.edu

B.Y. Lee
Georgia Institute of Technology
e-mail: blee30@gatech.edu

K.E. Kurtis
Georgia Institute of Technology
e-mail: kimberly.kurtis@ce.gatech.edu

replacement of cement), cement dilution results in an increase of the water-to-cement ratio (w/c) and a decrease in total cement content when water-to-solids ratio (w/s) is kept constant. The modification of particle size distribution due to chemically inert filler addition changes the system porosity, but the effect on cement hydration does not seem to be well documented in literature. The surface of the fine fillers has been shown to provide sites for nucleation of cement hydration products (C-S-H) and catalyzes the reaction by reducing the energy barrier [4]. The effectiveness of this catalysis depends on fineness and dosage of the filler [4]. In addition, other phenomena may occur, including water ab/adsorption by the nanoparticles, interactions with nanoparticle surface treatments, and reaction of materials previously presumed to be inert.

Most of the fine fillers previously examined react chemically to some extent in the cement hydration process. Limestone may be slightly reactive with the Portland cement forming a monocarbonate phase [3]. Silica fume reacts with calcium hydroxide by pozzolanic reaction. Considering the increasing interest in inert additives to cement, such as titanium dioxide (TiO_2), there needs to be a study that directly focuses on the effect of non-reactive filler on cement hydration. Titanium dioxide (TiO_2) is added as a filler to cement for its photocatalytic activity. Research has shown that the photocatalytic activity is superior in nano-crystalline TiO_2 and that it exhibits maximum efficiency in anatase phase compared to rutile or brookite phase [7, 8]. When added to Portland cement, TiO_2 is considered to act as inert filler and has not been believed to take part in the hydraulic reaction of Portland cement.

The nano-size of the TiO_2 particles could significantly affect the rate of hydration reaction especially in the early stages. Most of the previous researches on the effect of fine inert fillers on cement hydration were conducted using micrometer sized particles in the range of 0.5μm to 4μm. Jo et al. used nano-particles (of average particle size 40nm) of silicon dioxide (SiO_2) [9], but SiO_2 could react with cement during early hydration due to its high pozzolanic activity.

Thus, the objective of this research was to better understand the effect of addition of nano-sized and presumably non-reactive anatase TiO_2, on the early hydration of Portland cement. Specifically, the effects of variation particle size and percentage addition of TiO_2 nanoparticles were examined as a part of this research. Isothermal calorimetry was used to characterize the influence of these factors on early age cement hydration.

2 Research Methodology

2.1 Materials

Six blended cements were prepared and examined to study the effect TiO_2 dosage rate and particle size on the cement hydration process. The potential Bogue composition of the ordinary Portland cement used for making all TiO_2–blended cement was 51.30% C_3S, 19.73% C_2S, 8.01% C_3A and 9.41% C_4AF and 0.40% Na_2O_{eq}. Two TiO_2 powders (T1 and T2), of different surface areas were lab-blended with ordinary Portland cement, each at 5, 7.5 and 10% replacement by mass. The properties of the TiO_2 powders, obtained from Millennium Inorganic Chemicals are given in Table1.

Table 1 Properties of anatase TiO_2 added to cement (as obtained from the manufacturer)

	Crystal Size (nm)	Agglomerate Size (μm)	Surface Area (m^2/g)	pH	Purity (%)
T1	20-30	1.5	45-55	3.5-5.5	>97
T2	15-25	1.2	75-95	3.5-5.5	>95

2.2 *Sample Preparation*

A w/s of 0.50 was used for all the mixes. The mixing tools and materials were stored at a constant temperature of 23°C for 24 hours before mixing. TiO_2 powder was added to the water and mixed using a handheld mixer for 60 seconds to disperse the agglomerates. The cement was then added, mixed by hand using a stirrer for a maximum of 10 seconds and mixing continued with the handheld mixer. The entire mixing period was maintained within 60±5 seconds. The paste was then carefully poured into calorimeter capsules and the weight of the cement paste in the capsule is noted.

2.3 *Methods*

Isothermal calorimetry was performed on triplicate paste samples using an eight channel micro-calorimeter at 25°C. The capsules were placed inside the isothermal calorimeter within 240 seconds of addition of cement to water. The time of addition of cement to water is considered as the starting time for each sample. The data for initial 10 minutes of the tests was discarded to ensure that capsule reached thermal equilibrium with the calorimeter. The rate of hydration was measured every 60 seconds as power (mW) and was normalized per gram of cement. Since the difference between the triplicate samples was not significant, one of the triplicate samples per mix was chosen for the comparative studies.

3 Results and Discussion

Isothermal calorimetry was carried out on the laboratory blended TiO_2-cements to study the effect of TiO_2 replacement level and particle size on the early hydration reaction. The data for the first 48 hours, starting from the time of mixing of cement and water, was analyzed.

3.1 Rate of Hydration of TiO_2-Blended Cements

Figure 1 shows the variation of the rate of hydration of ordinary Portland cement and the TiO_2-blended cements at 5, 7.5 and 10% replacement levels. For the graph of ordinary Portland cement, the main peak of heat release corresponding to the reaction of C_3S can be observed from ~1 to ~8 hours. This main peak is followed by a secondary peak corresponding to C_3A hydration.

From Figure 1 it can be observed that all the mixes with TiO_2 addition showed accelerated hydration compared to ordinary Portland cement. For example, at 10% replacement, the peak of C_3A hydration was accelerated by around 80 and 180 minutes for the mixes with T1 and T2 respectively compared to the ordinary Portland cement. The increasing dosage of TiO_2 was found to accelerate the rate of hydration of the mixes and increase the peaks for C_3S and C_3A hydration. For instance, the increase in the peak of the C_3A hydration compared to the control mix was found to be 22.24%, 28.32% and 37.20% for the mixes with 5%, 7.5% and 10% replacement by T2. These results indicate that heterogeneous nucleation effect could be more dominant than dilution effect. Previous research using calorimetry has also shown that heterogeneous nucleation increases the rate of cement hydration when fine inert materials are added to Portland cement [4].

From Figure 1 it can also be observed that the rate of hydration for the cement mixes prepared by replacement with finer TiO_2(T2) was higher than all the mixes

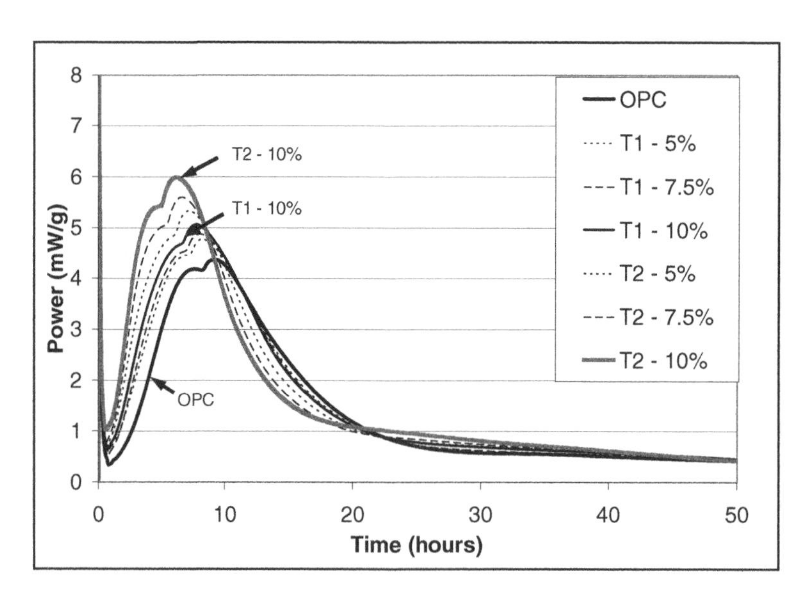

Fig. 1 Rate of hydration of TiO_2-blended cements

prepared with coarser TiO_2(T1). This shows that the rate of cement hydration in the presence of TiO_2 strongly depends on the size (Table1) of the nano-sized particles that are blended to the cement, with smaller particles accelerating the reaction much more than larger particles. This reinforces the conclusion that nucleation effect, which depends on the surface area of the particles, could be dominant than dilution effect in the case of addition of nano-TiO_2 particles to cement.

3.2 Cumulative Energy Released by TiO_2-Blended Cement

Figure 2 shows the variation of cumulative energy released with time of ordinary Portland cement and the lab-blended TiO_2 cements. For the graph of ordinary Portland cement it can be seen that after the initial dormant period there is a rapid increase in the total energy released, which corresponds to the peaks of C_3S and C_3A hydration.

From Figure 2 it can be seen that the total energy evolved during the hydration reaction increased as the percentage replacement of cement with TiO_2 was increased. The total energy released also increased with the increasing surface area of TiO_2 used, again reinforcing the significant effect of the addition of nano-sized particles in the nucleation reaction.

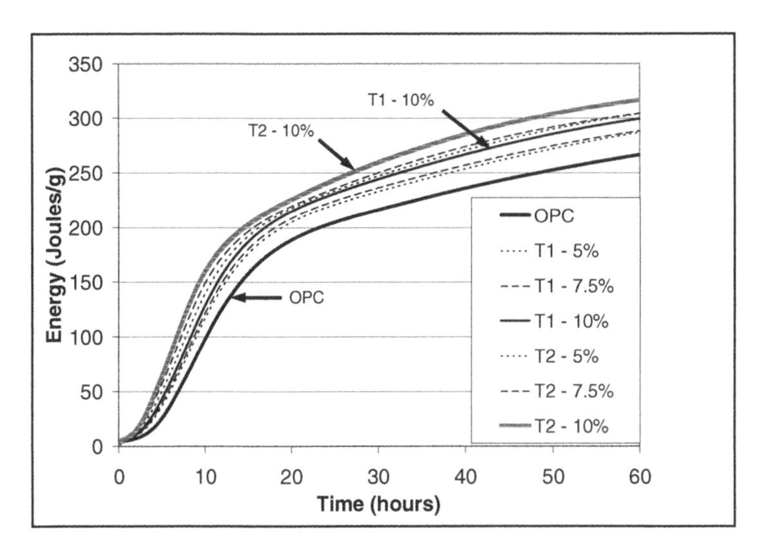

Fig. 2 Cumulative energy released by OPC and TiO_2-blended cements

As explained earlier, the addition of TiO_2 to cement paste affects the hydration rate by dilution effect and heterogeneous nucleation. With increased dosage of fine filler, the rate of hydration of cement and the total heat evolved will be decreased by the dilution effect and increased by nucleation effect. In all the tests that were conducted in this research it was observed that increasing the dosage and lowering of particle size of filler caused the rate of reaction to increase (Figure 1). Figure 2 shows that at all replacement rates by TiO_2 the cumulative heat evolved is greater than that of ordinary cement paste indicating an acceleration of cement hydration. This suggests that the heterogeneous nucleation effect is the dominant effect than dilution effect when nano-sized TiO_2 particles are added to cement.

It should be noted here that the TiO_2 used for these experiments were acidic, presumably due to surface treatments used in their manufacturing. The dissolution of surface groups (chlorides, sulphates or ammonium ions) could affect the rate of reaction of the cement particles which are in a highly basic environment during normal hydration. Research on the effect of the acidity and surface coating of TiO_2 particles on the rate of nucleation is needed. It is our understanding, however, that titania blended with portland cement in commercial use is not further processed to remove or alter the surface chemistry; that is, the TiO_2 examined herein should be quite similar to that used in practice.

4 Conclusions

The effect of the addition of nano-sized TiO_2 particles on the early hydration reaction of cement was studied as a part of this research. When TiO_2 of different particle sizes were added to cement, the hydration reaction was accelerated and the rate of hydration increased. The increase in the rate of reaction was proportional to the dosage of the TiO_2. Smaller particles of TiO_2 were found to accelerate the reaction more than larger particles. Heterogeneous nucleation effect was found to be dominant compared to the effect of dilution when inert TiO_2 particles were added to cement.

Acknowledgments. This research is supported by the National Science Foundation under Grant Nos. CMMI-0825373 and CMMI-0855034. Any opinions, findings, and conclusions or recommendations expressed in this material are those of the authors and do not necessarily reflect the views of the National Science Foundation.

References

1. Gutteridge, W.A., Dalziel, J.A.: The effect of the secondary component on the hydration of Portland cement: Part I. A fine non-hydraulic filler. Cement Concrete Res. 20(5), 778–782 (1990)
2. Kadri, E., Duval, R.: Effect of ultrafine particles on heat of hydration of cement mortars. ACI Mater J. 99(2), 138–142 (2002)
3. Lothenbach, B., et al.: Influence of limestone on the hydration of Portland cements. Cement Concrete Res. 38(6), 848–860 (2008)
4. Lawrence, P., Cyr, M., Ringot, E.: Mineral admixtures in mortars - Effect of inert materials on short-term hydration. Cement Concrete Res. 33(12), 1939–1947 (2003)
5. Zelic, J., et al.: The role of silica fume in the kinetics and mechanisms during the early stage of cement hydration. Cement Concrete Res. 30(10), 1655–1662 (2000)

6. Gutteridge, W.A., Dalziel, J.A.: Filler cement: The effect of the secondary component on the hydration of Portland cement: Part 2: Fine hydraulic binders. Cement Concrete Res. 20(6), 853–861 (1990)
7. Tanaka, K., Capule, M.F.V., Hisanaga, T.: Effect of crytallinity of TiO_2 on its photocatalytic action. Chem. Phys. Lett. 187(1-2), 73–76 (1991)
8. Bianchi, C.L., et al.: The role of the synthetic procedure of nano-crystalline TiO_2 on the photodegradation of toluene. In: International RILEM Symposium on Photocatalysis, Environment, and Construction Materials - TDP 2007. Rilem Publications, Florence (2007)
9. Jo, B.W., et al.: Characteristics of cement mortar with nano-SiO_2 particles. Constr. Build. Mater. 21(6), 1351–1355 (2007)

Nano-modification of Building Materials for Sustainable Construction

M. Kutschera, T. Breiner, H. Wiese, M. Leitl, and M. Bräu

Abstract. Nanostructured products or products which were developed by means of nanotechnology already exist in the field of construction chemistry or construction materials. Main driving force in the conservative construction industry to invent or adopt new technologies is to reduce energy (CO_2-footprint) during the construction process as well as during the utilization of buildings. Additional targets are increased service lifetimes of the constructions or new functionalities. E.g. photocatalytically active surfaces to reduce staining and increase air quality.The performance advantages of materials in the field of thermal insulation foams (nanofoams), nanocomposite colloidal particles (polymeric binders) and nanotechnologically improved inorganic binder systems have been investigated and will be discussed.

1 Introduction

Nanomaterials and nanotechnology have attracted a lot of attention both in the scientific field and in media communication. But up to now most announcements concerning nano-products still target future possibilities which will not be realized within the next decade. Thus nanotechnology seems to be a very trendy word but offers only little to no benefit for the average consumer.

However intentionally nanostructured products or products which only could have been developed by means of nanotechnology already exist.

Main driving force in the construction chemicals branch to invent or adopt new technologies is reducing energy during the construction process, reducing energy during the utilization of buildings and an increased service lifetime of the constructions. Also new functionalities e.g. photocatalytically active surfaces to

M. Kutschera and M. Leitl
BASF Construction Chemicals GmbH, Trostberg, Germany
e-mail: michael.kutschera@basf.com

T. Breiner, H. Wiese, and M. Bräu
BASF SE, Ludwigshafen, Germany

reduce staining and increase air quality have been developed. In all these areas solutions based on nanotechnology exist. Some of them are well known and recognized, others are not even noticed but still contribute to the construction performance.

Examples for reduced energy consumption and increased sustainability due to nanotechnology are:

Thermal insulation: Starting from conventional thermal insulation foams (Styropor® or Styrodur®) going on to technically improved solutions (like Neopor®) we will finally end up with nanostructured foams for top level performance. Polymer-foams with cell size in the nano scale are already available in the lab.
Coatings with high durability can contribute to sustainable construction and thus save energy. In this area organic-inorganic nanocomposites are one possible road to high performance coatings. Nanocomposite binders like Col.9® combine the advantages of organic acrylic material (film formation and elasticity) with the benefit of inorganic substances (low stickiness and moisture permeability).

Furthermore nanotechnology offers numerous possibilities to improve inorganic, cement based binders (concrete, mortar, etc.) with respect to decreased CO2 footprint and increased service life (carbonation, alkali-silicate reaction, freeze thaw resistance and adhesion to both old and new substrates).

In the following chapters technologies and selected examples in the areas nanofoams, nanocomposites and nanotech binders are investigated and discussed.

2 Nanofoams

Energy reduction or decreased CO2 emissions during the utilization of buildings is mainly archived by means of thermal insulation. State-of-the-art insulation materials in the construction area are mineral fibers, mineral foams and polymeric foams. All these systems reduce the thermal conductivity λ by minimizing the matrix contribution to thermal conductivity. This is done by replacing material by air voids (preferably in closed cells). The effect can be seen from equation (1).

$$\lambda_{\text{total}} = \lambda_{\text{matrix}} + \lambda_{\text{gas}} + \lambda_{\text{radiation}} \tag{1}$$

Thermal conductivity can be further decreased by blocking the losses caused by thermal radiation. This is done by using absorber pigments which are active in the infrared part of the spectrum (e.g. carbon black particles). Best effects are achieved when theses pigments are evenly distributed in the cell walls without destroying them. Radiation losses can be decreased by approx. 50% without deteriorating the foam quality.

Even higher leverage can be achieved by decreasing the heat transfer in the air voids caused by the movement of the gas molecules. This can be either done by directly decreasing the gas pressure (vacuum insulation panels) or using gas molecules with higher molecular weight (decreased speed of molecular movement).

Another possibility is to take advantage of the Knudsen effect. This is done be reducing the pore size of the foam down to length scales of the mean free path of the gas molecules thus substantially decreasing the thermal conductivity

Table 1 Typical contributions to thermal conductivity of different types of polymer foams

λ (mW/mK)	standard polymer foam	polymer foam with IR absorber	polymer nanofoam	Aerogel (high price !)
matrix	2	3	adjustable between normal polymer foams and Aerogels	4
gas	22	22		9
radiation	9	5		(depends)
total	34	30		13

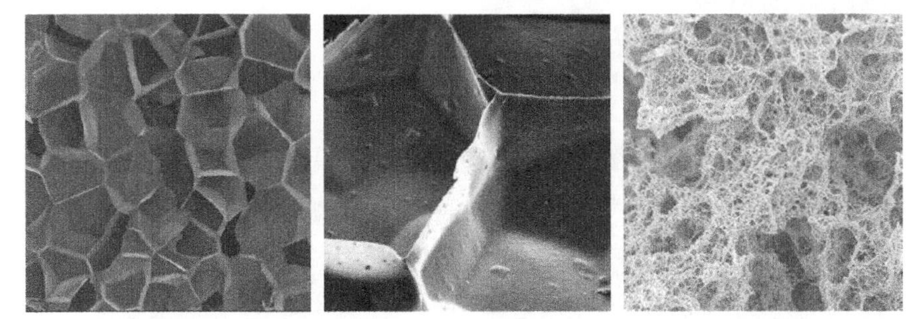

Fig. 1 Scanning electron microscopy images of polymer foams: conventional (left) IR-pigmented (mid) and nanofoam (right)

3 Nanocomposites

All typical construction materials share one common problem. If they are made of inorganic raw materials they show high strengths but appear to be brittle. They tend to fail by cracking which leads to low durability. On the other hand organic construction materials are flexible but often lack mechanical strength. In addition their surfaces often show significant dirt pickup.

Nature itself has solved the problem during the billion years of evolution by developing biominerals like bone, dental enamel or mother-of-pearl. These composite materials produced by nature combine the hardness of crystalline minerals like hydroxylapatite or aragonite with the flexibility of organic sub-stances such as collagen or chitin, making them some of nature's most stable materials.

A technical approach to mimic these materials is to synthesize nanocomposites materials in which inorganic material and organic "glue" bond together. One example for these type of material is the colloidal nanocomposite binder Col.9®. It is a dispersion of organic plastic polymer particles in which nanoscale particles of silica are incorporated directly during the synthesis of the organic polymer. This lead to a very evenly distribution of hard (inorganic) and flexible (organic) structuring on the nano scale. Only by this nanostructuring it is possible to combine the different advantages.

As a result the dried film (paint) composed from the colloidal nanocomposite particles shows the following advantages when compared to conventional polymeric colloidal films:

- significant lower dirt pickup
- hydrophilic surface leads to less moisture and less algae and fungus fouling
- stable color and no surface chalking
- crack free
- water vapor permeability but low water uptake

Fig. 2 Scanning force image of organic inorganic colloidal nanocomposites (left). Use of nanocomposite binders for exterior painting (right)

4 Nanotech Binders

The field of inorganic binders is very complex and manifold. On the other hand the majority of concrete and mortar systems sold today are based on different types of Ordinary Portland Cement (OPC), aluminate cements and mixtures thereof.

In the recent years new types of ultra high performance concretes (UHPC) have entered the market. They show significantly enhanced mechanical properties. The idea behind UHPC is to optimize the non-reactive filler material to ensure better packing density in the final hardened cement stone. This is supported by numerical calculations and a good part empiricism. Typical filler materials range from limestone flour, quartz flour, basalt powders and basalt fibers up to steel fibers. Additionally these materials often contain nano-scale SiO_2 in the form of agglomerated or disperse Microsilica of different qualities and sources [6, 7, 8]. Typical mechanical values for ordinary concrete and UHPC are given in Table 2. Newest

Table 2 Some typical average mechanical values for different types of concrete [2]

	standard concrete	quality concrete	UHPC
Compressive Strength (MPa)	~20	~30	100…180
Flexural Strength (MPa)	approx. 2…4	approx. 3…5	10

developments even use carbon nanotubes as internal reinforcement for ultra high performance concrete as well as for ultra high performance renders (e.g. for façade applications or in EIFS systems) and mortars [1, 3, 4, 5].

For the future we see an even higher potential by NOT adding nano-particles externally to the cementitious binder but understand the cement matrix itself as a nanostructured material. With the advent of modern, improved analytical tools deep insight into the processes during the cement reaction and hardening is gained. Some of the tools like modern microscopy (transmission electron, scanning electron, scanning force) or diffraction methods (x-ray, synchrotron) can just now be applied in new quality to cementitious systems with thriving results.

As a consequence we are now able to guide and control the same old cement material into new shape like different crystalline morphologies and habitus or new matrices with controllable nanostructure of voids, hydrate phases and aggregate distribution (Fig. 3). These materials show improved properties with respect to reaction speed, durability and adhesion. Example of these nanotechnologically optimized binders are the EMACO® Nanocrete repair mortars [9].

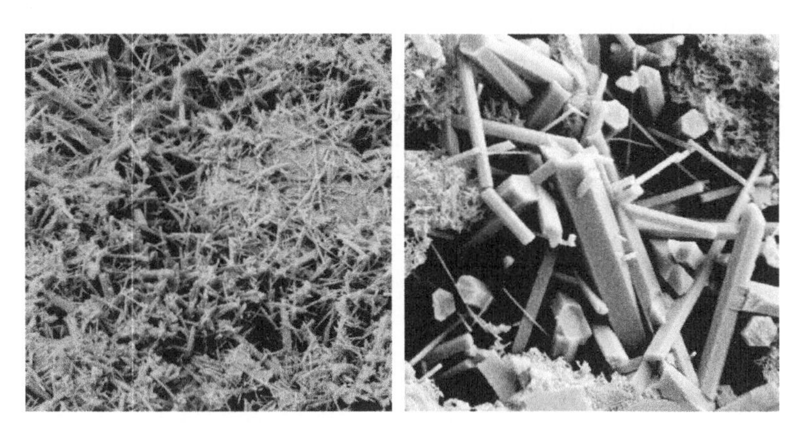

Fig. 3 Scanning electron microscopy images of different nano-structured cement matrices. The structuring is done by directly interfere into the hydration reaction by organic and inorganic additives

5 Discussion and Summary

Nanotechnology and nano-modification of building materials offer new possibilities for significantly improved materials. One focus with respect to energy reduction and preservation of natural resources is to create new nanostructured materials with extended durability.

Some examples of these new high tech materials are already available. Even more are still to be developed or to be invented. Nano is clearly one way to the future of construction.

References

1. Campillo, I., Dolado, J.S., Porro, A.: High-Performance Nanostructured Materials For Construction. In: Proceedings of the 1st International Symposium on Nanotechnology in Construction, United Pre-Prints, pp. 110–121 (2003)
2. Fehling, E., Schmidt, M., Stürwald, S.: Ultra high Performance Concrete (UHPC). Kassel university press, Kassel (2008)
3. Jiang, X., Kowald, T.L., Staedler, T., Trettin, R.H.F.: Carbon Nanotubes As A new Reinforcement Material For Modern Cement-Based Materials. In: Proceedings of the 2nd International Symposium on Nanotechnology in Construction, pp. 209–213. RILEM Publications s.a.r.l (2005)
4. Kowald, T., Trettin, R.: Influence of surface-modified Carbon Nanotubes on Ultra-High Performance Concrete. In: Proceedings of the International Symposium on Ultra High Performance Concrete, pp. 195–202. Kassel University Press, Kassel (2004)
5. Kowald, T., Trettin, R., Dörbaum, N., Städler, T., Jian, X.: Influence of Carbon Nanotubes on the micromechanical properties of a model system for ultra-high performance concrete. In: Second International Symposium on Ultra High Performance Concrete. Kassel University Press, Kassel (2008)
6. Liu, C., Shen, W.: Effect of crystal seeding on the hydration of calcium phosphate cement. J. Mater. Sci-Mater M 8(12), 803–807 (1997)
7. Patent WO2006111225 Hydraulic binders accelerated by nanoscale $Ca(OH)_2$
8. Patent WO2007128638 / WO2007128630 Accelerated binders by nanoscale TiO_2
9. http://www.emaco-nanocrete.com/ (accessed January 14, 2009)

Study of P-h Curves on Nanomechanical Properties of Steel Fiber Reinforced Mortar

S.F. Lee, J.Y. He, X.H. Wang, Z.L. Zhang, and S. Jacobsen

Abstract. Steel fiber reinforced mortars with w/b 0.3 and 0.5 with and without 10% silica fume by cement weight were investigated using a Hysitron Triboindenter® with Berkovich tip, indenting in the interfacial transition zone (ITZ) between steel fiber and matrix, and also on the steel fiber and aggregate using 5mN maximum force to obtain P-h (Load-Displacement) curves for elastic modulus and hardness analysis. Different P-h curves were generated at different points in the ITZ region, steel fiber and aggregate. The P-h curves in the ITZ reached the maximum force at larger displacement than those of aggregate and steel fiber, revealing that the microstructures in ITZ are loosely packed together. The unit structures in steel fiber are mainly bound together in regular way by covalent bond; therefore, it reached the maximum force earlier than that of the actual igneous granitic aggregate. Varying irregular P-h curves were observed, mostly in the ITZ, and reasons for this are discussed; voids in microstructure, weak zone, possible voids beneath the indented point, indenting in varying unhydrated and hydrated phases, possible leaching/washing out of binder during polishing of non-epoxy-reinforced samples.

1 Introduction

Steel fiber reinforced mortar consists of four phases: steel fiber, ITZ, matrix and aggregate. The ITZ, which maximum thickness ranges from 15μm up to 50μm [1, 2], is the region that has high porosity compared to the matrix [3]. It is formed due to the so called wall effect where the cement packs more loosely against the relatively large aggregate's and steel fiber's surface, and this also increases the local w/c ratio. Furthermore, it is also considered as a weakest link in the mechanical behavior of concrete [4].

S.F. Lee, J.Y. He, Z.L. Zhang, and S. Jacobsen
Department of Structural Engineering, Norwegian University of Science and Technology (NTNU), Trondheim, Norway

X.H. Wang
Department of Civil Engineering, Shanghai Jiaotong University, Shanghai, China

In the past few years, nanoindentation on the cement paste [5, 6] was carried out in order to understand properly the nanomechanical properties of the microstructures, so that the macroscopic properties of concrete can be controlled and improved.

In this paper, nanoindentation was performed on the ITZ, steel fiber and aggregate. The P-h curves, elastic modulus and hardness obtained were compared and related to the intrinsic property of each phase.

2 Experimental Procedures

2.1 *Materials*

Steel fiber reinforced mortars with w/b 0.3 and 0.5 with and without 10% silica fume (sf) by cement weight were made with 0.3 vol% straight high carbon steel fibers were added in each mix. The mix proportion is stated in ref. [7]. Norcem Anlegg cement (an Ordinary Portland cement in Norway), silica fume with >90% SiO_2, limestone filler, granitic sand with 4mm maximum size, polycarboxylate polymers superplasticizer, straight steel fibers with L13mm and D0.16mm were used. The fresh mortars were casted into 40x40x160mm moulds and vibrated on the vibrating table for 3 seconds. The mortars were then covered with plastic bags, demoulded after 24 hours and cured in water at 20°C for 28 days.

2.2 *Sample Preparation for Nanoindentation*

Small cubes with 16x16x16mm dimension were cut out from the centre of the mortar using a diamond saw at low speed with water as lubricant. The cubes were dried at room temperature before epoxy mounting, followed by grinding and polishing. A Struers grinding and polishing machine together with the MD-grinding discs and MD-polishing cloths were used. Ultrasonic cleaning in water to remove grit and an examination of the polished surface under the transmitted light microscope were performed at each step of grinding and polishing. The specimens were plane ground on the diamond discs of 68μm, 30μm and 14μm, fine ground with 9μm diamond suspension, polished with 3μm and 1μm diamond suspension, and finally done with oxide polishing, so that the surface flatness less than 1μm could be achieved. The forces and durations used in the grinding and polishing can be found in ref. [7].

2.3 *Nanoindentation*

A Hysitron Triboindenter® with a Berkovich diamond tip (a three-sided pyramidal diamond with included angle of 142.3°) was used to indent on the steel fiber, ITZ and aggregate. The maximum indentation load was 5mN. A series of P-h curves indenting on the steel fiber, ITZ and aggregate were collected and analyzed. An average of 10 indents was performed on the steel fiber and aggregate and nearly 40 indents were on the ITZ. The testing was repeated on two different areas of each specimen. Elastic modulus, E, and hardness, H, of each phase is calculated using the following equation:

$$\frac{1}{E_r} = \frac{(1-v^2)}{E} + \frac{(1-v_i^2)}{E_i} \tag{1}$$

$$H = \frac{P_{\max}}{A} \tag{2}$$

where E_r and A are the reduced modulus and the projected area of the elastic contact respectively, v is the Poisson's ratio of the phase, E_i and v_i are the elastic modulus and the Poisson's ratio of the tip with 1140GPa and 0.07, respectively. The reduced modulus, E_r, can be calculated as below:

$$S = \frac{dP}{dh} = \frac{2}{\sqrt{\pi}} E_r \sqrt{A} \tag{3}$$

where $S = dP/dh$ is the stiffness of the upper portion of the unloading curve [8].

3 Results and Discussion

In nanoindentation, the maximum load is determined so that the tip will stop indenting when the maximum load is reached, and thus, a P-h curve is obtained. This is different from the macromechanical test where the strength of the specimen is roughly estimated first before a machine is chosen. In nanoindentation, it is important to have a surface flatness less than 1μm for the ease of nanoindentation, and the nanomechanical properties calculated from the P-h curve for each phase could be compared effectively. In order to minimize the error caused by the washing out of binder during polishing all specimens were polished under the same preparation procedures in our study. The irregular P-h curves that possibly depicted material defect or tip slipping were discarded from being used so that the intrinsic property of each phase could be studied as close as possible.

Fig. 1 shows the typical P-h curves of steel fiber, aggregate, cement paste and some irregular P-h curves obtained during nanoindentation on the specimens. Fig. 1(b), an irregular P-h curve of steel fiber, shows a slight increase in load at the beginning of the displacement and followed by a very clean loading and unloading curve. This depicted that the feature indented had a well-arranged structure but with a local uneven surface. Fig. 1(d) shows that on unloading, some coarse grains might attach to the indenter, and with their irregular shapes, interlocking between coarse grains happened and stopped the load from decreasing smoothly with the

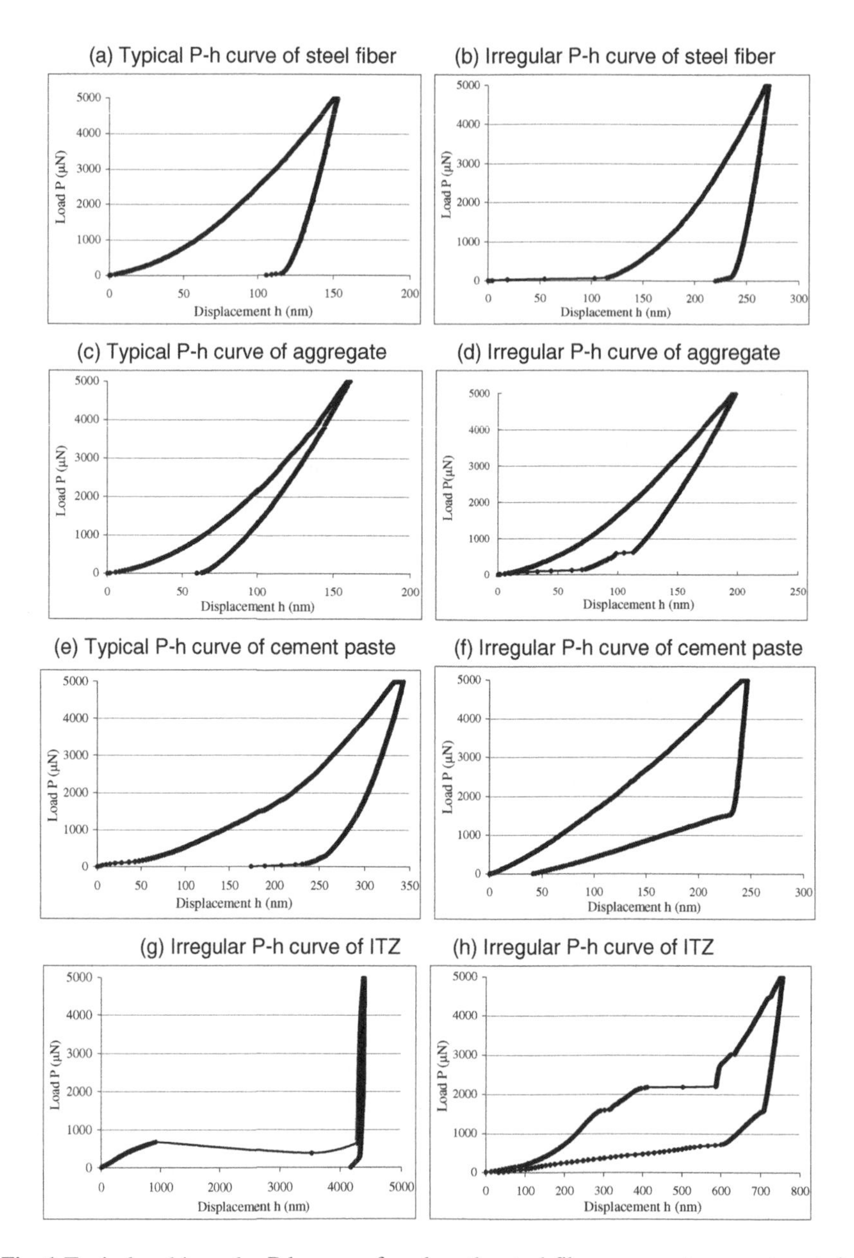

Fig. 1 Typical and irregular P-h curves found on the steel fiber, aggregate cement paste in the and ITZ

displacement. The same interlocking phenomenon shown in the nanoindentation on the microstructures in the cement paste, see Fig. 1(f). Fig. 1(g) and 1(h) reveals that there could be large voids underneath the microstructure as the load stopped increasing after a period of time of loading and remained the same even though the indenting depth increased. Possibly also leaching during polishing may have affected the results.

Straight high carbon steel fiber consists of mainly atom Fe, 0.7% C and some other compositions bound together in covalent bonds in a well-arranged structure. During loading, dislocation happened in the steel fiber. During unloading, the bonds between the layers of atoms are so strong that the dislocation was hardly disturbed by the uplifting indenter. Thus, the load dropped to zero at a small displacement on unloading. The similar unloading curves were shown in metals such as aluminum and tungsten [8]. However, if the surface of the steel fiber was not polished properly, the irregular P-h curve shown in Fig. 1(b) was commonly seen.

The igneous granitic aggregate consists of mainly coarse mineral crystals of quartz, feldspar and mica packed tightly together. From the P-h curves, see Fig. 1(a) and Fig. 1(c), it was found that the load reached the maximum at nearly the same displacement for the steel fiber and aggregate, and the hardness calculated was also nearly the same for both, see Table 1. This could attribute to the tightly-packed-together structure shown in both. Although the coarse minerals in aggregate have crystalline structures, they are packed together in week bonds. Therefore, the orderly arranged atomic structure with covalent bonds in the steel fiber could be responsible for its high elastic modulus when compared to the aggregate and microstructures in the ITZ.

The hydration products, such as calcium silicate hydrate (C-S-H), calsium hydroxide (CH), ettringite and monosulphate, found in the ITZ and bulk matrix have crystalline structures. However, see Fig. 1(e), a typical P-h curve of ITZ shows that the load reached the maximum at a displacement larger than that of steel fiber and aggregate, which means a lower elastic modulus and hardness in the ITZ than that of steel fiber and aggregate. The weak bonds between heterogeneous microstructures in the ITZ in fact were often found between the coarse minerals in the aggregate, however, the coarse minerals in the aggregate packed more tightly than the heterogeneous microstructures in the ITZ. The more porous characteristic of ITZ due to locally lower cement packing caused by wall effect and high w/c in ITZ than in bulk matrix, voids right under the indented surface and possible washing out of non-epoxy reinforced polished paste could be also additional reasons for the lower elastic modulus and hardness shown in the ITZ than in the aggregate and steel fiber. Hu et al. [3] revealed that with computer simulation, a higher volume fraction of hydration products were found in the ITZ than in the matrix, mainly due to the disproportional high rate of hydration in the ITZ, however, the packing discontinuity due to the wall effect caused the higher porosity in the ITZ than in the matrix for a matured concrete. Furthermore, irregular P-h curves with possible large voids were mostly found in the ITZ in our study.

Mondal et al. [5] performed nanoindentation on cement paste with w/c 0.5 and revealed that for unhydrated cement grain, the elastic modulus was 110GPa; for cement paste matrix was 21GPa, and for ITZ was 18GPa. Comparing our results, see Table 1, with Monda et al. [5] and Sorelli et al. [9], a high value of elastic modulus found in w/b 0.3 could be from the indentation either fully on the unhydrated cement grains or partially on the unhydrated cement grains and hydration products. In general, the hydrated microstructures in the ITZ have lower elastic modulus and hardness than the steel fiber and aggregate for both w/c studied.

4 Conclusions

The results of the nanoindentation revealing the microstructures in the ITZ had lower elastic modulus and hardness than steel fiber and aggregate greatly supports the wall effect and more porous characteristic in the ITZ of steel fiber and aggregate. This also possibly supports the assumption that ITZ is a weak link in the mechanical properties of the steel fiber reinforced mortar.

Table 1 Elastic modulus, E, and hardness, H, of steel fiber, aggregate and microstructures in the ITZ at a distance of 10 to 50μm from the steel fiber, and aggregate

	Steel fiber		ITZ (10-50μm)		Aggregate	
	E (GPa)	H (GPa)	E (GPa)	H (GPa)	E (GPa)	H (GPa)
w/b0.3, sf0%	285-310	7.2-8.5	14-50	0.3-1.8	48-85	6-10.5
			105-160	2.3-3.6		
w/b0.3, sf10%	250-310	7.8-9.8	2-85	0.1-4	60-115	6.5-15
w/b0.5, sf0%	240-280	7.8-9.8	6-38	0.2-1.3	65-95	6-12

References

1. Ollivier, J.P., Maso, J.C., Bourdette, B.: Interfacial transition zone in concrete – review. Adv. Cem. Based Mater. 2, 30–38 (1995)
2. Zheng, J.J., Li, C.Q., Zhow, X.Z.: Thickness of interfacial transition zone and cement content profiles around aggregates. Mag. Concrete Res. 57, 397–406 (2005)
3. Hu, J., Stroeven, P.: Properties of the Interfacial Transition Zone in Model concrete. Interface Sci. 12, 389–397 (2004)
4. Simeonov, P., Ahmad, S.: Effect of transition zone on the elastic behavior of cement-based composites. Cement Concrete Res. 25, 165–176 (1995)
5. Mondal, P., Shah, S.P., Marks, L.D.: Nanoscale characterization of cementitious materials. ACI Materials Journal 105, 174–179 (2008)
6. DeJong, M.J., Ulm, F.: The nanogranular behavior of C-S-H at elevated temperatures (up to 700ºC). Cement Concrete Res. 37, 1–12 (2007)
7. Wang, X.H., Jacobsen, S., He, J.Y., Zhang, Z.L., Lee, S.F.: Application of nanoindentation testing to study of the interfacial transition zone in steel fiber reinforced mortar. Cement Concrete Res. (2008) (submitted)
8. Oliver, W.C., Pharr, G.M.: An improved technique for determining hardness and elastic modulus using load and displacement sensing indentation experiments. J. Mater. Res. 7, 1564–1583 (1992)
9. Sorelli, L., Constantinides, G., Ulm, F., Toutlemonde, F.: The nano-mechanical signature of ultra high performance concrete by statistical nanoindentation techniques. Cement Concrete Res. 28, 1447–1456 (2008)

Evolution of Phases and Micro Structure in Hydrothermally Cured Ultra-High Performance Concrete (UHPC)

C. Lehmann, P. Fontana, and U. Müller

Abstract. Thermal curing of Ultra-High Performance Concrete (UHPC) has a strong influence on its mechanical properties. By applying a water vapor saturation pressure additional to the increased temperature the curing conditions are strongly enhanced and lead to a significant improvement of the degree of hydration of the cement paste. Increasing temperature accelerates the formation of crystalline calcium silicate hydrates by dehydrating the cement paste and ends in the formation of gyrolite, truscottite and xonotlite at 200 °C and 15 bars. Thereby the micro structure undergoes an obvious change. Cement paste consists of close networked crystal fibers with dimensions up to 1 μm. By filling cracks and small pores with crystalline C-S-H phases, flaws in the matrix are healed and generate a more homogeneous micro structure. Additionally, autoclaving encourages dissolution processes at quartz grains, which produces a better cohesion between fillers and the fine crystalline cement paste. As a consequence, autoclaving enhances compressive and flexural strength significantly but, compared to simple heat treatment at 1 bar, with very low scatter of the test results.

1 Introduction

Concrete technology turns its attention more and more to materials with enhanced properties such as high strength and durability as well as increased ecological performance. One of the more recent research topics in the field of improving the concrete composition is Ultra-High Performance Concrete (UHPC). The exceptional strength of UHPC of 150 MPa and more as well as its remarkably increased durability is mainly based on its dense micro structure which is a result of the high

---

C. Lehmann, P. Fontana, and U. Müller
Federal Institute for Materials Research and Testing (BAM), Berlin, Germany
e-mail: Christian.Lehmann@bam.de, Patrick.Fontana@bam.de,
Urs.Mueller@bam.de
www.bam.de

cement content, the very low water cement ratio, the use of highly reactive silica fume and the granulometric adjustment of fillers [4]. Due to the addition of highly effective plasticizers a good workability of the fresh concrete is maintained. Previous studies showed that thermal curing of UHPC can have a strong influence on its mechanical properties [1,3,5]. One effect of heat treatment is the development of a denser micro texture with formation of crystalline calcium silicate hydrate (C-S-H) phases, what usually results in increased compressive strengths [2]. The enhancement of the curing conditions by additional application of water vapour saturation pressure may increase significantly the flexural strength too [3]. The main interest of our study was focussed in how the micro structure and phase equilibrium is changed by autoclaving UHPC and how this relates to the mechanical properties. In addition the reaction rate of the used fly ash was of particular interest, since a higher rate could help to reduce the cement and silica fume content significantly.

2 Experimental Setup

2.1 Materials and Curing Regimes

The mix design of the UHPC was based on commercial raw materials to achieve results with practical relevance. In addition to a white cement CEM I 42.5-R a micro fly ash (median particle diameter 0.2 μm) and silica fume were used as reactive components. The chemical compositions of the cement and the fly ash are shown in Table 1. The maximum size of the quartz aggregate was 2 mm. In order to optimize the particle size distribution a quartz filler with a median size of 50 μm was added. The water cement ratio was 0.26. The total water binder ratio was 0.22. The composition of the UHPC is given in Table 2. The self-compacting properties of the fresh UHPC were adjusted using a polycarboxylate-based superplasticizer. The slump-flow was 260 mm (small cone according to EN 1015-3).

First, the solid components were dry mixed in a high shear mixer to homogenize the material. Then, water and superplasticizer were added and the material was mixed thoroughly. The fresh concrete was casted in prismatic steel moulds (160 x 40 x 40 mm³) and demoulded after 1 day. Thereafter the specimens were cured under six different conditions (Table 3).

Table 1 Chemical composition of cement and fly ash measured by X-ray fluorescence analysis in Wt.-%

Material	MgO	Al_2O_3	SiO_2	P_2O_5	SO_3	K_2O	CaO	MnO	TiO_2	Cr_2O_3	Fe_2O_3	CO_2
Cement	0.60	4.73	20.73	-	3.14	0.95	66.42	-	0.11	0.06	0.38	2.88
Fly ash	1.00	18.77	58.47	0.66	0.41	1.63	2.49	0.02	0.70	-	3.60	12.22

Table 2 Composition of UHPC

Material	content
Cement (kg/m³)	745.0
Silica fume (kg/m³)	72.4
Fly ash (kg/m³)	74.5
Quartz filler (kg/m³)	243.1
Quartz aggregate 0-0.5 mm (kg/m³)	238.2
Quartz aggregate 0.5-1.0 mm (kg/m³)	357.1
Quartz aggregate 1.0-2.0 mm (kg/m³)	357.1
Water (kg/m³)	168.0
Superplasticizer (g)	53.5

Table 3 Curing conditions

	Curing condition	Curing time
Series 1 – reference	23 °C / 1 bar	6 days
Series 2 – heat treated	90 °C / 1 bar	2 days
Series 3 – heat treated	150 °C / 1 bar	2 days
Series 4 – heat treated	200 °C / 1 bar	2 days
Series 5 – autoclaved	150 °C / 5 bar	8 hours
Series 6 – autoclaved	200 °C / 15 bar	8 hours

After 7 days the specimens were dried at 40 °C and 40 mbar to stop the hydration.

2.2 *Analytical Techniques*

For phase-analysis a combination of several techniques was used to achieve reliable results. Firstly the samples were analyzed using a Philips PW 1710 X-ray diffractometer with Cu Kα radiation. All samples were scanned over a 2θ-range from 3 to 65° using a step size of 0.02° and a measuring time of 4 s each step. Additionally detailed scans followed over a 2θ-range from 3 to 20° and a step size of 0.005°. To optimize the identification of C-S-H phases, and in particular the puzzolanic consumption of portlandite, differential thermo analysis and thermal gravimetry was used (Netzsch – STA 449 Jupiter). Finally, for precise chemical and textural analysis in micro- and nanometer range, a scanning electron microscope (Leo Gemini 1530 VP) was employed. On all series compressive strength and flexural strength were tested. Mercury intrusion porosimetry was performed to examine the evolution of pore diameters under the different curing conditions.

3 Results

3.1 Microstructure and Mechanical Properties

Heat curing or autoclaving the UHPC generated an obvious change in the micro structure. The hydration of clinker phases and the fly ash was improved. Simple heat treatment produced a slightly denser micro structure and a remarkable increase of pore volume with a median pore diameter of 12 nm. Autoclaved samples exhibited an obviously higher degree of hydration. The texture of autoclaved UHPC showed a homogeneous, dense cement paste which consisted of close networked crystal fibers with a length up to one micrometer in the specimen cured at 200 °C and 15 bars. Cracks and small pores were filled with crystalline C-S-H (Fig. 1). The pore volume decreased compared to the heat treated specimen and the median pore diameter was reduced to 5 nm.

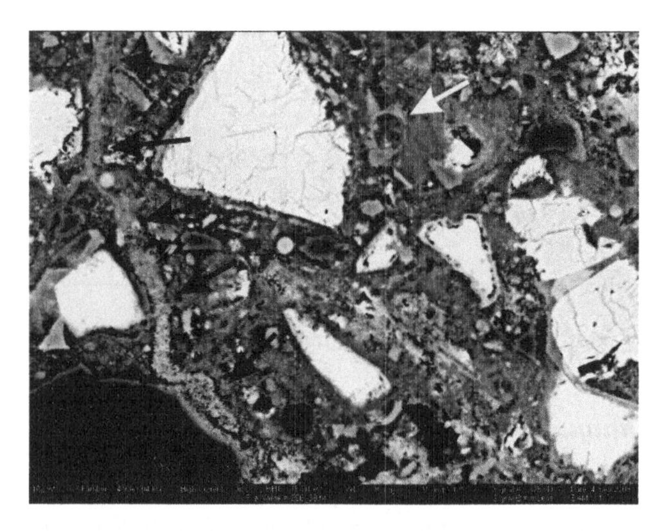

Fig. 1 SEM image of UHPC autoclaved at 200 °C / 15 bars. A crack is filled with crystalline C-S-H (black arrows). The white arrow points to a completely hydrated grain of fly ash, which indicates a high degree of puzzolanic reaction

The autoclaved series exhibited dissolution processes around quartz grains, which produced a better cohesion between fillers and the fine crystalline cement paste (Fig. 2).

The mechanical tests showed a general increase of compressive strength proportional to the curing temperature. Indeed, the results of the autoclaved samples showed a smaller scatter and reached a higher final strength. The flexural strength increased clearly by autoclaving, while it decreased slightly by heat curing.

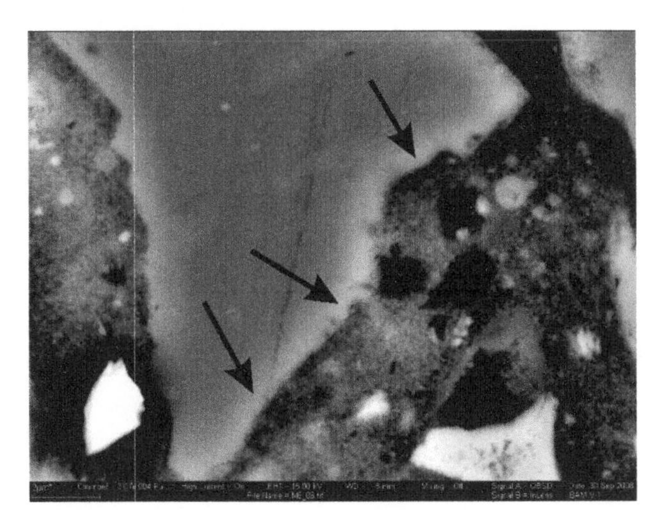

Fig. 2 SEM image of UHPC autoclaved at 200 °C / 15 bars. Dissolution rims on a grain of the quartz filler producing strong cohesion with crystalline cement paste (black arrows)

3.2 Microchemistry and Phase Composition

With increasing temperature the amorphous C-S-H phases in the sample changed to crystalline C-S-H phases. The level of dehydration was proportional to the curing temperature. In simply heat treated samples the Si/Ca atom ratio increased slightly from 0.65 to 0.75. The C-S-H first crystallized to tobermorite and jennite (90 °C). Heat curing at 150 °C effected the formation of foshagite from jennite and quartz (1) or jennite and tobermorite (2). Additionally jennite decomposed to afwillite in a small amount (3).

$$4\,jennite + 3\,quartz \rightarrow 9\,foshagite + 35\,H_2O \tag{1}$$

$$3\,jennite + tobermorite \rightarrow 8\,foshagite + 30\,H_2O \tag{2}$$

$$jennite \rightarrow 3\,afwillite + 2\,H_2O \tag{3}$$

The formation of xonotlite and gyrolite from tobermorite was visible at 200 °C (4). Afwillite and portlandite react to hillebrandite (5), which subsequently reacted with portlandite to jaffeite (6).

$$8\,tobermorite \rightarrow 4\,xonotlite + gyrolite + 18\,H_2O \tag{4}$$

$$afwillite + portlandite \rightarrow 2\,hillebrandite + 2\,H_2O \tag{5}$$

$$2\,hillebrandite + 2\,portlandite \rightarrow jaffeite + H_2O \tag{6}$$

Hydrothermal curing led to total absence of portlandite, which indicates a high degree of puzzolanic reaction of the fly ash. The Si/Ca atom ratio increased significantly from 0.75 to 1.1. The specimens autoclaved at 150 °C and 5 bars exhibited the same coexisting phases as the samples simply heat treated at the same temperature. Additionally afwillite and portlandite turned into hillebrandite (5) and jennite reacted with quartz to xonotlite (7). Finally tobermorite decomposed to xonotlite and gyrolite at 200 °C and 15 bars (4). Gyrolite itself reacted with quartz to the Si-rich truscottite (8). Also hillebrandite was stable in absence of portlandite.

$$2\,jennite + 6\,quartz \rightarrow 3\,xonotlite + 19\,H_2O \tag{7}$$

$$7\,gyrolite + 24\,quartz \rightarrow 8\,truscottite + 78\,H_2O \tag{8}$$

4 Conclusions

The homogenous cement paste matrix, consisting of close networked C-S-H crystal fibers, which develop a more stable structure than amorphous C-S-H phases, and the "healing" of flaws by filling them with C-S-H crystals are the main reasons for the improved mechanical properties of autoclaved UHPC. In addition autoclaving results in increased cohesion between cement paste and fillers by partly dissolution of quartz grains, and distinctive reduction of pore sizes.

The puzzolanic reaction of fly ash is significantly accelerated, so autoclaving might be a useful tool to reduce the high cement and silica fume content in UHPC by using secondary cementitious materials.

The thermal treatment of C-S-H phases results generally in the development of crystalline phases. Hereby the water content of the phases decreases with an increase in temperature. Pure heat treatment at 1 bar leads to formation of foshagite and xonotlite and Ca-rich phases like jaffeite. However, portlandite is still present at 200 °C in this series but the amount is strongly reduced.

Due to autoclaving the reaction rate and the final degree of hydration of cementitious materials are substantially enhanced. Portlandite is not observed anymore in autoclaved samples and the Si/Ca atom ratios in the cement paste are higher than in simply heat treated samples what is shown by the formation of Si-rich truscottite. Furthermore xonotlite, hillebrandite and foshagite are detected as final phases.

References

1. Cheyrezy, M., Maret, V., Frouin, L.: Microstructural analysis of RPC (Reactive Powder Concrete). Cem. Concr. Res. 25, 1491–1500 (1995)
2. Dehn, F.: Ultrahochfeste Betone. In: König, G., Tue, N., Zink, M. (eds.) Hochleistungsbeton – Bemessung, Herstellung und Anwendung. Ernst & Sohn, Berlin (2001)

3. Müller, U., Kühne, H.-C., Fontana, P., Meng, B., Neme ek, J.: Micro texture and mechanical properties of heat treated and autoclaved Ultra High Performance Concrete (UHPC). In: Schmidt, et al. (eds.) Proc. Int. Symp. Ultra High Performance Concrete, 2nd edn., Kassel, Germany, March 5-7, 2008, pp. 213–220 (2008)
4. Richard, P.: Reactive Powder Concrete: A new ultra-high strength cementitious material. In: 4th Int. Symp. on utilization of high strength concrete, pp. 1343–1349 (1996)
5. Sauzeat, E., Feylessoufi, A., Villieras, F., Yvon, J., Cases, J.M., Richard, P.: Textural analsis of Reactive Powder Concretes. In: Proc. 4th Int. Symp. Utilization of High-Strength/High-Performance Concrete (1996)

Interparticle Forces and Rheology of Cement Based Suspensions

D. Lowke

Abstract. Rheological properties of cement based suspensions are affected by the surface properties of the particles, the properties of the solvent and the adsorbed polymers. To understand the interaction between these parameters and the rheology of cement based suspensions the surface forces of the colloidal powder particles were considered. Three surface forces are taken into consideration – the attractive van der Waals forces, the repulsive double layer forces and the polymer induced steric forces. A theoretical basis for the evaluation of these forces in cement based suspension is given. Within the experimental program the superplasticizer adsorption was determined for cement and ground limestone suspensions. Rheological measurements were performed with these suspensions to determine yield stress and thixotropy. The results indicate a strong correlation between polymer adsorption, rheological properties of the suspensions and evaluated forces.

1 Introduction

Fresh properties of High Performance Concretes (SCC, UHPC) like flowability, segregation resistance and formwork pressure are determined by viscosity, yield stress and thixotropy. Thus the control of rheological properties is the key to successful application of these concretes. The rheological properties are affected by the surface properties of the particles, the properties of the solvent and the adsorbed superplasticizer polymers. In particular, the use of superplasticizers to adjust rheological properties of fresh modern high performance concrete has gained in importance during recent years. To understand the interaction between properties of the raw materials and the rheology of cement based suspensions the surface forces of the colloidal powder particles have to be considered. This paper focuses

D. Lowke
Technische Universität München, Centre for Building Materials (cbm)
e-mail: lowke@cbm.bv.tum.de
www.cbm.bv.tum.de

on the effect of superplasticizer adsorption on yield stress and thixotropy explained by the interparticle potential energy.

2 Materials and Methods

Portland cement and ground limestone were used as powder materials in the investigations. The density determined by helium pycnometry, the mineralogical composition determined by X-ray diffraction, the surface area determined by nitrogen adsorption and the mean Particle diameter of the powder materials determined by laser diffraction are shown in Table 1. Quartz sand with a 0/2 mm grading and a density of 2.6 g/cm³ was used as aggregate. The mixes were prepared with a polycarboxylate ether superplasticizer (SP) with 35% solids in aqueous solution and a mean molecular mass of 44,500 g/mol. The mean hydrodynamic $R_{h,avg}$ was determined at 9.5 nm.

Table 1 Characteristics of the powder materials

	Alite [wt.%]	Belite [wt.%]	C_3A [wt.%]	C_4AF [wt.%]	ρ [g/cm³]	A_s [m²/cm³]	d_{50} [μm]
Cement	62.3	10.5	8.0	6.7	3.2	6.38	15.1
Ground Limestone	-	-	-	-	2.8	3.39	7.0

Five Mortar mixes consisting of 150 L/m³ cement, 150 L/m³ limestone, 400 L/m³ sand and 315 L/m³ water were used in the investigations. The volumetric fraction of water and powder V_w/V_p was 1.06. For the reference mix the dosage of superplasticizer was adjusted to 0.51 wt.% with respect to cement to yield a slump flow of 260 ± 10 mm. To investigate the effect of superplasticizer content on yield stress and thixotropy four mortars with varying amounts of superplasticizer were prepared. The amount of superplasticizer was changed by up to ± 0.05 wt.%, Table 2. The rheological measurements for the determination of yield stress and thixotropy were performed 10 min after water addition using a rheometer with a rotating ball with a radius r of 10 mm. Before commencing the measurements, the mortar was subjected to shear stress for 30 s in order to break up agglomerates enabling the subsequent observation of structure formation. The force of resistance F_{max} needed to move the ball through the mortar suspension was measured after waiting periods of 5, 30, 90 and 120 s during which the mix was at rest. Each measurement was performed at the very low rotational speed of $5 \cdot 10^{-4}$ m/s ($\dot{\gamma} \approx 0.05\ s^{-1}$). Owing to the low shear rates in the investigations, the measured resistance force F_{max} is mainly due to the yield stress of the mortar. Neglecting the shear force due to the rotation motion, static yield stress τ_{max} was calculated from the results of the rheological investigations using $\tau = Y_G\ F/(2\pi r^2)$,

$Y_G = 0.14334$, according to [1]. The first derivative of the yield stress as a function of time (between 5 and 120 s) is a measure of the thixotropy T_{120}.

Furthermore, three pastes with the same composition of cement, limestone and water as the mortars and varying superplasticizer contents of 0.37, 0.51 and 0.95 wt.% were prepared to determine the superplasticizer adsorption. Pore Solution was extracted of the pastes 15 min after water addition. The total organic carbon (TOC) content of the pore solution and the superplasticizer solution was determined by high-temperature oxidation of the organic ingredients. The amount of adsorbed superplasticizer was calculated as the difference between the TOC of the added superplasticizer solution and the pore solution of the mortar.

3 Results

The superplasticizer adsorption of the pastes with superplasticizer contents of 0.37, 0.51 and 0.95 wt.% are shown in table 2. There was a strong linear relationship between the amount of added superplasticizer and the adsorbed superplasticizer. Thus the adsorption for the pastes with a superplasticizer content of 0.46, 0.50, 0.52 and 0.57 were calculated by a linear regression.

The effect of superplasticizer on the development of yield stress and thixotropy of mortar is shown in Fig 1. It is apparent that larger amounts of superplasticizer lower the yield stress as well as the thixotropy significantly. An increased SP-content of 0.05 wt.% from the reference mix leads to an reduction of the initial yield stress $\tau_{max,5}$ after 5s from 14 to 6 Pa and a reduction of thixotropy T_{120} from 0.16 to 0.09 Pa/s.

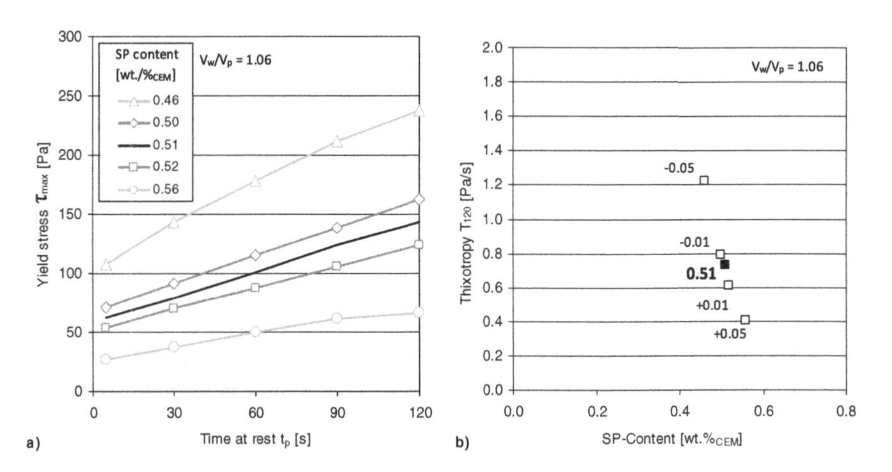

Fig. 1 Yield stress as a function of time at rest at varying superplasticizer contents (a) and effect of superplasticizer content on thixotropy (b)

4 Discussion - Interparticle Forces, Yield Stress and Thixotropy

The effect of superplasticizer in fresh mortar may be understood in terms of the effect of polymer adsorption on the interaction forces between the individual particles of the powder suspension which may be divided into attractive van der Waals forces, electrostatic forces and repulsive steric forces. In the case of cementitious suspensions and its usual surface potential the electrostatic forces are thought to be negligible compared with the van der Waals forces and polymer induced forces [2]. According to [3], the van der Waals interparticle potential energy G_{vdw} between two spheres of radius a, separated by a distance h (between their surfaces along the axis through the centre points) is given by

$$G_{vdw} = -\frac{H}{6}\left(\frac{2}{(h/a+2)^2-4}+\frac{2}{(h/a+2)^2}+\ln\left(\frac{(h/a+2)^2-4}{(h/a+2)^2}\right)\right) \tag{1}$$

where H is the Hamaker constant for the interaction. According to [4], the interaction energy G_{ster} between adsorbed polymer layers can be calculated as follows.

$$G_{ster} = \pi a \frac{N_A}{\upsilon_3}\left(\frac{\Gamma}{\rho\delta}\right)^2 k_B T\left(\frac{1}{2}-\chi_1\right)(2\delta-h)^2,\ h<2\delta \tag{2}$$

Here N_A is Avogadro's number, v_3 the molar volume of the solvent, Γ the specific mass of the adsorbed polymer, ρ the density of the polymer, χ the polymer segment interaction parameter and δ the thickness of the polymer layer. The combination of the repulsive steric interaction with the attractive van der Waals interaction yielding in the total interaction energy G_{tot} between the particles is shown in Fig. 3a as a function of particle separation. As the particles approach, the van der Waals attraction increases rapidly until the adsorbed polymer layers on the particles meet ($h = 2\delta$) and the steric repulsion superimposes on the van der Waals energy, e.g. point C in Fig. 3a. For polymer concentrations encountered in flowable concretes, the polymer layer is effectively a wall by inducing a very steep repulsion and prevents a further approach when $h \leq 2\delta$. In this case the exact size of v_3, Γ, ρ and χ has little effect on the interaction so that the thickness of the polymer layer is the paramount. Thin polymer layers mean that the particles can approach to smaller distances and the interparticle attraction is stronger, point A in Fig. 3a.

The thickness of the polymer layer was determined on the basis of absorptiometry, molecular weight and size. Assuming a spherical conformation of the polymer in solution (compare [5]), the volume of a polymer molecule was estimated with the experimentally determined hydrodynamic radius (Fig. 2).

In a good solvent the side chains are stretched well in all directions around the backbone which is situated in the centre of the polymer bundle. Due to the

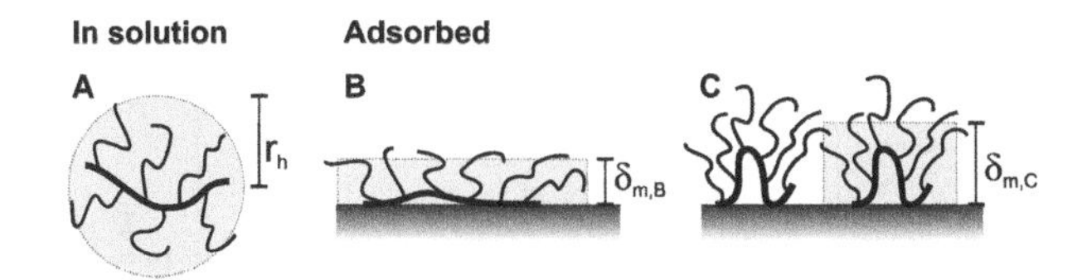

Fig. 2 Polymer conformation in solution and adsorbed on a surface and effect of adsorbed polymer concentration on polymer layer thickness

negative charged carboxyl groups of the backbone, polycarboxylate ether adsorbs with the backbone directly onto the surface. Thus the volume of the adsorbed polymer bundle is about half the volume of the free polymer in the solvent. The mean layer thickness δ_m is determined by the available surface area for the adsorbed polymer. This means that, the higher the amount of adsorbed polymer the higher is the thickness of the polymer layer (Fig. 2). The calculated mean polymer layer thicknesses δ_m of the mortars with varying superplasticizer contents are in a range of 7.2 to 8.2 nm, see Table 2. They are in good agreement with values experimentally determined by [6, 7] using AFM.

Table 2 Variation of superplasticizer content, superplasticizer adsorption, calculated mean thickness of the polymer layer and minimum of total interparticle energy

SP content [wt.%$_{CEM}$]	0.37	0.46	0.50	0.51	0.52	0.57	0.95
Variation in SP [wt.%]	-0.14	-0.05	-0.01	±0.00	+0.01	+0.05	+0.44
SP adsorption [mg/m²$_{Solid}$]	0.409	0.461[1]	0.484[1]	0.490	0.496[1]	0.522[1]	0.728
Calc. mean polymer layer thickness δ_m [nm]	-	7.2	7.7	7.8	7.9	8.2	-
Minimum total interparticle energy G_{min}/kT [-]	-	-58.1	-53.8	-52.8	-52.3	-50.0	-

[1] Interpolated

The total interparticle potential energies between two particles with a radius of 1 µm and different superplasticizer contents are shown in Fig. 3a. The depth of the minimum in the total energy curves G_{min} (Table 1) define the maximum attraction between the particles and determines the rheological properties of the mortars at static conditions, like yield stress and thixotropy. The minimum depends on the thickness of the polymer layer. With a decreasing layer thickness the minimum decreases (Fig 3). According to [8] particles coagulate when the minimum energy becomes smaller than -5kT. This causes a structure formation and thus yield stress and thixotropy. A strong correlation between the minimum of the total interparticle potential energy G_{min}, yield stress $\tau_{max,5}$ and thixotropy T_{120} was found for the investigated mortars, see Fig 3b. The lower G_{min} the higher are yield stress $\tau_{max,5}$ and thixotropy T_{120}.

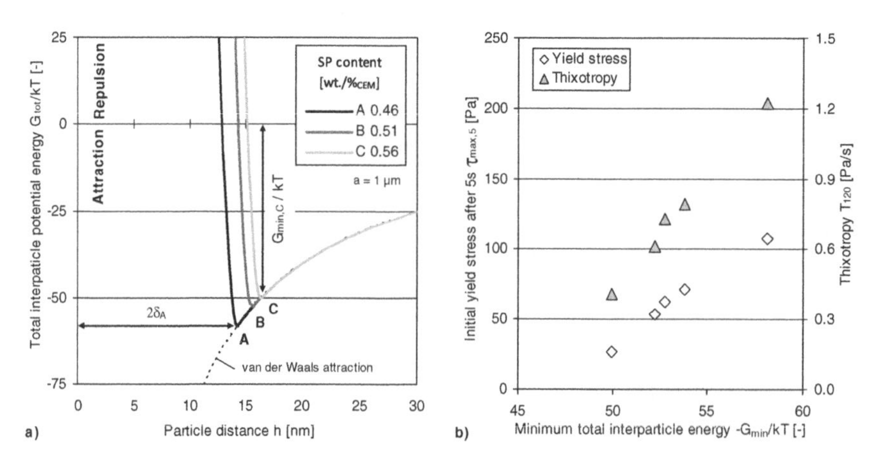

Fig. 3 Total interparticle potential energy (a) and correlation between minimum of total interparticle potential, yield stress and thixotropy (b)

5 Conclusion

The experimental investigations focused on the effect of superplasticizer content on yield stress and thixotropy of fresh mortar. An increase in superplasticizer content led to a reduction in yield stress and thixotropy. The effect of superplasticizer content on rheological properties can be understood in terms of the effect of polymer adsorption on the interaction energy between the individual particles of the powder. An increase in superplasticizer content results in a thicker adsorbed polymer layer and consequently weaker van der Waals attraction between the particles so that less forces is needed to disperse the particles – yield stress and thixotropy decreases.

It was shown that thixotropy and yield stress of the investigated mortars can be explained by the interactions between the powder particles. The lower the minimum of the total interparticle potential energy G_{min}, this means the higher the attraction between the particles, the higher are yield stress and thixotropy.

Owing to the complex nature of the interactions (e.g. heterogeneous composition of clinker particles, irregular particle shape, hydration reactions) the calculation of the interparticle interactions contain many assumptions and simplifications so their accuracy is limited. However, this approach may be used to improve the understanding of the mechanisms responsible for the rheological behaviour of fresh mortar and concrete suspensions.

Acknowledgments. The author would like to thank the German Research Foundation (DFG) for the financial support.

References

1. Beris, A.N., Tsamopoulos, J.A., Armstrong, R.C., Brown, R.A.: Creeping motion of a sphere through a Bingham plastic. J. Fluid Mech. 158, 219–244 (1985)
2. Kjeldsen, A.M., Geiker, M.: Modelling inter-particle forces and resulting agglomerate sizes in cement-based materials. In: SCC 2005, pp. 105–111 (2005) ISBN 0-924659-64-5
3. Yoshioka, K., Sakai, E., Daimon, M., Kitahara, A.: Role of steric hindrance in the performance of superplasticizers for concrete. J. Am. Ceram. Soc. 80, 2667–2671 (1997)
4. Flatt, J.R.: Interparticle forces and superplasticizers in cement suspensions. Dissertation. Lausanne (1999)
5. Gay, C., Raphaël, E.: Comb-like polymers inside nanoscale pores. Adv. Colloid Interf. 94, 229–236 (2001)
6. Kauppi, A., Andersson, M., Bergström, L.: Probing the effect of superplasticizer adsorption on the surface forces using the colloidal probe AFM technique. Cem. Con. Res. 35, 133–140 (2005)
7. Laaraz, E., Kauppi, A., Andersson, K., Kjeldsen, A.M., Bergström, L.: Dispersing multi-component and unstable powders in aqueous media using Comb-type anionic polymers. J. Am. Ceram. Soc. 89, 1847–1852 (2006)
8. Hesselink, F.T., Vrik, A., Overbeek, J.T.G.: On the theory of the stabilization of dispersions by adsorbed macromolecules. J. Phys. Chem. 75, 2094–2103 (1971)

Nanocomposite Sensing Skins for Distributed Structural Sensing

J.P. Lynch, K.J. Loh, T.-C. Hou, and N. Kotov

Abstract. The operational safety of civil engineered structures can be jeopardized by structural deterioration (*e.g.*, corrosion) and damage (*e.g.*, yielding, cracking). Structural health monitoring has been proposed to provide engineers with sensors and algorithms that can detect structural degradation in a timely manner for cost-effective correction. In this paper, a thin film material engineered at the nano-scale is proposed for distributed sensing of metallic structures. Assembled from single wall carbon nanotubes (SWNT) and polymers, the thin film's electrical properties are designed to change in response to external stimulus such as strain or tearing. Electrical impedance tomographic (EIT) conductivity imaging is adopted to make a measurement of the film conductivity over its complete surface area. The end result is a true distributed sensor offering engineers with impressive two-dimensional maps from which strain and damage can be observed in fine detail.

1 Introduction

During the past decade, an infusion of expertise and high-technologies from related engineering domains have dramatically improved the state-of-art in sensors and actuators used for monitoring and controlling large-scale civil structures. Concurrent to these advances has been an ongoing revolution in the emerging field

J.P. Lynch
Department of Civil and Environmental Engineering, Universtiy of Michigan
e-mail: jerlynch@umich.edu

K.J. Loh
Department of Civil and Environmental Engineering, Universtiy of California Davis
e-mail: kjloh@ucdavis.edu

T.-C. Hou
Department of Civil and Environmental Engineering, University of Michigan
e-mail: tschou@umich.edu

N. Kotov
Department of Chemical Engineering, University of Michigan
e-mail: kotov@umich.edu

of nanotechnology [1]. The transcendent advances of the nanotechnology field introduce a paradigm shift in how the next generation of smart structures will be designed. Specifically, it is now possible to design materials with specific macroscopic mechanical, electrical and chemical properties by controlling structure and assembly at the nano-scale [2].

This paper explores the design of a novel thin film assembled at the nano-scale using single wall carbon nanotubes (SWNT) and polyelectrolytes (PE) to create a homogenous SWNT-PE composite with superior mechanical strength and with electrical conductivities that change in response to stimulus (in this paper, strain and tearing). This multifunctional material can therefore sense the stimulus everywhere the material is; hence, it is capable of true distributed sensing. To unleash the distributed sensing capabilities of the thin film, electrical impedance tomography (EIT) is explored. EIT inversely solves for the distribution of film conductivity using only electrical measurements made along the film boundary. To highlight the film's functionality as a distributed sensor, laboratory experiments are made using the skin to detect strain fields and cracking in steel plates.

2 Multifunctional Nanocomposite Films

While individual carbon nanotubes (single and multi wall) undoubtedly offer impressive mechanical and electrical properties, they must be processed to offer macroscopic materials endowed with similar properties. To date, the processing of SWNT to attain desired macroscopic material properties (*e.g.*, strength, bulk conductivity) has been a major challenge. Early approaches used vacuum filtration of an aqueous solution of suspended SWNT to form a thin carbon nanotube mat on filtration paper. Unfortunately, this "buckypaper" is brittle and incapable of high strain due to the weak van der Waals interaction between the nanotubes [3]. In response to this limitation, polymer-carbon nanotube composites have been proposed. These composites have only shown moderate strength enhancements when compared to other carbon fiber composite materials [4-6] due to a lack of uniform nanotube connectivity throughout the polymer matrix.

In this study, sequential layering of chemically-modified SWNT and polymers is proposed to fabricate a homogenous polymer-carbon nanotube composite [7]. The layer-by-layer (LBL) approach offers SWNT-polyelectrolyte (SWNT-PE) composites with outstanding phase integration and homogeneity [7; 8]. LBL assembly entails the dipping of a solid substrate (*e.g.*, glass or silicon) in solutions of the individual components (in this study, SWNT and polyelectrolytes). First, SWNT are dispersed using non-covalent methods. A high molecular weight polyelectrolyte, poly(sodium 4-styrene-sulfonate) (PSS) is non-covalently bonded to the surface of individual SWNT providing them with an overall negative charge. Similarly, a positively charged solution of poly(vinyl alcohol) (PVA) is prepared as a conjugate polymeric material for the LBL assembled composite. A substrate is then prepared by cleaning its surface using a piranha solution. LBL

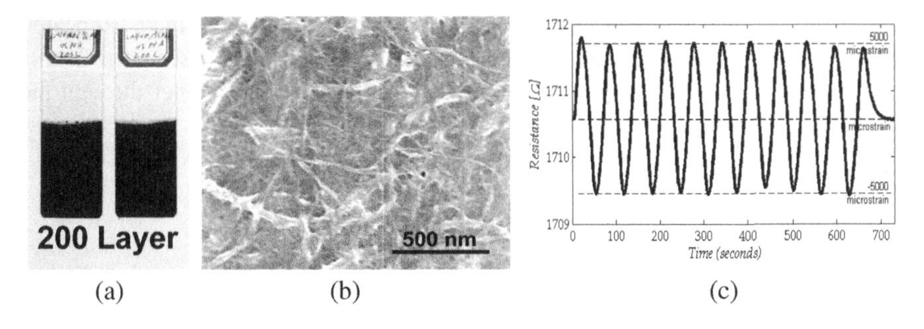

Fig. 1 (a) SWNT-PE film on glass slides; (b) SEM image of a 200-layer film; (c) resistance time history under a saw-tooth +/- 5000 $\mu\varepsilon$ load pattern (100 cycles) [9]

assembly begins by dipping the treated substrate for 5 minutes in the PVA solution. Next, the substrate is rinsed using deionized water and dried. After a PVA monolayer is deposited, the substrate is dipped in the SWNT-PSS solution to deposit negatively charged SWNT to the surface of the positively charged PVA monolayer. This process is repeated to build up films (Fig. 1a) of any number of bilayers. The resulting film is referred to as $(SWNT\text{-}PSS/PVA)_n$ where n is the number of bi-layers. The film can be either left in place or released into its free-standing form using etchants such as hydrofluoric acid. When viewing the film using a scanning electron microscope (SEM), it is evident that SWNT are uniformly distributed in the film with a well interdigitated morphology (Fig. 1b).

The mechanical properties of SWNT-PE thin films have been extensively tested in the laboratory to characterize their stress-strain properties. The homogenous distribution of SWNT in polyelectrolyte matrices allows the strength of the SWNT to be transferred to the composite. Tensile strain testing of multiple $(SWNT\text{-}PSS/PVA)_n$ thin film specimens reveal tensile strengths of more than 250 MPa with Young's modulus of 10 GPa or greater [9]. In addition to incredible strength, the film is also inherently piezoresistive. In general, SWNT-composite materials exhibit a piezoresistive behavior under applied strain which motivates their use as strain sensors [10; 11]. Prior work fundamentally explored a methodical approach to optimizing the fabrication parameters of the SWNT-based film to derive high strain, high gage factor strain sensors [12]. Parameters such as the SWNT dispersive agent, polyelectrolyte conjugate pair, dipping time, annealing process, among other parameters have been varied to produce SWNT-PE LBL films exhibiting linear changes in conductivity under high strain levels ($\epsilon > 4\%$) with high gage factors (GF) in excess of 5. For example, Fig. 1c presents the measured change in resistivity of a SWNT-PSS/PVA thin film axially loaded in tension; the resistivity change is reversible and nearly linear.

The beauty inherent to sensing-based multifunctional materials is that measurement of the material conductivity (which in this case is correlated to strain) can be made anywhere the material is. Therefore, such materials intrinsically offer the capability of *distributed* sensing. Distributed sensing is realizable if the material can be probed repeated over its full area to develop a conductivity map. Repeated probing is labor intensive and rules out its use in an automated system. In this study, conductivity mapping will be conducted in an indirect manner through the use of electrical impedance tomography.

3 Electrical Impedance Tomography

The development of EIT evaluation represents a major step-forward in the further development of nanoengineered materials for distributed sensing. EIT is essentially an inverse problem intended to estimate the spatial distribution of conductivity of a body based on boundary electrical measurements (*i.e.*, voltages) made during stimulation (*i.e.*, regulated current injection) of the boundary. When current is applied to a conductive thin film material, the flow of electrical current can be described by the 2D Laplace vector equation:

$$\nabla \cdot [\sigma(x, y)\nabla\phi(x, y)] = -I(x, y) \tag{1}$$

where σ is the material conductivity, ϕ is the electrical potential (voltage), and I is the applied current at a point source. The two in-plane dimensions of the thin film are designated by the position variables, x and y. Conductivity, σ, measures how easy it is for electrical current to flow normal to two faces of a unit volume of material. Two boundary conditions are specified for the Laplace equation including the Neuman (the sum of current crossing the film boundary, S, is zero) and Dirichlet (electric potential v along S is equal to the internal potential, ϕ, at the boundary) conditions [13]. If the conductivity distribution, $\sigma(x,y)$, is known, the internal electrical potential, $\phi(x,y)$, can be found from a known current across the film boundary; this approach is often termed the forward problem [13]. In contrast, the inverse problem attempts to find a mapping of conductivity, $\sigma(x,y)$, based upon voltage measurements at the boundary based on a regulated applied current (DC or AC). However, this inverse problem is ill-posed and requires a set of boundary measurements corresponding to multiple applied current distributions.

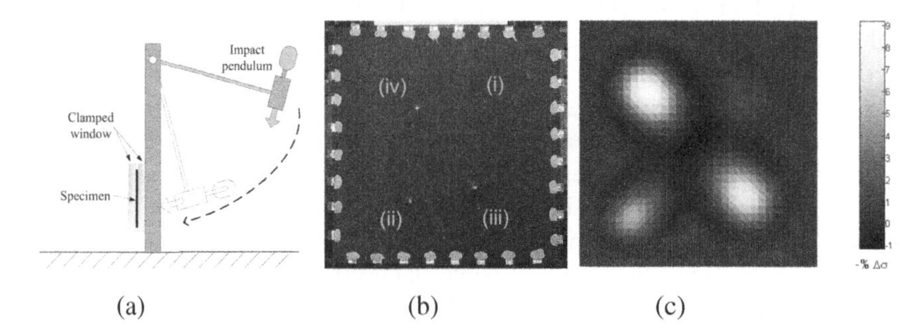

Fig. 2 (a) Impact test apparatus; (b) four impacts upon the sensing skin coated plate element; (c) corresponding EIT conductivity map with percentage change in conductivity shown [9]

In recent years, a number of researchers have proposed effective solutions to the non-linear inverse problem providing the tools necessary to perform accurate EIT. In the study reported herein, the inverse problem is solved using a finite element (FE) model of the thin film for repeated solution of the forward problem [14]. An iterative Gauss-Newton algorithm is used for modification of finite element conductivity until convergence. In this case, convergence is defined by minimization of the difference of the experimentally obtained boundary voltage, v_{exp}, and the boundary voltage predicted by each solution of the FE forward model, ϕ_{FE}.

4 Experimental Validation

To validate the distributed sensing properties of the nanocomposite thin film, a simple set of experiments in which impacts are used to introduce damage are described. A100-layer $(SWNT\text{-}PSS/PVA)_{100}$ thin film is deposited on a primer-coated thin steel plate (110 mm by 110 mm and 0.75 mm thick) to produce a large structural specimen to which damage can be introduced. After deposition of the sensing skin, 32 copper electrodes are attached to the plate along its boundary with 8 electrodes on each of the four sides. The sensing skin-coated steel plate is then clamped into an impact apparatus in which a pendulum is used to impact the plate. The apparatus (Fig. 2a) can control the amount of energy imparted to the plate by changing the height from which a sharp-tipped weight is dropped. Damage is introduced in two forms: permanent residual deformation and plate/skin penetration.

The plate is impacted four time with controlled energy input of 0.09, 0.38, 0.81, 1.17 J. The four impacts are sequentially numbered in terms of the energy as shown in Fig. 2b with the lowest energy given the index "*i*". Evident in Fig. 2b is the residual deformation of the plate with locations *ii*, *iii*, and *iv* clearly dented. At location *iv*, the plate is mildly cracked due to excessive deformation. As shown in Fig. 2c, the EIT-derived conductivity map clearly identifies the location and extent of "damage" introduced in the plate. The percentage change in conductivity (when comparing before and after) is correlated to the degree of damage.

5 Conclusions

A powerful, new nanocomposite assembled from SWNT and PE has been proposed as a distributed sensing skin. Controlled of the assembly of the composite at the nano-scale allows for the development of a multifunctional material with impressive mechanical properties and self-sensing functionality. This work largely explored the piezoresistive properties of the SWNT-PE sensing skin. In particular, EIT was adopted to provide a means of automated spatial conductivity mapping. Conductivity maps are capable of presenting the spatial distribution of strain and deformation in a structure as illustrated using a simple impact test on a steel plate. With sensing skins in their infancy, more work is needed to refine the skin for field use. Current work is also exploring the embedment of other sensing mechanisms including the sensing of corrosion using the SWNT-PE sensing skin.

Acknowledgments. This research is supported by the National Science Foundation (NSF) under grant number CMS-0528867 (program manager: Dr. S. C. Liu). The authors would also like to express their gratitude to NSF and Prof. Jeffrey Schweitzer (University of Connecticut) for sponsoring the travel of the authors to the NICOM3 conference.

References

1. Chong, K.P.: Nanotechnology in civil engineering. Adv. Struct. Eng. 8, 325–330 (2005)
2. Nalwa, H.S.: Nanostructured Materials and Nanotechnology. Academic Press, San Diego (2002)
3. Kang, I., Heung, Y.Y., Kim, J.H., Lee, J.W., Gollapudi, R., Subramaniam, S., Narasimhadevara, S., Hurd, D., Kirikera, G.R., Shanov, V., Schulz, M.J., Shi, D., Boerio, J., Mall, S., Ruggles-Wren, M.: Introduction to carbon nanotubes and nanofiber smart materials. Compos. Part B-Eng. 37, 382–394 (2006)
4. Shaffer, M.S.P., Windle, A.H.: Fabrication and characterization of carbon nanotube/poly(vinyl alcohol) composites. Adv. Mat. 11, 937–941 (1999)
5. Haggenmueller, R., Gommans, H.H., Rinzler, A.G., Fischer, J.E., Winey, K.I.: Aligned single-wall carbon nanotubes in composites by melt processing methods. Chem. Phys. Lett. 330, 219–225 (2000)
6. Watts, P.C.P., Hsu, W.K., Chen, G.Z., Fray, D.J., Kroto, H.W., Walton, D.R.M.: A low resistance boron-doped carbon nanotube-polystyrene composite. J. Mater. Chem. 11, 2482–2488 (2001)
7. Mamedov, A.A., Kotov, N.A., Prato, M., Guldi, D., Wicksted, J., Hirsch, A.: Molecular design of strong SWNT/polyelectrolyte multilayers composites. Nat. Mat. 1, 190–194 (2002)
8. Decher, G.: Fuzzy nanoassemblies toward layered polymeric multicomposites. Science 277, 1232–1237 (1997)
9. Loh, K.J.: Development of multifunctional carbon nanotube nanocomposite sensors for structural health monitoring. Ph.D. Thesis, University of Michigan, Ann Arbor, MI (2008)
10. Dharap, P., Li, Z., Nagarajaiah, S., Barrera, E.V.: Nanotube film based on single-wall carbon nanotubes for strain sensing. Nanotechnology 15, 379–382 (2004)
11. Kang, I., Schulz, M.J., Kim, J.H., Shanov, V., Shi, D.: A carbon nanotube strain sensor for structural health monitoring. Smart Mater Struct. 15, 737–748 (2006)
12. Loh, K.J., Kim, J.H., Lynch, J.P., Kam, N.W.S., Kotov, N.A.: Multifunctional layer-by-layer carbon nanotube-polyelectrolyte thin films for strain and corrosion sensing. Smart Mater. Struct. 16, 429–438 (2007)
13. Barber, D.C.: A review of image reconstruction techniques for electrical impedance tomography. Med. Phys. 16, 162–169 (1989)
14. Hou, T.C., Loh, K.J., Lynch, J.P.: Spatial conductivity mapping of carbon nanotube composite thin films by electrical impedance tomography for sensing applications. Nanotechnology 18, 315501 (2007)

Utilization of Photoactive Kaolinite/TiO_2 Composite in Cement-Based Building Materials

V. Matějka, P. Kovář, P. Bábková, J. Přikryl, K. Mamulová-Kutláková, and P. Čapková

Abstract. Titanium dioxide (TiO_2) is the most studied photocatalyst with application potential in many branches of industry. Building industry represent the sector, where the photoactive TiO_2 have been already successfully utilized. Concretes, plasters, paints are building materials where the photoactive TiO_2 is widely tested. However the amount of TiO_2 in these materials is limited with respect to their final properties. If the TiO_2 replaces the certain amount of cement in concretes, the resulting compressive strength decreases when this photocatalyst is added in non-adequate content. The surface of kaolinite particles can serve as a matrix for nanosized TiO_2 growing what results in photoactive composite – kaolin/TiO_2 formation. After the calcination of this composite the process of kaolinite dehydroxylation is responsible for metakaolinite formation and composite metakaolinite/TiO_2 with latently hydraulic properties originates. If the metakoline/TiO_2 is used for partial cement replacement the compressive strength of resulting samples is notably increased and its surface shows photodegradation ability against rhodamine B.

1 Introduction

Self-cleaning and antibacterial properties, as well as photodegradation of environmental pollutants are the added values which make the materials with TiO_2 perspective for applications in building industry. The increasing number of

V. Matějka, K. Mamulová-Kutláková, and P. Čapková
CNT, VŠB-Technical university of Ostrava, Ostrava, CR
e-mail: vlastimil.matejka@vsb.cz

P. Kovář and J. Přikryl
ČTC AP a.s., Přerov, CR
e-mail: pavel.kovar@precheza.cz

P. Bábková
CPIT, VŠB-Technical University of Ostrava, Ostrava, CR
e-mail: petra.babkova@vsb.cz

experimental works dealing with photodegradation of NO_x e.g. [1,2] or VOCs e.g. [3,4] emphasized the importance of photocatalysis by TiO_2 in control of environmental pollution. Limitation of massive utilization of photocatalytic technologies arises mainly due to the higher price of building materials with photoactive TiO_2. Practically, TiO_2 can be added to the bulk of building material or can be applied as an ingredient of thin surface layer. The addition of TiO_2 to the cement based building materials has to be reasonably considered mainly in respect to the final strength.

Kaolin is a sub-group mineral and includes 4 different polymorphs: kaolinite, dickite, nacrite and halloysite. If the kaolinite is heated, its dehydroxylation occurs and metakaolinite is formed. Metakaolinite belongs to the group of material with latent hydraulic properties. For metakaolinite hydraulicity activation, alkali activators as hydroxides of alkali metals and water glass are often used. In building materials based on cement binder, hydraulic properties of metakaolinite are activated with $Ca(OH)_2$ which originate during the process of cement hydratation. With respect to this fact metakaolinite can partially replace cement binder without loosing of strength of final product.

This work is focused on the kaolin/TiO_2 (*KATI*) composite application in cement based building materials. As prepared composite KATI shows photodegradation activity against organic dyes which serves as model pollutants. After the burning of KATI at the 600 °C the sample KATI600 is obtained. KATI600 combine latent hydraulic properties of metakaolinite and photoactivity of nanosized TiO_2. The increase in compressive strength of cement based testing samples containing KATI600 is comparable with this increase obtained at the samples containing commercially available metakaolin. Photodegradation activity of testing samples with KATI600 is approved with discoloration of Rhodamine B applied on the surface of samples.

2 Experimental

Composite $KTiO_2$ preparation and characterization

Kaolin SAK47 – *K* (LB minerals) and titanyl sulphate – *$TiOSO_4$* (Precheza a.s.) were used as received without any purification, for hydrolysis distilled water was used. The process of KATI composite preparation is schematically described in Fig. 1.

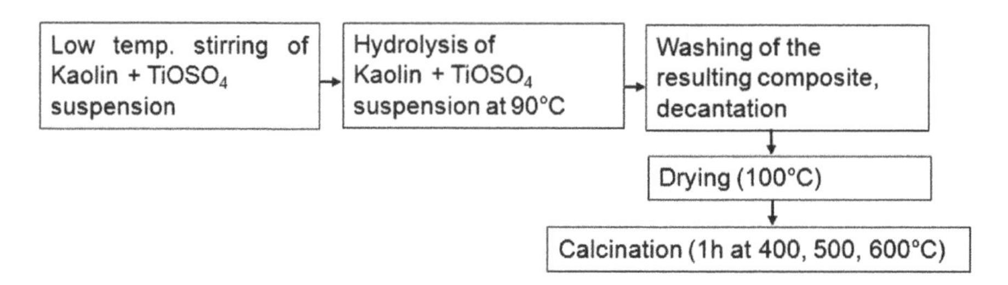

Fig. 1 Process of KATI composite preparation

The amount of TiO_2 in prepared composites was analyzed using X-ray fluorescence spectroscopy - XRFS (Spectro XEPOS), the phase composition was studied using X-ray powder diffractometry - XRPD (Bruker D8 Advance). Photodegradation activity of prepared composites was evaluated on the basis of discoloration of methylene blue – *MB* (Fluka) solution after the 1h irradiation with UV lamp (UVP Ltd) emitted maximum light at 365nm.

Testing samples preparation

Compositions of testing samples expressed in weight fractions (wt. %) of all the ingredients in prepared mixtures are shown in Tab. 1. In this table the letter *M* is used for assignment of K, KATI or MEFISTO admixtures, respectively. Prepared mixtures were schematically assigned as M(*T*)_*w* (where *T* shows the temperature used for K or KATI calcination (400, 500 and 600 °C respectively), *w* represents weight fraction (wt. %) of K, KATI or MEFISTO which replace the appropriate amount of cement binder. Commercially available metakaolinite MEFISTO was used as received without any thermal treatment. Ordinary portland cement (*OPC*) CEM I 42.5R (Cement Hranice a.s.) was used as hydraulic binder. The weight fraction of aggregates (three fractions of silica sands) and water was kept constant. Testing samples were prepared according to ČSN EN 196-1 [5], ČSN EN 450-1+A1 [6] respectively, their compressive strength after the 28-days hydration was tested also according to ČSN EN 196-1 [5].

Table 1 Composition of prepared mixtures

Sample	W (aggregate)			w (water)	w (OPC)	w (M)	w(M)/ w(OPC)*100	w/(c+M)
	0.1-0.6 mm	0.1-1.0 mm	0.3-1.6 mm					
	wt. %							
Ref	22.2	22.2	22.2	11.1	22.3	0	0	0.5
M(T)_5	22.2	22.2	22.2	11.2	21.2	1.1	5.2	0.5
M(T)_10	22.2	22.2	22.2	11.3	20.1	2.2	11.0	0.5
M(T)_15	22.2	22.2	22.2	11.4	19.3	3.0	15.5	0.5
M(T)_20	22.2	22.2	22.2	11.5	18.4	3.8	20.7	0.5

The photodegradation ability of surface of prepared samples was tested using modified Italian standard UNI 11259:2007, utilizing photodegradation of rhodamine B [7].

3 Results and Discussion

Using XRFS the amount 22 wt. % of TiO_2 was analyzed in composite KATI, original kaolin contain 1 wt. % of TiO_2. With respect to the TiO_2 content the testing sample in which 20 wt. % of cement was replaced with KATI contain 0.5 wt. % of photoactive TiO_2.

X-ray powder patterns of KATI before and after the calcination at selected temperatures are shown in Fig. 2. The sample KATI calcined up to 400 °C consist of kaolinite and anatase, quartz represent typical admixture in raw kaolin. After the calcination of KATI on temperatures higher than 500 °C the basal 001 diffraction peak of kaolinite disappears what signalize transformation of kaolinite to metakaolinite, the anatase particles become better defined, what is apparent from the constriction of 101 diffraction peak of anatase.

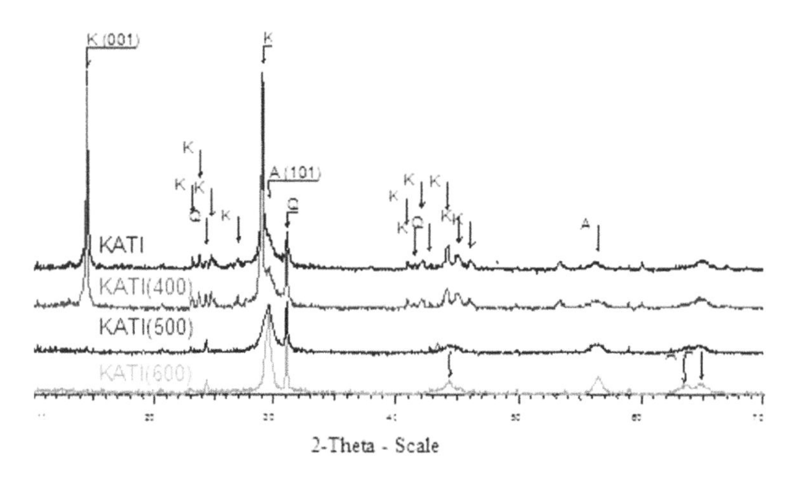

Fig. 2 XRPD patterns of KATI, KATI(400), KATI(500), KATI(600)

Photodegradation ability of KATI freshly prepared and after the calcination at 400, 500 and 600 °C is shown on the Fig. 3. Calcination of composite up to 600 °C doesn't decrease photodegradation activity of KATI and reach approx. 85 %, what means that approx. 85 % of amount of MB is removed after the 1h UV irradiation.

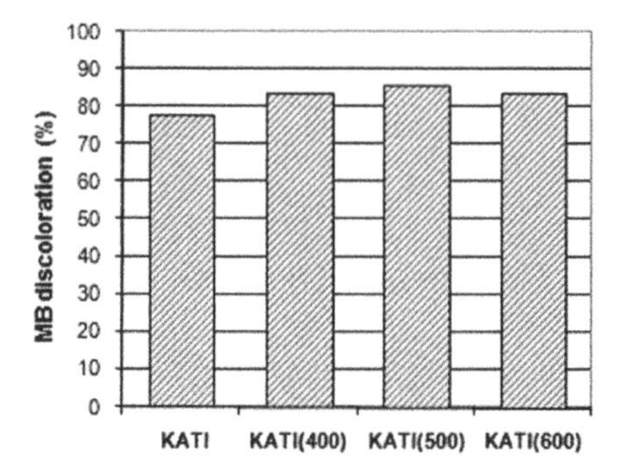

Fig. 3 Photodegradation ability of KATI, KATI(400), KATI(500), KATI(600) against MB after 1h irradiation

The influence of cement replacement with K and KATI on compressive strength is shown in the Fig. 4. The compressive strength of the samples is related to compressive strength of reference sample *Ref* (for *Ref* composition see table 1) after the 28-days hydration, compressive strength of this sample reached 40.2 MPa. Addition of both calcined K and calcined KATI composite increased compressive strength of prepared samples. The obtained values of compressive strength at samples with KATI(600) are comparable with those obtained for samples containing MEFISTO. The values of compressive strength obtained for K(600) are lower in comparison to samples with MEFISTO and KATI(600).

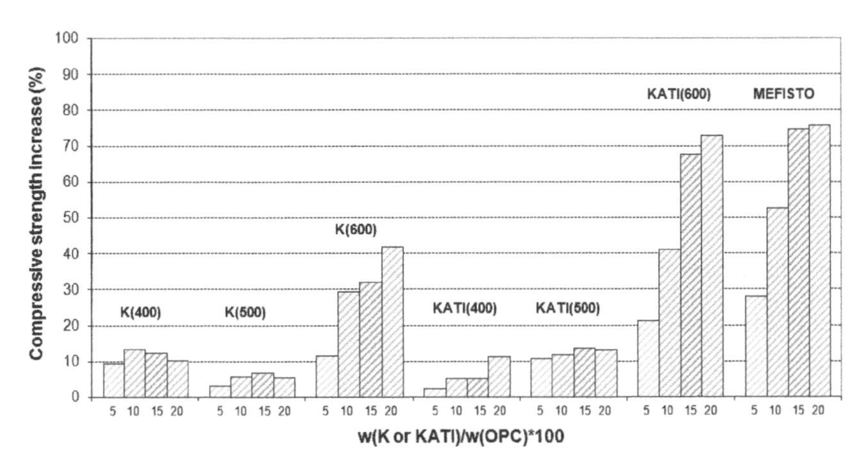

Fig. 4 Influence of calcined kaolin and calcined KATI composite on compressive strength (after the 28 days curing) of cement-based testing samples

Photodegradation ability of the surface of testing samples against rhodamine B is well documented on the Fig. 5.

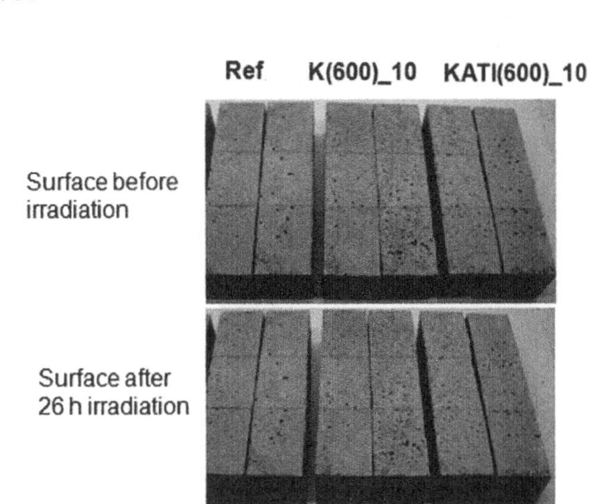

Fig. 5 The pictures of the surfaces of testing samples painted with rhodamine B

Rhodamine B applied on the surface of testing sample KATI(600)_10 (the sample in which 10 wt.% of cement was replaced with KATI(600)) is in significantly higher extent removed after the 26 h irradiation.

4 Conclusions

Partial replacing of portland cement with photoactive composite KATI(600) in cement-based testing samples increases significantly their compressive strength and the surface of prepared samples exhibits photodegradation ability against rhodamine B. Composite KATI calcined at 600 °C, at which synergistic effect of latent hydraulic properties of metakaolinite and photoactivity of nanosized TiO_2 is achieved, represents promising ingredient for cement-based building materials. Further research will be employed to explain the effect of the presence of TiO_2 at composite KATI(600) on the compressive strength of cement mortars which is significantly higher in comparison with compressive strength of cement mortars with metakaolinite obtained after the 1h calcination of kaolin at 600 °C.

Acknowledgments. This research has been funded by the Czech Ministry of Industry and Trade research project FT-TA4/025 and the Ministry of Education of the Czech Republic project MSM 6198910016.

References

1. Poon, C.S., Cheung, E.: NO removal efficiency of photocatalytic paving blocks prepared with recycled materials. Constr. Build Mater. (2007) doi: 10.1016/ j.conbuildmat. 2006.05.018
2. Maggos, T., Plassais, A., et al.: Photocatalytic degradation of NOx in a pilot street canzon configuration using TiO_2-mortar panels. Environ. Monit. Assess (2008) doi: 10.1007/s10661-007-9722-2
3. Diamanti, V.M., Ormellese, M., Pedeferri, M.P.: Characterization of photocatalytic and superhydrophilic properties of mortars containing titanium dioxide. Cement Concrete Res. (2008) doi:10.1016/j.cemconres.2008.07.003
4. Demeestere, K., Dewulf, J., et al.: Heterogenous photocatalytic removal of toluene from air on building materials enriched with TiO_2. Build Environ. (2008) doi: 10.1016/ j.buildenv.2007.01.016
5. CSN EN 196-1: Methods of testing cement - Part 1: Determination of strength (2005)
6. SN EN 450-1+A1: Fly ash for concrete - Part 1: Definition, specifications and conformity criteria (2008)
7. UNI 11259:2007 Determinazione dell'attività fotocatalitica di leganti idraulici - Metodo della rodammina

Nanomechanical Properties of Interfacial Transition Zone in Concrete

P. Mondal, S.P. Shah, and L.D. Marks

Abstract. This research provides better understanding of the nanostructure and the nanoscale local mechanical properties of the interfacial transition zone (ITZ) in concrete. Nanoindentation with in-situ scanning probe microscopy imaging was used to compare the properties of the bulk paste with the properties of the ITZ between paste and two different types of aggregates. ITZ was found to be extremely heterogeneous with some areas as strong as the bulk matrix. Higher concentration of large voids and cracks along the interface was observed due to poor bonding. Nanoindentation results on relatively intact areas of the interface disagreed with the notion of increasing elastic modulus with distance from the interface. Depending on the aggregate type, average modulus of the ITZ was 70% to 85% of the average modulus of the paste matrix. The main problem the ITZ poses on the overall mechanical properties of concrete was concluded to be due to extreme heterogeneity within the interface and poor bonding between aggregate and paste. It was noted that the connectivity of the weaker areas such as large voids and cracks along the interface governs failure.

1 Introduction

In concrete, paste matrix works as the glue that holds aggregates together to behave as a whole. However, this composite action depends on a thin interface layer (ITZ) that exists between the aggregates and the paste matrix. This is a region of gradual transition of properties, where the effective thickness of the region varies with the microstructural feature being studied, and with degree of hydration [1]. In many studies, it was concluded that in ordinary Portland cement concrete, the ITZ

P. Mondal
University of Illinois at Urbana Champaign, IL, USA

S.P. Shah
ACBM Center, Northwestern University, IL, USA

L.D. Marks
Northwestern University, IL, USA

consists of a region up to 50 μm around each aggregate with fewer unhydrated particles, less calcium-silicate-hydrate, higher porosity, and greater concentration of calcium hydroxide and ettringite.

Simeonov et al. [2] reported the influence of the ITZ on the overall elastic properties of mortar and concrete. In normal strength concrete, it is considered to be the weakest link in the mechanical system [1]. Although it is widely accepted that the properties of the ITZ have to be taken into account in modeling the overall mechanical properties of concrete [2-5], it is difficult to determine its local mechanical properties because of the complexity of the structure and the constraints of the existing measurement techniques [6]. Most of the time, the modulus of the ITZ is assumed to be uniform and less than that of the paste matrix by a constant factor. This factor is assumed to have a value between 0.2 and 0.8, although there is not enough theoretical or experimental evidence to support this assumption [5]. In some recent studies, attempts were made to determine the local mechanical properties of the ITZ using microindentation or microhardness testing [7-9]. Asbridge et al. reported about 20 percent reduction in Knoop microhardness in the ITZ than the bulk matrix at w/c of 0.4 and 0.5 [9], however similar Knoop microhardness at w/c 0.6. In their study, the width of indentation was reported as 10-15 μm, which is comparable with the width of the ITZ itself. Therefore, though microindentation revealed some information, effects of adjacent phases on hardness results make microhardness test not suitable for determining the local properties of the ITZ. Furthermore, there is still very little information available about the nanoscale mechanical properties of the ITZ. Nanoindentation has been used successfully in the recent past to investigate the local mechanical properties of cement paste and concrete [10-13]. This study is one of the first efforts which strive to characterize the nanomechanical properties of the ITZ using nanoindentation.

2 Experimental Details

Since different factors such as w/c ratio, age of sample, aggregate type and size affect properties of the ITZ, it was decided to keep most of them constant and vary one or two factors at a time: w/c ratio and age of the sample were respectively 0.5, and 1 month. Two types of model concrete samples were made, one with round gravel and the other one with limestone aggregates. Samples were cured under water at 25° C. For nanoindentation, samples were prepared following the method described in a different paper by the authors [12]. Figure 1 shows scanning electron microscopy image of polished model concrete sample with gravel. In most of the areas, there were large voids adjacent to an aggregate or a crack running along the interface. This proves weak bonding between aggregates and cement paste. Even in areas with no obvious interfacial zone (no voids or cracks), fewer unhydrated cement particles were found. Using a Hysitron Triboindenter, nanoindentation was done away from large voids and cracks. This is due to the practical constraints that nanoindentation can not be done on such a large void or a

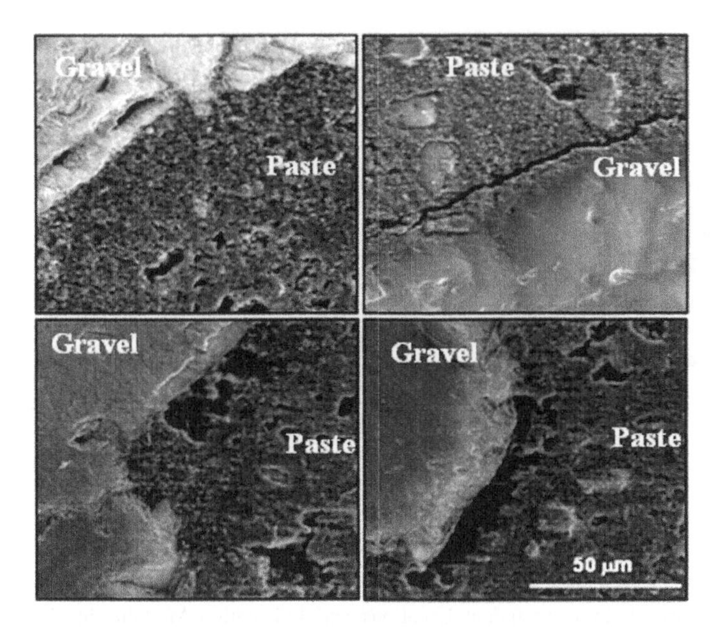

Fig. 1 Scanning Electron Microscopy Images of Different Areas of the Interfacial Transition Zone in the Model Concrete Sample

crack. In addition, these features are associated with the bonding issues between aggregates and paste rather than nanoscale properties of materials present in ITZ. AFM like imaging capability of the Triboindenter was indispensable to find the narrow area of the interface and position the indenter.

3 Results and Discussion

Figure 2 shows the SEM image of cement paste adjacent to gravel and results of nanoindentation on the same area. From the SEM image (Figure 2 (a)), no obvious differences in properties were found at the interface. To determine nanomechanical properties, exactly the same area was identified using the AFM imaging mode of the indenter. Indentation was performed on the selected locations as shown in Figure 2 (b). Figure 2 (c) shows the modulus in GPa obtained from each indent. No difference in the modulus was observed in this area adjacent to the gravel.

In a different area of the same sample, cement paste close to the gravel showed higher porosity in SEM and AFM image. Local elastic modulus determined from nanoindentation on this area reflected the effect of higher porosity. The indentation modulus obtained was lower than the previous case. The average modulus in the interfacial zone was found to be 85% of that of the paste matrix. This is close to the value assumed by a few resent researchers for modeling purposes [5]. The

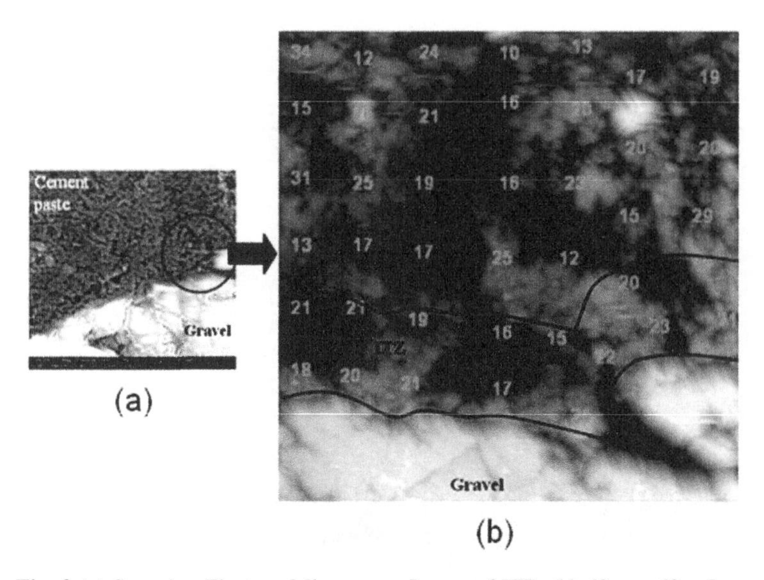

Fig. 2 (a) Scanning Electron Microscopy Image of ITZ, (b) 60 μ × 60 μ Image of the ITZ between Paste and Gravel Showing Indent Locations and Indentation Modulus in GPa Written on each Indent Locations

modulus obtained from indentation all around one gravel plotted against distance from the gravel showed no trend of increasing modulus of cement paste with increasing distance from the gravel. This is in contrast to what has been reported by Zhu et al. [8], although one has to keep in mind that Zhu et al. used microindentation to study the ITZ between paste and steel bar.

To eliminate any uncertainty that might be present due to the mixed chemical composition of round gravel, limestone aggregates were used for further study. To determine the variation of modulus with distance from limestone, the experiment was repeated on eight different locations around the same limestone aggregate.

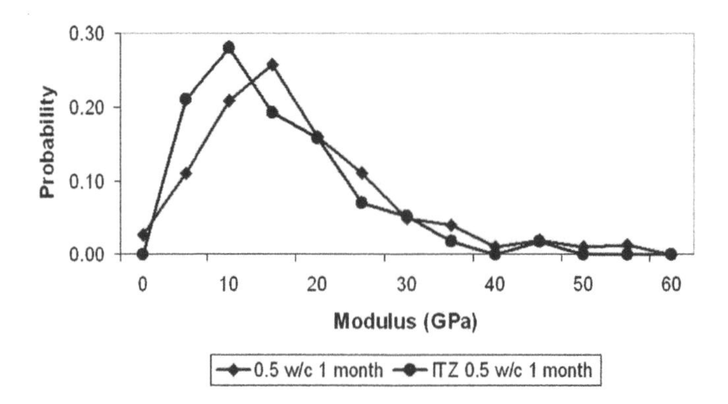

Fig. 3 Comparison between the Modulus of the ITZ and the Bulk Paste

Analyzing results of over 200 indents, no trend of increasing modulus with distance from aggregate was found. Figure 3 shows a comparison between the distributions of modulus obtained from nanoindentation on the interfacial zone with modulus of bulk cement paste. Comparisons showed that the average modulus of the ITZ is 30% lower than the average modulus of the bulk cement paste. Furthermore, the percent of data within the modulus range of 5 to 15 GPa is much higher in the case of the ITZ compared to that of the bulk paste. This implies higher porosity in the interfacial zone.

4 Conclusions

Nanoindentation with imaging was proved to be indispensable to identify the narrow area around aggregate and position the indenter in the same area. Interfacial transition zone was found to be extremely heterogeneous with some areas as strong as the bulk matrix. A higher concentration of large voids and cracks along the interface was observed due to poor bonding. Nanoindentation was performed on relatively intact areas of interface. Results from this study disagree with the notion of increasing elastic modulus with distance from the interface. However, results show higher porosity in the ITZ and the average modulus of the ITZ is 70% to 85% of the average modulus of the bulk paste depending on the aggregate type. Still, extreme heterogeneity within the interface and poor bonding between the aggregate and paste remain as the main problem that affects the overall mechanical properties of concrete. It is important to note that the connectivity of the weaker areas such as large voids and cracks along the interface will govern failure. Thus, modeling concrete as three phase material considering the ITZ as a weak zone around aggregate with average property some percentage less than the paste matrix may not be sufficient to predict overall strength.

References

1. Scrivener, K.L., Crumbie, A.K., Laugesen, P.: The interfacial transition zone (ITZ) between cement paste and aggregate in concrete. Interfac. Sci. 12, 411–421 (2004)
2. Simeonov, P., Ahmad, S.: Effect of transition zone on the elastic behavior of cement-based composites. Cement Concrete Res. 25, 165–176 (1995)
3. Bentz, D.P., Garboczi, E.J., Schlangen, E.: Computer simulation of interfacial zone microstructure and its effect on the properties of cement-based composites. Mat. Sci. Concrete 4, 44 (1995)
4. Lutz, M.P., Monteiro, P.J.M., Zimmerman, R.W.: Inhomogeneous interfacial transitionzone model for the bulk modulus of mortar. Cement Concrete Res. 27, 1113–1122 (1997)
5. Sun, Z., Garboczi, E.J., Shah, S.P.: Modeling the elastic properties of concrete composites: Experiment, differential effective medium theory, and numerical simulation. Cement Concrete Comp. 29, 22–38 (2007)

6. Ramesh, G., Sotelino, E.D., Chen, W.F.: Effect of transition zone on elastic stresses in concrete materials. J. Mater. Civil Eng. 10, 275–282 (1998)
7. Zhu, W., Bartos, P.J.M.: Application of depth-sensing microindentation testing to study of interfacial transition zone in reinforced concrete. Cement Concrete Res. 30, 1299–1304 (2000)
8. Zhu, W., Sonebi, M., Bartos, P.J.M.: Bond and interfacial properties of reinforcement in self-compacting concrete. Mater Struct. 37, 442–448 (2004)
9. Asbridge, A.H., Page, C.L., Page, M.M.: Effects of metakaolin, water/binder ratio and interfacial transition zones on the microhardness of cement mortars. Cement Concrete Res. 32, 1365–1369 (2002)
10. Constantinides, G., Ulm, F.J.: The effect of two types of C-S-H on the elasticity of cementbased materials: Results from nanoindentation and micromechanical modeling. Cement Concrete Res. 34, 67–80 (2004)
11. Hughes, J.J., Trtik, P.: Micro-mechanical properties of cement paste measured by depthsensing nanoindentation: A preliminary correlation of physical properties with phase type. Mater. Charact. 53, 223–231 (2004)
12. Mondal, P., Shah, S.P., Marks, L.D.: Nano-scale characterization of cementitious materials. ACI Mater.J. 105, 174–179 (2008)
13. Nemecek, J., Kopecky, L., Bittnar, Z.: Size effect in nanoindentation of cement paste. In: Proceedings of the International Conference held at the University of Dundee. Thomas Telford, Scotland (2005)

Mitigation of Leachates in Blast Furnace Slag Aggregates by Application of Nanoporous Thin Films

J.F. Muñoz, J.M. Sanfilippo, M.I. Tejedor, M.A. Anderson, and S.M. Cramer

Abstract. The reutilization of slag materials as aggregates is seriously limited by the production of contaminant leachates rich in heavy metals and sulfur when these materials are contacted by water. A unique type of thin-film nanotechnology was used to ameliorate this problem. The surface of the slag was altered by depositing a thin-film comprised of nanoporous oxides. The deposition was performed by coating the aggregates with a suspension containing nanoparticles. Once the water evaporated, a nanoporous thin-film (<0.5 μm) remained firmly attached to the surface of the slag. Different leachate experiments under semi-anoxic conditions were performed using three different nanoparticles oxides films including silica, and titanium. These films were compared against a control. The preliminary results demonstrated that samples coated with one layer of these oxides can decrease the amount of sulfur and calcium in the leachate by 70 and 80%, respectively.

1 Introduction

The combination of different factors such as the aim to reduce greenhouse gas emissions, the achievement of a sustainable development, and simply economical

---

J.F. Muñoz
University of Wisconsin-Madison
e-mail: jfmunoz@wisc.edu

J.M. Sanfilippo
University of Wisconsin-Madison
e-mail: jmsanfilippo@wisc.edu

M.I. Tejedor
University of Wisconsin-Madison
e-mail: tejedor-anderson@engr.wisc.edu

M.A. Anderson
University of Wisconsin-Madison
e-mail: nanopor@facstaff.wisc.edu

S.M. Cramer
University of Wisconsin-Madison
e-mail: cramer@engr.wisc.edu

constraints, have motivated a growing interest to explore new applications for materials that in the recent past have been considered as waste. This is the case of blast furnace slag emanating from iron and steel production that, depending on its gradation, is used as traditional supplementary cementitious materials or as a potential substitute for natural aggregates. However, when the material is contacted by water, the application of slag materials as aggregates is seriously limited by the production of contaminant leachates rich in heavy metals and sulfur (H_2S) [1-4]. The hydration of the high amounts of lime present in the aggregates triggers a rise in pH that leads to a degradation of the amorphous silicates that otherwise binds the sulfur and the other contaminants to the aggregate.

The objective of this research is to explore the capacity of nanoporous thin-films comprised of nanoparticulate oxides, judicially located on the surface of the aggregates, to eliminate or ameliorate the leachate produced when air cooled blast furnace (ACBF) slag aggregates are used as base layer materials.

2 Materials and Sample Preparation

The slag materials selected for this research were ACBF slag aggregates obtained from iron production with ¾ inch size gradation. These aggregates were coated with two types of basic nanoparticle coatings using a "dip coating" method. One of thin-film coatings was a nanoporous silica dioxide and the second a nanoporous titanium dioxide. Both oxide solutions were prepared using standard sol-gel processes [5, 6]. The coating of the slag has been done by immersing baskets of slag aggregate into the sols and later draining the sol from the aggregate at a constant speed. The slag was coated with one layer of either materials and left to dry. As the last steep in the process, the coated samples were heated at 300 ºC for 3 hours. The sintering temperature was selected taking into account the thermal stability of the aggregates and the capability for the sintering particles to themselves and to the slag. The thermal stability of the slag was determined from thermogravimetric and differential thermal analysis of the slag heated in air.

3 Methodology

The methodology initially chosen to study the leaching of these slag materials was based on the standards 1027 and 212-02T from the Ohio State and the Indiana Departments of Transportations, respectively. These two methods were selected since they are commonly used to evaluate the leaching capacity of the ACBF slag aggregates. In order to save time and material, it was decided to scale-down specimen size. In our modified method, 500 g of slag aggregate was used. In these leaching experiments, the samples were kept immersed in deionized water inside 500 ml propylene bottles under almost anoxic conditions as little overhead air remained in the container. These experimental conditions were chosen in order to minimize the lost of the water phase through evaporation. Aliquots of the leachate were extracted and filtered after 24 and 48 hours from preparing the initial samples. Three different analyses were performed on the samples: i) color evaluation of the filtrate and the solid retains in the filter paper, ii) pH measurements of the filtrate, and lastly, iii) a measurement of the calcium and sulfur concentration in the filtrates using inductively coupled plasma (ICP).

From an analysis of the results, we could see some limitations of these particular initial analytical protocols. As a result, a more thorough methodology for the analysis of leached calcium and sulfur was developed in order to quantitatively determine the capability of the thin film coatings to ameliorate the leaching of the ACBF slag aggregates. A scheme of the final methodology used to quantify the leachates is represented in Figure 1. The results predicted by the corresponding chemical equilibrium diagrams were corroborated by applying the analytical protocol to a solution of 0.25 M of $Ca(NO_3)_2$ and K_2SO_4. The analyses done by ICP indicated that the recovery of calcium and sulfate by using the proposed methodology is very close to 100%.

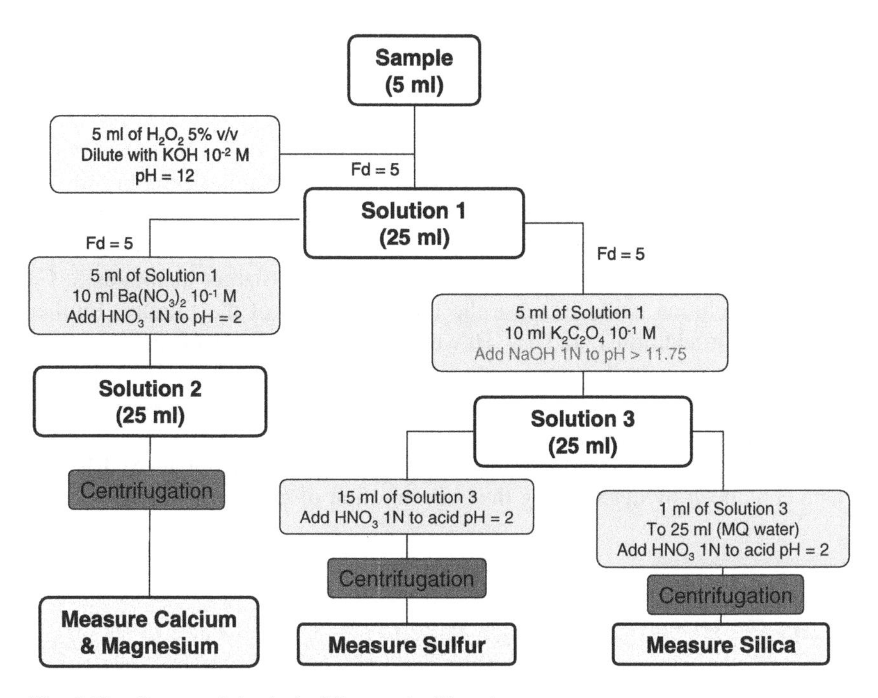

Fig. 1 New Proposed Analytical Protocol of Leachates

4 Results and Discussion

4.1 Analysis of Leachates Using Standards 1027 and 212-02T

The results obtained from the three types of analysis are summarized below. None of the filtrates showed any color. The color of all solids retained by the filter matched a light brown color labeled as 10YR 8/2 in the rock-color chart. Exceptions were the samples coated with titanium that exhibited a darker brown color (10YR 5/4).

The measured pH values of the filtrates were different for coated and uncoated samples as can be seen in Figure 2. The pH in control samples oscillated from basic (~ 10) to acid (~ 2) values during the first 48 hours of leaching. A similar trend but over a different pH range and also smaller in magnitude was observed in the samples coated with silica. The samples with a coating of titanium showed a different behavior. The pH shifted to more basic values during the first 48 hours but the shift was very small when compared with the other two systems. The low values of pH in most of the filtrates can be explained by the leaching of sulfur as sulfide or polysulfide. This is to be expected as the leaching test was performed under rather anoxic conditions. During these extraction and filtration procedures, sulfur species were exposed to atmospheric oxygen. Under these conditions, the sulfides can easily oxidize to sulfates, as it is expressed in equation 1.

$$H_2S + 4H_2O \leftrightarrow HSO_4^- + 9H^+ + 8e^- \tag{1}$$

The oxidation of one mole of sulfides produces nine moles of protons that explains the acidification observed in the filtrate. The fact that the 24 hours filtrate sample has a very basic pH can have several explanations: a smaller leaching of sulfur than in the rest of the systems; most of the sulfur being retained on the filter as colloidal polysulfates; and even a third explanation that larger quantities of Ca leaching into solution could increase the buffer capacity of the filtrate. Further hypothesis await additional studies. However, one thing seems clear, the coated slag produces a more similar pattern of pH values in the filtrates than do uncoated slag samples. Thus, it can be concluded that nanoporous coatings on slag result in quite different leaching behaviors.

This first evaluation of the potential of the coatings to ameliorate ACBF slag leachate concluded by determining the concentration of calcium and sulfur in the

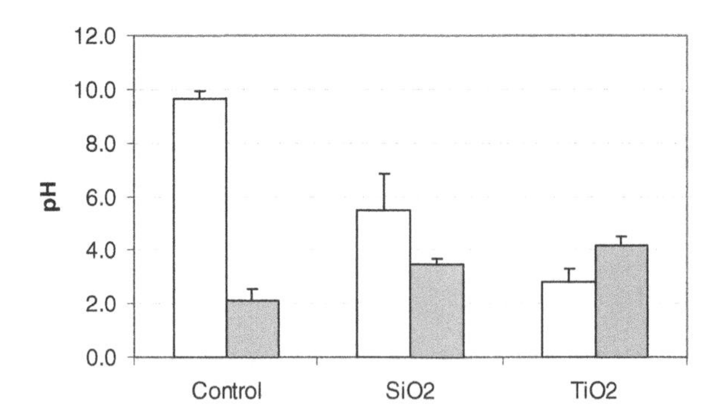

Fig. 2 pH Values of the Filtered Water Measured at 24 (☐) and 48 (☐) Hours

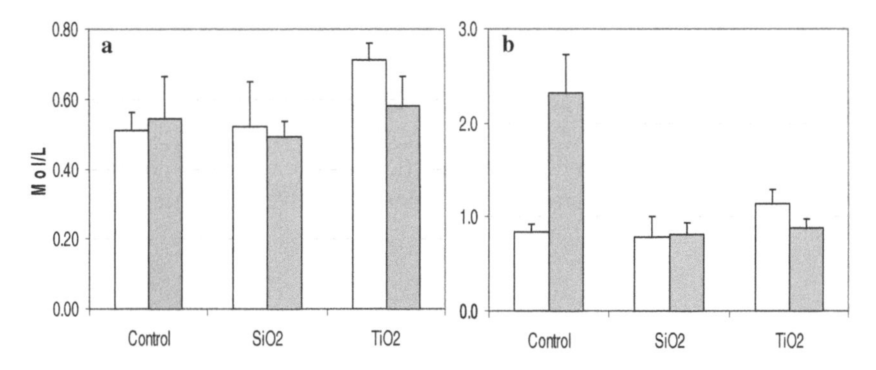

Fig. 3 Concentration of Calcium (a) and Sulfur (b) at 24 (☐) and 48 (■) Hours in Slag Leachate

filtrates. These two elements were chosen as tracers of the leaching activity of the slag aggregates. The results are represented in Figure 3. The values obtained for the concentration of calcium did not show any significant difference with respect to leaching time of coated versus uncoated slag. On another hand, Figure 3b shows a *more than 50% reduction* in the concentration of sulfur in the filtrates associated with silica and titanium dioxide coated slag after 48 hours of leaching.

In the leaching mechanism proposed by Schwab et al. [7], the amount of sulfur liberated is directly dependant of the amount of soluble calcium originating during the hydration of the lime. Originally, the sulfur is trapped inside some of the inter-granular amorphous silica matrix. The hydration process of the lime triggers the resulting basic pH of the system to dissolve this matrix. Therefore, a lower amount of leached sulfur from the coated samples should be accompanied with a lower amount of calcium. This correlation was not observed in the analysis of the filtrates of Figure 3. The homogeneity in the values of calcium concentration could be explained if the soluble calcium was controlled by the solubility product of some calcium salt present as a solid phase in the leachate. In this case, the leached calcium will be the sum of the calcium in the filtrate and the calcium on the filter. Therefore, an analysis of leachated solutes in the filtered will not allow one to evaluate the total leached calcium. A similar problem can be encountered when measuring sulfate in the filtrates, as some of the sulfates can be present in the leachate as an insoluble phase. Under the anoxic conditions of the test, the sulfur is as sulfide that could easily be in the formation of polysulfides. The polysulfide particles are colloidal in nature and could be retained and/or adsorbed by the paper filter. Furthermore, the sulfates can form calcium sulfates, which is not very soluble. Despite the limitations of these analytical protocols, there are some encouraging signs indicating a different behavior for coated and uncoated slag with respect to leaching.

4.2 Analysis of Leachates Using the New Analytical Protocol

The new analysis protocol was applied to leachates taken from the three systems mentioned above, 60 days after mixing the slag with water. It is worthwhile to mention that, after 2 months of near anoxic conditions, only the bottles of the control displayed a characteristic green color, indicative of higher presence of polysulfides.

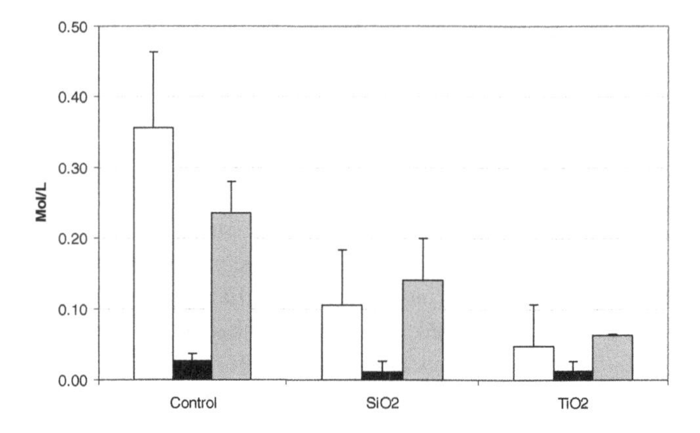

Fig. 4 Leaching Concentration under Anoxic Conditions of Calcium (□), Magnesium (■), and Sulfur (□) Measured at 60 Days

Results of these analyses, shown in Figure 4, indicated that coating the slag with a thin-layer of either oxide significantly *decreased* the amount of *calcium and sulfur leached* under anoxic conditions. The amount of calcium leached with SiO_2 and TiO_2 is only 28% and 14% of the one leached in the control system. The same trend is true for the quantity of leached sulfide. The results illustrate the higher capacity of the titanium oxide versus the silica coating to ameliorate leaching.

5 Conclusions

These results clearly suggest that the thin-film nanotechnology has the potential to stop or significantly decrease the production of environmental unfriendly leachates of ACBF slag aggregates. The experiments have proved that the nanoporous coatings of metal oxides can be used as an effective barrier to avoid this diffusion and ultimately decrease the leaching in ACBF slags. Therefore, it is worthwhile to study this subject matter in more detail, for example, the influence of the number of coatings, different oxide coatings, etc.

This new way of applying nanoparticles in concrete processing opens the door to the possibility of managing and manipulating the physical-chemical properties of aggregates depending on specific needs. In other words, this technology could be applied in concrete to improve the flexural and tensile strengths, permeability of concrete in addition to its resistance to alkali silica reaction.

References

1. Kosson, D.S., van der Sloot, H.A., Sanchez, F., Garrabrants, A.C.: An integrated framework for evaluating leaching in waste management and utilization of secondary materials. Environ. Eng. Sci. 19, 159–204 (2002)
2. Barna, R., Moszkowicz, P., Gervais, C.: Leaching assessment of road materials containing primary lead and zinc slags. Waste Manage. 24, 945–955 (2004)
3. Rastovcan-Mioc, A., Cerjan-Stefanovic, S., Curkovic, L.: Aqueous leachate from electric furnace slag. Croat. Chem. Acta. 73, 615–624 (2000)
4. Mayes, W.M., Younger, P.L., Aumonier, J.: Hydrogeochemistry of alkaline steel slag leachates in the UK. Water Air Soil. Poll. 195, 35–50 (2008)
5. Anderson, M.A., Gieselmann, M.J., Xu, Q.Y.: Titania and Alumina Ceramic Membranes. J. Membrane Sci. 39, 243–258 (1988)
6. Chu, L., TejedorTejedor, M.I., Anderson, M.A.: Particulate sol-gel route for microporous silica gels. Microporous Materials 8, 207–213 (1997)
7. Schwab, A.P., Hickey, J., Hunter, J., Banks, M.K.: Characteristics of blast furnace slag leachate produced under reduced and oxidized conditions. J. Environ. Sci. Heal A 41, 381–395 (2006)

Possible Impacts of Nanoparticles on Children of Thai Construction Industry

W. Musikaphan and T. Kitisriworaphan

Abstract. A possible impact of nanoparticles on human health becomes a concerned issue especially among children who probably lack of self protection. For Thai construction workers, their pre-school children are more likely to expose such the fine particles due to they have to spend their lives in construction site. This study points out the health problems related to nanoparticles exposition among pre-school children of Thai construction workers. The finding indicated that children who reside and play in construction site are more likely to expose to chemical particles and left behind toxic materials during pre and post construction process than others. Thus, urgent policy is strongly recommended for this vulnerable group since all children are very important as the main source of the national productivity in the future, especially in the aging society.

1 Introduction

Thailand is a developing country in which most population work in agricultural sector. Since 1962, Thailand launched the first National Economic Development Plan which provided the country and people to enjoy with economic growth. An influence of economic growth for Thailand is urbanization. The extension of cities, especially the capital; Bangkok, has pulled a lot of unskilled labors from rural areas to work in many sectors including construction sites [1]. Urban sprawl is still going along with urbanization pulling more and more number of rural people to work in construction industry. Almost 2 million rural-urban migrants in Thailand participate in the construction industry and these majority workers are unskilled

W. Musikaphan
National Institute for Child and Family Development, Mahidol University, Thailand
e-mail: wimontip79@yahoo.com
http://www.cf.mahidol.ac.th

T. Kitisriworaphan
Institute for Population and Social Research, Mahidol University Thailand
e-mail: nanosoctk@gmail.com
http://www.ipsr.mahidol.ac.th

[2]. Normally, these workers are vulnerable groups of people due to they are low or uneducated as well as having no power to ask for safe health condition in working site.

As we all know that technological change is one condition for keeping economic growth for any country. Among new technologies that is releasing from their laboratory to the market, nanotechnology is an outstanding technology that most scholar expected to change the globalized industry [3]. With technological change, a coming up of nanomaterials in Thai construction industry generates the higher quality buildings as well as the higher number of industry's benefit. The construction companies can save their cost with more powerful nanomaterials. Nowadays, there are some estimation from National Nanotechnology Center, Ministry of Science and Technology that it is about more than 100 nanoproducts in Thai construction industry [4]. Anyway, any changing of innovation is significantly needed new matching skill and proper knowledge to deal or apply with due to the new one always comes with both benefit and harmful effects, especially the very tiny particle of nanomaterials.

In fact, the terms "nanotechnology" or "nanoproduct" is known only among the well-to-do people due to they have strong potentiality to access information. But for those 2.0 million workers who are low educated and work for little earning in construction business, nanotechnology or nanoproduct is meaning nothing. The harmful effects and possible consequences of nanoparticles on biological systems are mentioned in many various forums. The very large surface area of ultra-small particles can result in the direct generation of harmful oxyradicals (ROS): these can cause cell injury by attacking DNA, proteins and membranes [5]. Furthermore, the ability of the nanoparticles to penetrate the body and cells (e.g., via skin and respiratory system) is possible. A study of T.K.Joshi in the topic on Impact of Nanotechnology on Health mentioned that nanotechnology is likely to become a source for human exposures by different routes: inhalation (respiratory tract), ingestion, dermal (skin) and injection (blood circulation) [6]. In addition, even few studies done to investigate the pulmonary toxicity of nanoparticles in rats but their result showed that lung exposures to ultrafine or nanoparticles produce greater adverse inflammatory and fibrotic responses when compared with larger-sized particles among rats [7].

As far as one concerns, there are about 200,000 workers working in construction sites in Bangkok and periphery. These workers have about 400,000 children aged 0-3 years old on average who are allowed to play and do their activities in construction sites and nearby areas. These children are accepted as one of the vulnerable groups due to they have no ability to protect themselves from unseen toxic particles. So it is believed that they have more chance to expose toxicity available in nanoproducts applied in the sites more than other children in the older ages and of other occupations.

The paper is intended to examine possible health impact of nanoparticles available in environment among pre-school age children of construction workers in construction sites in Bangkok and periphery aimed to point out the possible bad effect to those who are vulnerable groups. Finding of the paper will be beneficial for concerned agencies to take more action for dealing with unseen upcoming toxicity. We also set up the hypothesis that there are some positive relationship of lower respiratory symptom and skin symptom that caused from unspecific particle among preschool children in construction site.

2 Methodology

As Nakorn Pathom is a one of fifth migration destination especially for Northeastern construction worker due to number of increasing in construction area. Data for analysis are collected from Puttamonthon hospital, a district hospital located in Nakornpathom province; a suburb province of Bangkok in 2007. Child patients in preschool ages (aged 0-3 years old) are studied comparing with those 4- 12 years old who face with suspected respiratory symptoms and 0-3 years old child patients are studied and compared with those aged 4-60 years old who face irritation skin. Besides, the data was selected by consideration on climate effect that all cases only selected from summer time (April to July) that will reduce some effects from patient who might get cold because of season change during rainy and winter times.

Table 1 Availability of variables

Variables	Respiratory symptom	Skin irritation
Dependent variable		
- facing with respiratory symptom	0= upper respiratory symptom	0= other skin symptoms
	1= suspected respiratory	1= suspected symptoms
	2= lower respiratory symptom	
Independent variable		
- sex	0= female	0= female
	1= male	1= male
- age	0= 4-12 years old	0= 4-60 years old
	1= 0-3 years old	1= 0-3 years old
- parents' occupation	0= not work/live in construction site	0= not work/live in construction site
	1= work/live in construction site	1= work/live in construction site
- medical expense for 0-99 baht	0= pay more than 99 baht	0= pay more than 99 baht
	1= pay 0-99 baht	1= pay 0-99 baht
- medical expense for 100-199 baht	0= pay more 199 or lower than 100 baht	0= pay more 199 or lower than 100 baht
	1= pay 100-199 baht	1= pay 100-199 baht
- medical expense more than 200 baht	0= pay less than 200 baht	0= pay less than 200 baht
	1= pay more than 200 baht	1= pay more than 200 baht

For respiratory symptom, Multinomial logistic regression is employed for data analysis due to dependent variable composes of being upper respiratory symptom, suspected respiratory and lower respiratory symptom. Concept for dividing the respiratory symptoms into above 3 groups is from "type J" symptom and disease identified by International Classification of Disease (ICD-10) which shows diseases of clear cause and some unclear. For those symptoms and diseases of unclear causes, as we assume them partly occurring from nanoparticles. For skin irritation, Binary logistic regression is employed for data analysis due to dependent variable is categorized as dummy variable of 0= being other skin symptoms and 1= being suspected symptoms. As same as the respiratory grouping concept, ICD-10 in "type L" is applied for grouping vivid cause of diseases and those of unclear one. Details of variables are shown in table 1 above.

3 Results and Discussion

3.1 Respiratory Symptoms

According to the assumption that child patients who their parents are construction workers being more likely to expose toxic nanoparticles than those of other occupations. Thus, table 2 below is basically designed for expressing the number of samples identified by their parents' occupations.

Table 2 Parents' occupation of child patients identified by types of symptom

Symptom	Living condition		Total
	Not work/live in construction site	Work/live in construction site	
Upper respiratory	127 (29.3)	306 (70.7)	433 (100.0)
Suspected	135 (51.5)	127 (48.5)	262 (100.0)
Other lower respiratory	26 (47.3)	29 (52.7)	55 (100.0)
Total	288 (38.4)	462 (61.6)	750 (100.0)

3.2 Finding on Respiratory Symptom

Table 3 presents results from multivariate analysis, which is mainly conducted to examine parents' occupation of construction worker on children's respiratory problem with nanoparticles dispersed in construction site. The coefficients in the form of odds ratios are presented. An odds ratio greater than 1 indicates that the independent variable increase the log odds, all else being equal, while odds ratio less than 1 indicates that the independent variable decreases the log odds. Results are discussed as follow.

The first contrast shows the log odds of being suspected symptom relative to being upper respiratory symptom among child patients. It appears that age has positive effect to child patients' probability of being suspected respiratory symptom, given that the child is upper respiratory symptom. Compared to a child who aged 4-12, the log odds of a child aged 0-3 years old on being suspected respiratory symptom versus being upper respiratory symptom is about seven times more. Result shows that the younger children are more likely to get risk than those of older age children due to the younger they are the lower protection they have. Interestingly, compared to a child who his/her parents do not working or living in construction site, the log odds of a child whose being suspected respiratory symptom is reduced by 84 per cent, given that a child of construction-worker parents. With this finding, it means that toxicity can be found everywhere and it can expose to children no matter whether children live or play in construction site or not.

Table 3 Odd ratios of respiratory symptoms among child patients

Independent variables	Suspected symptom vs. upper respiratory		Other lower respiratory vs. upper respiratory		Other lower respiratory vs. Suspected symptom	
Age of child patients						
- Patient age 4-12 (ref.)						
- Patients age 0-3	7.360***	(0.217)	2.080*	(0.332)	0.283***	(0.340)
Sex						
-Female (ref.)						
-Male	0.751	(0.179)	3.873***	(0.384)	5.155***	(0.393)
Parents' occupation						
-Not work/live in construction site (ref.)						
-Work/live in construction site	0.160***	(0.217)	0.347**	(0.329)	2.171*	(0.338)
Medical expense						
> 200 baht (ref.)						
0-99 baht	0.669*	(0.193)	3.205**	(0.387)	4.790***	(0.402)
100-199 baht	5.344*	(0.329)	17.533***	(0.486)	3.281**	(0.460)
Constant	0.716	(0.205)	0.025***	(0.502)	0.035***	(0.510)
Model chi2	231.44					
Df	10					
N	750					
p-value	0.000					

*** $p < 0.001$, ** $p < 0.01$, * $p < 0.05$

For making clear of finding, simulation is employed for showing the level of being suspected respiratory symptom among child patients. In terms of parents' occupation, there are about 22% of children of construction worker being suspected respiratory symptom compared with 52% of children of other occupations. The result from simulation through adjusted proportional distribution technique shows similar finding with multinomial logistic regression technique which confirms that toxic substance can harm children even they are in construction site or not.

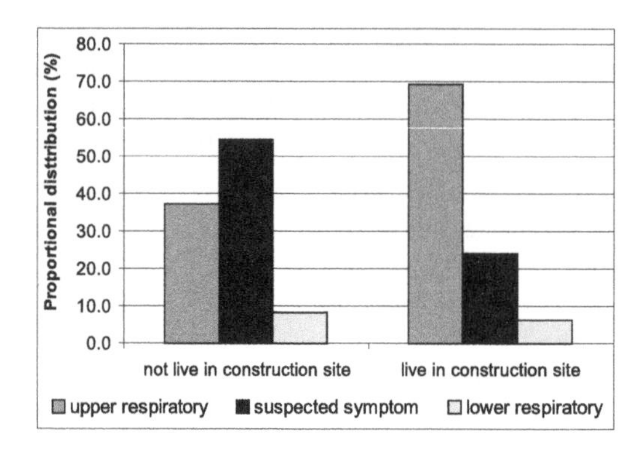

Fig. 1 Proportional distribution for being respiratory symptoms between children of construction worker and children of other occupation

3.3 Skin irritation Symptoms

This part is done under the hypothesis that child patients whose their parents are construction workers being more likely to expose toxic nanoparticles than those of other occupations. There was a study done to evaluate whether metallic nanoparticles smaller than 10 nm could penetrate and permeate the skin. This study found that nanoparticles were able to penetrate the hair follicle [8]. Thus, skin irritation from unclear cause might be one symptom relating to toxic exposing.

3.4 Finding on Skin Symptom

Table 4 presents results from binary logistic regression analysis, which is mainly conducted to examine parents' occupation of construction worker on children's skin problem with nanoparticles dispersed in construction site. *Model 1*, the first model shows that patients who live in construction site are 2.5 times more for being suspected skin symptom. In *Model 2*, two independent variables covering age of patients and sex are added into the model and finding shows that a person living/working in construction site is 2.6 times more for being suspected skin symptom. More interestingly, compared with patients aged 4-60 years old, pre-school patients are three times more for being suspected skin symptom. The last model (*Model 3*) shows that medical expense directly relates with being suspected skin symptom as patients who pay 0-99 baht and 100-199 baht are12 times and 4 times more for being suspected skin symptom than those who pay more than 200 baht. This finding means that suspected skin symptom is more likely to appear in low income people than those of higher income. Even this finding might be concrete, but it can tell us who are more likely to be victim if nanoparticle can generate unpredictable harmful effect.

Table 4 Odd ratios of suspected skin symptoms among patients

Variable	Model1		Model2		Model3	
Constant	0.688***	(0.105)	0.909	(0.153)	0.229***	(0.238)
Not live in construction site (ref.)						
Live in construction site	2.543***	(0.164)	2.66***	(0.170)	2.774***	(0.188)
Patient age 4-60 (ref.)						
Pre-school patients (0-3)			3.092***	(0.262)	1.401	(0.294)
Female (ref.)						
Male			0.470***	(0.170)	0.557**	(0.186)
Medical expense > 200 baht (ref.)						
Medical expense 0-99 baht					12.299***	(0.272)
Medical expense 100-199 baht					4.832***	(0.225)
Model chi2	33.40		77.20		183.37	
Df	1		3		5	
N	650		650		650	
p-value	0.000		0.000		0.000	

*** $p < 0.001$, ** $p < 0.01$, * $p < 0.05$

4 Conclusions

Even though this study is done under nano-related information constrain, but it is a starting point for the next step of studying the effect of nanotechnology on people in society, especially for vulnerable people like pre-school aged children. These children have no power to out-cry as well as no chance to protect their rights themselves. Thailand is going to be aging society which means that the number and quality of children is very important for running the country development further.

References

1. National Economic and Social Development Board (NESDB), The Ninth National Economic and Social Development Plan. Office of Prime Minister (2002)
2. National Statistic Office (NSO), Labor force survey round 1-3. National Statistical Office. Ministry of Interior (2006)
3. Compañó, R., Hullmann, A.: Forecasting the development of nanotechnology with the help of science and technology indicators. Nanotechnology 13, 243–247 (2002)
4. Taepakum, S.: Application of Nanotechnology in Construction Industry. In: International Conference on Nanotechnology in Thailand (in Thai) (2008)
5. Brown, D.M., Wilson, M.R., MacNee, W., Stone, V., Donaldson, K.: Size-dependent flammatory effects of ultrafine polystyrene particles: a role for surface area and oxidative stress in the enhanced activity of ultrafines. Toxicol. Appl. Pharmacol. 175, 191–199 (2007)
6. Joshi, T.K.: Impact of nanotechnology on Health. National Conference on nanotechnology and Regulatory Issues, January 9-10, 2009, Calcutta University, India (2009)
7. Warheitet, D.B., et al.: Health effects related to nanoparticle exposures: Environmental, health and safety considerations for assessing hazards and risks. Pharmacol. Therapeut., 35–42 (2008)
8. Baroli, B., et al.: Penetration of Metallic Nanoparticles in Human Full-Thickness Skin. J. Invest. Dermatol. 127, 1701–1712 (2007)

Characterization of Alkali-Activated Fly-Ash by Nanoindentation

J. Němeček, V. Šmilauer, and L. Kopecký

Abstract. Nanoindentation was employed for the characterization of reaction products, mainly N-A-S-H gel, within alkali-activated fly ash samples. Heat and ambient-cured samples from ground fly ash were indented in a grid of hundreds of indents. The intrinsic Young's modulus of N-A-S-H gel was found around the mean value 17.70 GPa, regardless on the curing procedure. Such finding elucidates intrinsic stiffness of mature N-A-S-H gel with different origin. Partly-activated slag, slag and fly-ash particles were further distinguished by histogram deconvolution.

1 Introduction

Alkali-activated fly ash (AAFA) is a new promising material forming stable inorganic binder. AAFA provides high potential in a partial replacement of ordinary concrete due to improved durability, acid and fire resistance, low calcium content, low drying shrinkage, no alkali-silica reaction, good freeze/thaw performance or lower creep induced by mechanical load [11]. The potential utilization of fly ash, as a by-product of coal power plants, brings attention of several researchers [6, 8, 11, 12].

Chemically, the main reaction product of fly ash is an amorphous aluminosilicate gel (denoted further as N-A-S-H gel) and/or C-S-H gel forming in the

J. Němeček
Czech Technical University in Prague
e-mail: jiri.nemecek@fsv.cvut.cz
http://mech.fsv.cvut.cz

V. Šmilauer
Czech Technical University in Prague
e-mail: vit.smilauer@fsv.cvut.cz
http://mech.fsv.cvut.cz

L. Kopecký
Czech Technical University in Prague
e-mail: lubomir.kopecky@fsv.cvut.cz
http://mech.fsv.cvut.cz

presence of calcium and low alkalinity activator [1]. The chemical composition of the N-A-S-H gel is similar to crystalline natural zeolitic materials but the microstructure is of amorphous nature. The N-A-S-H gel consists of three-dimensional structure, built from SiO_4 and AlO_4 tetrahedra connected by shared O atoms and forming polymeric chains [0, 0]

$$M_n[-(SiO_2)_z - AlO_2]_n \cdot wH_2O \tag{1}$$

where M stands for sodium, potassium or calcium supplied with alkali activator and fly ash, n is the degree of polymerization, z quantifies the amount of SiO_2 monomer units in the gel, typically within the range from 1 to 3 and w is the amount of binding water.

Several experimental techniques can be applied to characterize mechanical behavior of individual components of the AAFA composite. Nanoindentation plays an important role among the experimental techniques working at submicron length scale. Nanoindentation is based on the direct measurement of the load-displacement (P-h) relationship using a very small tip (typically diamond) pressed into the material. Standard processing of the measured P-h relation is based on the analytical solution of a contact problem involving an indenter and a semi-infinite solid body and provides the hardness and Young's modulus. The Oliver-Pharr [7] solution assumes perfectly flat surface and isotropic elasto-plastic material. Results from a similar cementititous material can be found in [2, 3].

The objectives of this paper aim at the characterization of intrinsic N-A-S-H gel properties in the heterogeneous microstructure of AAFA on the scale of micrometers. Ambient and heat-cured samples were prepared from the same composition to explore the differences in the curing procedure.

2 Experimental

2.1 Materials

The raw fly ash (RFA) originates from Chvaletice, Czech Republic, with the Blaine specific surface 210 m^2kg^{-1}. The average chemical composition of this RFA is given in Tab. 1 with SiO_2/Al_2O_3 mass ratio 1.58. RFA was ground in a small-scale ball mill in the quantity of 8 kg for 45 minutes. Activating solution was prepared by dissolution of $NaOH$ in a tap water with the addition of sodium soluble water glass in the proportions specified in [10]. The cylindrical moulds of 22 mm in diameter and 40 mm in length were filled, vibrated for 5 minutes and sealed. Curing was performed either at 80°C for 12 hours or at ambient temperature conditions at 20°C for 170 days. AAFA remained sealed before cutting and polishing for nanoindentation.

Table 1 Average chemical composition of the raw fly ash (main components)

Component	SiO_2	Al_2O_3	Fe_2O_3	CaO	MgO	Na_2O	K_2O	TiO_2	As_2O_3	V_2O_5	Cr_2O_3	ZnO	C	Rest	Total
Weight (%)	51.9	32.8	6.3	2.7	1.1	0.33	2.12	1.89	0.03	0.067	0.29	0.024	0.2	0.5	100

2.2 Methods

Before nanoindentation procedure, samples were polished on a series of emery papers, polishing cloth and cleaned in an ultrasonic bath. Than, three representative areas from each sample were selected. Nanoindentation was performed as a series of grids of about 10 x 10 = 100 imprints in each area. The distance between individual indents varied in order to cover heterogeneity of the sample and was set in the range between 10 and 50 μm. Nanohardness tester CSM was used for all of the tests. All together, around 700-800 imprints have been carried out for each AAFA sample. All experimental nanoindentation measurements were performed in a load control regime. Trapezoidal loading diagram was prescribed for all tests. Linear loading 4 mN/s (lasting for 30 s) was followed by the holding period (30 s) and unloading 4 mN/s (30 s). Maximum load was prescribed 2 mN for all indents. The applied load led in a maximum penetration depths ranging from 100 nm to 400 nm (average 260 nm) depending on the hardness of the indented material phase. The effective depth captured by the tip of nanoindenter can be estimated as four times of the penetration depth. It yields the effective depth around 1 μm for this particular case.

The environmental scanning electron microscope XL30 ESEM FEI PHILIPS was employed at gathering pre- and post-indentation images.

3 Results and Discussion

3.1 ESEM

Heat and ambient-cured polished samples were observed by ESEM in back scattered electrons, Fig. 1. The light luminous points are the iron rich particles (*Fe-Mn* oxides). The light gray compact spheres are alumina- silica rich glass particles. Only a small part of porous fly ash particles and slags remain intact by the alkali activation process. A great portion of the dark gray matter is N-A-S-H gel arising preferentially from activation of slags and, to a lesser extent, from amorphous silica from spherical fly ash particles.

The grinding process of RFA has a positive impact on the opening of internal structure of highly porous slag particles. The sickle-like crushed thin shells of non-activated fly ash particles are observable in the figure. The degree of alkali activation is estimated by image analysis around 50 %.

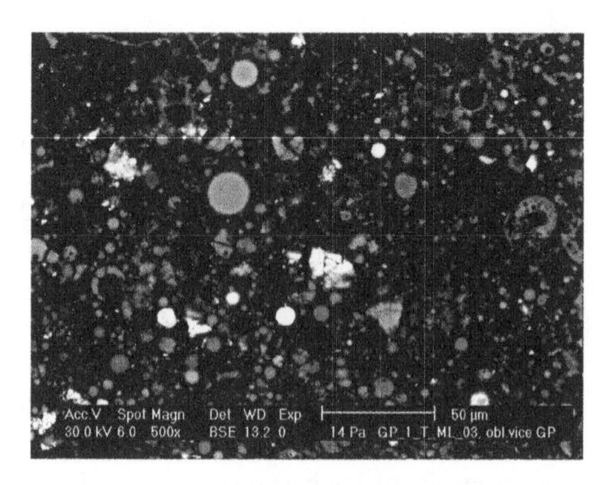

Fig. 1 Typical ESEM (back-scattered electrons) image of heat-cured AAFA

3.2 Nanoindentation

For all indents, elastic moduli were evaluated according to Oliver & Pharr [7] methodology from experimental P-h curves. Poisson's ration was assumed 0.2 for all measurements. Examples of P-h curves belonging to individual material phases are shown in Fig. 2 in which N-A-S-H phase is the most compliant one while non-activated fly-ash particle exhibits the stiffest response.

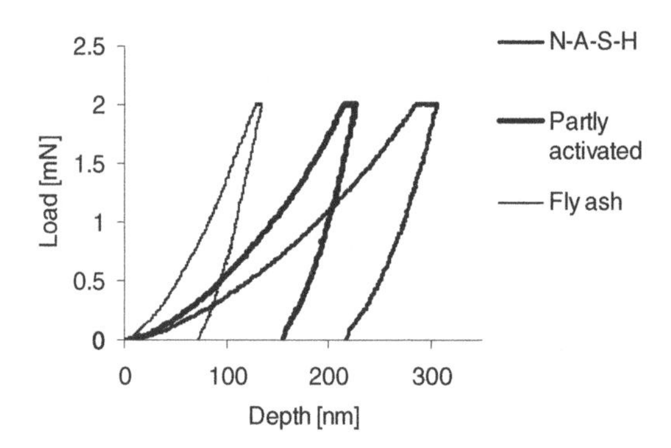

Fig. 2 Typical indentation load-depth diagrams of distinguished phases in AAFA

Preliminary ESEM observation led to the conclusion that AAFA heterogeneity occurs not only on a micrometer range but also on the scale of hundreds of µm, far exceeding the size of fly ash particles. This hypothesis was confirmed experimentally by nanoindentation. Several uniform grids from different AAFA locations yield different histograms of elastic properties on heat cured samples. From these measurements containing approximately 100 indents each may be derived that some areas are rich in a soft N-A-S-H gel while other areas shift toward higher moduli in the area of less activated fly ash. As opposed, ambient curing seems to produce homogeneous AAFA on the scale of hundreds of micrometers. The results are averaged through all grids from each AAFA sample.

Overall results from the measurements (approximately 700 indents for each sample) are merged and plotted in Figs 3 and 4. The mutual comparison shows higher frequency of low elastic modulus for ambient cured sample. The explanation lies probably in different reaction kinetics between ambient and heat-cured sample. Previous microcalorimetry measurement determined the ratio of reaction kinetics between heat and ambient cured sample as 406 [10], favoring more homogeneous formation of N-A-S-H gel due to ion equilibration over large distances in an ambient-cured sample.

In order to identify individual phase properties, statistical deconvolution was applied to both histograms of E modulus. Gaussian distributions were assumed for the deconvolution. In order to identify several material phases, we suggested to apply deconvolution of histograms into four phases, namely N-A-S-H phase (well activated), partly activated phase (higher stiffness), non-activated particles (mainly slag and high stiffness) and fly-ash particles (the highest stiffness) consisting dominantly from amorphous SiO_2.

Heat-cured samples exhibit two important peaks for the activation products, Fig. 3. The first peak can be attributed to N-A-S-H gels while the second one to a partly activated slag. Third and fourth peaks correspond to non-activated particles; probably slags and fly ash. As opposed, ambient-cured samples in Fig. 4 almost lack the second peak which points to a better activation with regard to the heat-cured sample. Also the third and fourth peaks of non-activated particles are smaller.

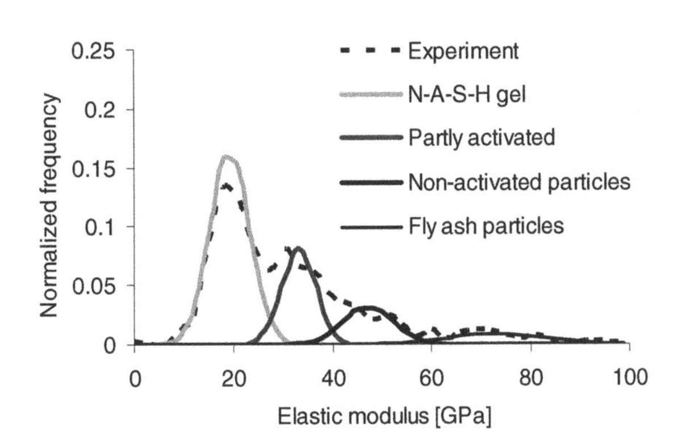

Fig. 3 Deconvolution into four phases for heat-cured samples

Tables 2 and 3 summarize mean values and standard deviations for individual components. The N-A-S-H gel phases have almost identical properties for both heat and ambient-cured samples but frequency of the occurrence in the statistical set is different. Higher frequency was obtained for ambient-cured samples which again satisfy the assumption of higher portion of the well activated fly ash. The elastic properties of minor phases are similar for heat and ambient-cured samples but again their frequencies are different.

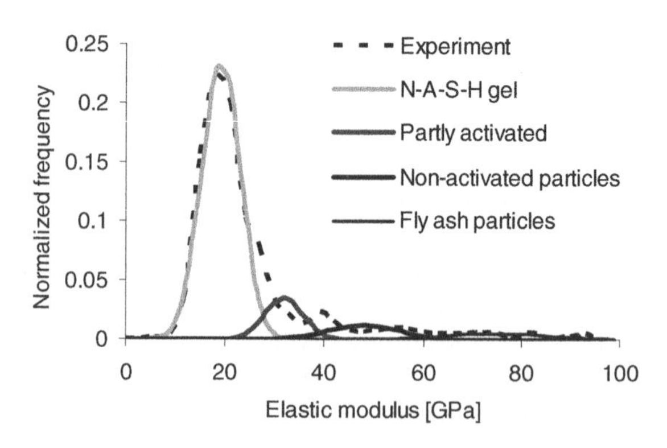

Fig. 4 Deconvolution into four phases for ambient-cured samples

Table 2 Elastic properties of individual material phases of heat-cured samples

	N-A-S-H	Partly activated	Non-activated	Fly-ash
Elastic modulus [GPa]	17.65±3.92	31.50±3.37	45.54±5.03	71.49±9.53
Frequency of occurrence [%]	55.3	24.0	13.5	7.2

Table 3 Elastic properties of individual material phases of ambient-cured samples

	N-A-S-H	Partly activated	Non-activated	Fly-ash
Elastic modulus [GPa]	17.75±3.77	30.50±3.61	46.63±6.45	74.01±10.05
Frequency of occurrence [%]	77.9	10.8	6.8	4.5

4 Conclusions

Nanoindentation was used to characterize dominant phases in the alkali-activated brown low-calcium fly ash. The main reaction product, N-A-S-H gel, seems to exhibit an intrinsic Young's modulus irrespective on the curing procedure on the tested scale of 1 μm. Such finding is important in the view of yet fully unexplained N-A-S-H gel structure [9]. In the parallel comparison with C-S-H gel studies [2, 3], one can speculate about similarly arranged building block with the same solid fraction in the indentation volume. The nanoindentation technique is therefore indispensable as a tool for the characterization on various length-scales.

Acknowledgments. The presented research has been supported by the Ministry of Education, Youth and Sports of the Czech Republic under grant MSM6840770003 and by the Czech Science Foundation under projects 103/08/1639 and 103/09/1748.

References

1. Alonso, S., Palomo, A.: Calorimetric study of alkaline activation of calcium hydroxide-metakaolin solid mixtures. Cem. Concr. Res. 31, 25–30 (2001)
2. Constantinides, G., Ulm, F.-J.: The effect of two types of C-S-H on the elasticity of cement-based materials: results from nanoindentation and micromechanical modeling. Cem. Concr. Res. 34, 67–80 (2004)

3. Constantinides, G., Ulm, F.-J.: The nanogranular nature of C–S–H. J. Mech. Phys. Sol. 55, 64–90 (2007)
4. Davidovits, J.: Chemistry of geopolymeric systems terminology. In: Geopolymer 1999 International Conference, France (1999)
5. Fernández- Jiménez, A., Palomo, A., Criado, M.: Microstructure development of alkali-activated fly ash cement: a descriptive model. Cement and Concrete Research 35, 1204–1209 (2004)
6. Hardjito, D., Rangan, B.: Development and properties of low-calcium fly ash-based geopolymer concrete, Research report GC 1, Curtin University of Technology, Perth, Australia (2005)
7. Oliver, W.C., Pharr, G.M.: An improved technique for determining hardness and elastic modulus using load and displacement sensing indentation experiments. J. Mat. Res. 7, 1564–1583 (1992)
8. Rangan, B.V.: Fly ash-based geopolymer concrete, Research Report GC 4, Curtin University of Technology, Perth, Australia (2008)
9. Sherer, G.: Structure and properties of gels. Cem. Concr. Res. 29, 1149–1157 (1999)
10. Škvára, F., et al.: Material and structural characterization of alkali activated low-calcium brown coal fly ash. Journal of Hazardous Material (2009) (submitted)
11. Wallah, S., Rangan, B.: Low-calcium fly ash-based geopolymer concrete: Long term properties, Research Report GC 2, Curtin University of Technology, Perth, Australia (2006)
12. Williams, P.J., et al.: Microanalysis of alkali-activated fly ash - CH pastes. Cem. Conc. Res. 32, 963–972 (2002)

Multi-scale Performance and Durability of Carbon Nanofiber/Cement Composites

F. Sanchez, L. Zhang, and C. Ince

Abstract. This paper reports on recent work that is directed at understanding the fundamental controlling mechanisms of multi-scale, environmental weathering of nano-structured cement-based materials through an integrated experimental and computational program. The effect of surface treatment and admixture addition on the incorporation of carbon nanofibers (CNFs) in cement composites was studied. Silica fume and surface treatment with nitric acid facilitated CNF dispersion. The CNFs were found as individual fibers anchored in the hydration products throughout the cement pastes and as entangled networks in cavities. The presence of the CNFs did not modify the compressive or tensile strength of the composite but did provide it with a fair level of mechanical integrity post testing. Preliminary results on durability indicated a residual effect of the CNFs after decalcification of the composites as manifested by a slow load dissipation after peak load under compression. Molecular dynamics modeling of the reinforcing structure-cement phase interface demonstrated that manipulation of the interface characteristics may provide a method to control the composite properties.

1 Introduction

Nano-level modifications of the structure of cement-based materials have the potential of greatly enhancing the material mechanical properties and durability and of opening the door for new applications in civil engineering infrastructure. A promise of nanotechnology is the use of carbon nanofibers and nanotubes as nano-reinforcement, or nano-rebar, to replace the steel rebar, a main cause of concrete degradation. High specific strength, good chemical resistance, and electrical and thermal conductivity are several properties that make carbon nanofibers/nanotubes interesting as cement reinforcement [1, 2]. However, understanding the evolution and performance of the nano-reinforcement interface is of critical importance.

F. Sanchez, L. Zhang, and C. Ince
Vanderbilt University
e-mail: florence.sanchez@vanderbilt.edu

Vapor grown carbon nanofibers (CNFs) are multiwall, highly graphitic structures with diameters ranging from 70 to 200 nm and lengths up to a few hundred microns. CNFs present numerous exposed edge planes along their surface, which in turn constitute potential sites for advantageous chemical or physical interaction. In addition, these fibers are well characterized, offer similar benefits as carbon nanotubes at a lower cost, and are already produced in ton per year quantities [3].

There is a complex, time-dependent and multi-scale interaction that occurs between an aging material and its surrounding environment. Exposure to weathering forces moves components into and out of the material causing internal chemical changes and stresses that affect the reinforcing fiber-cement interface. The properties of nanofiber reinforced, cement-based materials exist in, and the degradation mechanisms occur across multiple length scales (nano to macro). The nano-scale ultimately affects the properties and performance of the bulk material.

This paper reports on recent work [4-6] that is directed at developing CNF/cement composites that have long-term performance and durability. The objective is to understand the fundamental controlling mechanisms of multi-scale, environmental weathering of nano-structured cement-based materials through an integrated experimental and computational program focusing on how molecular level, chemical phenomena at internal interfaces influence long-term, bulk material performance. The performance of CNF/cement composites is discussed in terms of microstructural, physical, and mechanical properties.

2 Experimental Approach

Commercially available vapor grown CNFs (Pyrograf®-III PR-19-LHT, Applied Sciences, Inc., Cedarville, OH, USA) were used for the study. The CNFs were used "as received" and after surface treatment with 70% nitric acid. The CNFs were added to Portland cement (PC) pastes and PC pastes with 10 wt% silica fume (SF cement). The following materials were prepared: (i) plain reference PC paste, (ii) PC paste containing 0.5 wt% of "as received" CNFs, (iii) PC paste containing 0.5 wt% of surface treated CNFs with nitric acid, (iv) reference SF cement paste, and (v) SF cement paste containing 0.5 wt% of "as received" CNFs. A water to cementitious material (cement + SF) ratio of 0.33 was used for all mixes. After a minimum curing time of 28 days, some specimens were conditioned for 95 days under a concentrated solution of ammonium nitrate (590 g/L NH_4NO_3) to accelerate decalcification. A variety of tests were conducted on the non-degraded and degraded composites, including compression and splitting tensile tests, scanning electron microscopy (SEM) observation of the fracture surface, x-ray diffraction, BET analyses, and thermal analyses. A summary of the main findings is provided below. Details of the experimental techniques can be found in [4, 5].

3 Results and Discussion

3.1 Microstructure of CNF/Cement Composites

For all composites examined, SEM observations of the fractured surface revealed entangled networks of CNFs filling cavities created in the cement paste. Van der Waals interactions between "as received" CNFs presented a significant barrier to fiber dispersion. The current challenge to improving the composite properties is the break-up of the initial clumps of fibers. In general, a certain level of break-up was observed to occur with the addition of SF and after surface treatment of the CNFs with nitric acid [5, 6]. For these two cases, the CNFs were found as individual fibers well anchored inside the hydration products throughout the cement pastes (Fig. 1) in addition to the entangled networks (clumps of intertwined CNFs) in cavities. These results clearly demonstrated the potential for CNFs to intimately interact with the cement phases.

Fig. 1 SEM of the fracture surface of CNF/cement composites with nitric acid surface treated CNFs, showing individual CNFs anchored in the paste

3.2 Macroscopic Properties of CNF/Cement Composites

For all mixes tested, the splitting tensile strength of the CNF/cement composites was comparable to the reference cement pastes (Fig. 2). Subjected to compressive loads, though no significant change in the strength was observed, the CNF/cement composites retained a certain mechanical integrity post testing (Fig. 3). The propagation of cracks may have been limited by (i) the entangled clumps of CNFs inside the cavities, (ii) the well anchored fibers at cavity edges bridging the paste and the CNF networks, and/or (iii) the individually dispersed fibers (SF and surface treated CNF composites only). While static compression and tensile tests are an incomplete measure of the mechanical properties, these results are encouraging because no attempt to optimize the dispersion was made. Performance enhancements may be expected from on-going work using chemical functionalization of the surface, optimum physical blending, and/or the use of surfactants. Aspects of this work are being guided by the use of molecular dynamics modeling.

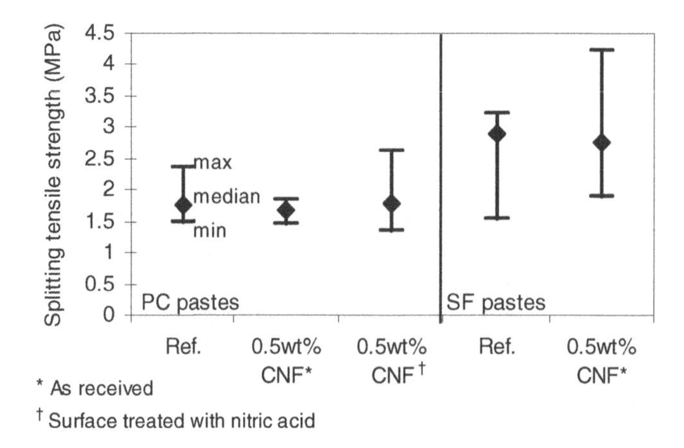

Fig. 2 Splitting tensile strength of CNF/cement composites

Fig. 3 CNF/ cement composites post compression testing. a) Reference PC paste and b) PC paste with 0.5 wt% nitric acid surface treated CNFs

3.3 Durability of CNF/Cement Composites

Many types of concrete degradation are closely associated with decalcification of the cement paste. It has been shown that calcium can be used as a good indicator of the chemical deterioration of concrete [7]. The load-displacement curves of the PC pastes with 0 wt% and 0.5 wt% CNF loading obtained before and after exposure to accelerated decalcification using ammonium nitrate solution for 95 days are presented in Fig. 4 . These initial results showed no evident difference in compressive strength after exposure to ammonium nitrate between the reference PC paste and the corresponding paste with CNFs. Decalcification resulted in ca. 50% reduction in compressive strength. With decalcification, the compressive strength behavior evolved from a more brittle to a more ductile behavior with a slow load dissipation after failure. This was more pronounced for the PC paste with CNFs than for the reference PC paste, indicating a residual effect of the CNFs.

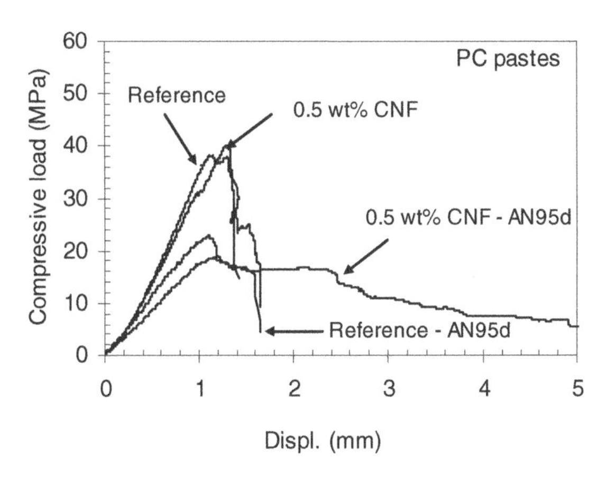

Fig. 4 Comparison of compressive load displacement curves of CNF/cement composites for PC pastes before and after decalcification for 95 days (AN95d)

3.4 Fiber-Cement Interaction

Molecular dynamics simulations were performed to investigate the interactions between PC pastes and surface treated carbon fibers [8]. A model derived from a model for the 9 Å tobermorite structure was used to represent the C-S-H phase of cement. Standard models were used for graphite surfaces with several different attached, reactive moities and a plain surface with no attached moities. In the development of CNF/cement composites, they offer insight into the local interactions among individual atoms, groups of atoms, and phases. The results indicated that significant improvement in interfacial interaction is possible through appropriate surface functionalization of the graphite surface. H-bonds and calcium counter ions played a significant role in bridging the structure across the interface. Careful control of the type and amount of functionalization is necessary to optimize the strength of the H-bond network and other ionic interactions.

4 Conclusions

Silica fume and surface treatment with nitric acid facilitated CNF dispersion and improved the interfacial interaction between the CNFs and the cement phases. Though the ultimate load failure during static compression and tensile testing were unchanged, improvements were observed post failure with a fair level of mechanical integrity observed for composites containing CNFs. Additionally, preliminary results on durability indicated that after decalcification the CNF composite was more ductile, retaining some residual strength post peak load. Molecular dynamics modeling was found to be a useful and promising technique for understanding the interfacial interaction between the cement phases and the reinforcing structure.

Acknowledgments. Funding from the National Science Foundation under NSF CAREER CMMI 0547024 is gratefully acknowledged.

References

1. Makar, J.M., Beaudoin, J.J.: Carbon nanotubes and their applications in the construction industry. In: Proceedings of the 1st International Symposium on Nanotechnology in Construction, Paisley, Scotland, June 23-25 (2003)
2. Chong, K.P., Garboczi, E.J.: Smart and designer structural material systems. Prog. Struct. Mat. Eng. 4, 417–430 (2002)
3. Kang, I., Heung, Y.Y., Kim, J.H., et al.: Introduction to carbon nanotube and nanofiber smart materials. Compos. Part B-Eng. 37, 382–394 (2006)
4. Sanchez, F., Ince, C.: Effect of carbon nanofiber (CNF) loading on the macroscopic properties and microstructure of hybrid CNF/ Portland cement composites. Compos. Part A-App. S (submitted) (May 2008)
5. Sanchez, F., Ince, C.: Microstructure and macroscopic properties of hybrid carbon nanofiber/silica fume cement composites. Compos. Sci. Tech. (submitted) (July 2008)
6. Sanchez, F.: Carbon nanofiber/cement composites: challenges and promises as structural materials. Int. J. Materials and Structural Integrity (submitted) (December 2008)
7. Thomas, J.J., Chen, J.J., Allen, A.J., et al.: Effects of decalcification on the microstructure and surface area of cement and tricalcium silicate pastes. Cement. Concr. Res. 34, 2297–2307 (2004)
8. Sanchez, F., Zhang, L.: Molecular dynamics modeling of the interface between surface functionalized graphitic structures and calcium-silicate-hydrate: Interaction energies, structure, and dynamics. J. Colloid. Interf. Sci. 323, 349–358 (2008)

Nano-structured Materials in New and Existing Buildings: To Improved Performance and Saving of Energy

F. Scalisi

Abstract. Improving well-being in buildings, in relation to energy conservation, represents a great challenge. In southern Italy a basic problem is that of keeping buildings cool in the summer months. This problem affects not only newly-erected buildings, but also the large number of existing buildings, some of which are of historical importance. Nano-technology represents an excellent opportunity to harness the salvage of existing buildings to the living requirements of contemporary society. The use of nano-structured materials in newly-erected buildings will lead to improved performance and a considerable saving of energy. Above all, the use of nano-structured materials in existing buildings will provide the possibility of intervention in these buildings and help improve, for example, insulation or lighting, without invasive intervention and consequent damage to the building itself.

1 Introduction

Nanotechnology is about the manipulation of matter at the nanoscale. A nanometre is a billionth of a metre (m=10-9 m). It is an 80.000th of a diameter of a hair. Nanotechnology opens up new possibilities in material design. On this level material behaves differently to how it does on the macro-level; objects can change colour and shape much more easily and fundamental properties such as force, surface/mass relationship, conductibility and elasticity can be improved in order to create material that can provide a better performance than present ones. The possibilities provided by nanotechnology embrace the most disparate sectors, from electronics to medicine, from energy to aeronautics, to name but a few; building is one of these and is considered a promising area of application for nanotechnology. The considerable modifications in materials and, consequently, building processes

---

F. Scalisi
University of Palermo, Department of Progetto e Costruzione Edilizia – DPCE
e-mail: francescascalisi@gmail.com

indicates that nanotechnology can provide radical and systematic innovation in architecture; the extent to which, and the manner in which architects, engineers, researchers, builders and producers embrace this innovation will determine the future of architectural operations.

2 Nanostructured Materials for the Energy Efficiency of Buildings

In the architectural sphere the advent of nanostructured materials is considered decisive for the energy efficiency of buildings. Nanotechnology provides new technological means with which to tackle climatic changes and contribute to reducing gas emissions in the near future. The first phase of the Kyoto Protocol will end in 2012 and CO_2 emissions throughout the world will have to be halved by 2050. Energy efficiency in buildings is therefore indispensible, especially since constructions are one of the major producers of CO_2 emissions. Architects are called to find innovative solutions in order to slow down climatic change, combining the requirements of dwelling-areas with energy efficiency.

One of the basic problems linked to energy consumption in buildings is represented by winter heating and summer cooling. Heat-loss and gain are closely connected to the presence of glass surfaces and to the insulating capacity of the outer cladding. As regards glass surfaces, nanotechnology is reducing heat-loss and gain by using glass covered with layers of thin thermo-chromatic, photo-chromatic and electro-chromatic film. Thermo-chromatic technology is capable of varying its own light absorption in function of its external surface temperature, becoming opaque above a certain critical temperature and then becoming transparent again with a fall in temperature.

Photo-chromatic technology autonomously modifies its light transmission in function of the amount of incident light on its surface. Lastly, electro-chromatic cladding gradually varies its own transmission in function of an electric signal; in order for the glass to become transparent again a new backward electrical impulse signal is required. All these applications are intended to reduce the use of energy for heating and cooling buildings and might contribute to helping diminish energy consumption in buildings.

Another category of material that has received a great boost from the arrival of nanotechnology is that of cladding/coating. Insulation coating represents a field of notable importance for the application of nanotechnology; it heralds the creation of materials with a greater insulating action than conventional insulation, but of a lesser thickness.

These performances characterise Vacuum Insulation Panels (VIP), which are capable of guaranteeing the same thermic transmittance as traditional insulation with a thickness that is ten times inferior; they are made up of a nucleus of

Fig. 1 Vacuum Insulation Panel (VIP)

material of low thermic conductibility, which can be subjected to high pressure, whilst the cladding is made of plastic or extremely flexible and resistant metals. Research has highlighted the need, apart from great resistance to compression and low thermic conductibility, for the central nucleus material to be characterised by a high degree of porosity, in order to facilitate the passage of air; therefore, importance must be given to the size of the pores, which must be less than 100 nanometres, in order to avoid phenomena of thermic gas conductibility.

"Aerogel is an ultra-low density solid, a gel in which the liquid component has been replaced with gas. Aerogel has a content of 5 percent solid and 95 percent air, and can support over 2,000 times its own weight. Aerogel panels are available with up to 75 percent translucency, and their high air content means that a 9cm (3.5") thick aerogel panel can offer an R-value of R-28, a value unheard of in a translucent panel. One of the greatest potential energy-saving characteristics of nanocoatings and thin films is their applicability to existing surfaces for improved insulation. Adding thermal insulation to existing European buildings could cut current building energy costs and carbon emissions by 42 percent or 350 million metric tons" [2].

Nanotechnology promises to render insulation more efficient, less dependent on non-renewable resources and less toxic. Producers estimate that insulation materials deriving from nanotechnology will be about 30% more efficient than those from conventional materials.

One of their most important characteristics of insulation nano-coating is its applicability to existing surfaces to improve their insulation; it can be applied directly to the surfaces of existing building, whilst the post-construction addition of conventional insulation materials such as cellulose, glass-fibre, polystyrene is extremely invasive.

Its application to existing structures could lead to huge savings in energy and it does not seem to pose a threat to the environment and health in the way that glass-fibre and polystyrene do.

Nanotechnology promises to render insulation more efficient, less dependent on non-renewable resources and less toxic.

Fig. 2 Silica aerogel

Table 1 Example of masonry in a building in Sicily

Masonry	Thickness
External plaster of lime and gypsum	mm 30
Extruded polystyrene foam	mm 40
Brick (250x120x50)	mm 120
vertical layer of air	mm 60
Brick (250x120x250)	mm 120
Internal plaster of lime and gypsum	mm 20

Transmittance is 0531 W/m^2K with a thickness of 390 mm, with an insulating nanostructured could have a better transmittance with a lower thickness.

Fig. 3 Aerogel with glass

3 Conclusion

It should be pointed out that buildings are responsible for a quarter of carbon emissions in the European Union, 70% of which stems from heating requirements. By saving on the heating of spaces through better insulation, the European Union could reduce carbon dioxide emissions by 100 million tonnes per year, and by so doing ensure that Europe alone might reach its goal of reducing carbon emissions by 25% by 2010. In spite of its enormous potential, there are several factors that might impede the adoption of nanotechnology on a large scale: above all the high cost of nano-products compared to conventional ones. Nanotechnology does represent a relatively recent accomplishment and prices are destined to fall, as is usually the case, over the course of time, with all new technology. Secondly, the building market is extremely conservative and therefore tends to proceed cautiously in adopting new technologies; those in the trade seem to know very little about nanotechnology and its potential implications for the building sector. Knowledge and skills are still too fragmentary to enable it to spread extensively in the building sector. Moreover, from the point of view of demand, there will be a certain reluctance regarding the introduction of nanotechnological materials until convincing documentation is produced regarding its functionality and the long-term effects. Finally, there is considerable anxiety about the general public's seeming reluctance to accept nanotechnology.

References

1. Alagna, A.: Energie & tecnologie in architettura, ricerche per una possible casa passive in Sicilia. DPCE, Palermo (2004)
2. Elvin, G.: Nanotechnology for Green Building. Green Technology Forum (2007)
3. Hegger, M., Fuchs, M., Stark, T., Zeumer, M.: Atlante della sostenibilità. UTET, Torino (2008)
4. Leydecker, S.: Nanomaterials in Architecture, Interior Architecture and Design. Birkhäuser, Basel (2008)
5. Mann, S.: Nanotechnology and Construction. Nanoforum (2006)
6. Sala, M.: Recuperoedilizio e bioclimatica. Sistemi Editoriali, Napoli (2001)
7. Scalisi, F.: I materiali nanostrutturati nel settore edilizio. In: Sposito, A. (ed.) Agathòn, vol. 2, Offset Studio, Palermo (2008)
8. Scalisi, F.: Nanotechnology in construction: the new means for the sustainable development. In: Fabris, L.M.F. (ed.) Enviroscape a manifesto. 2nd blu+verde International Congress, Maggioli, Rimini (2008)

Stability of Compressed Carbon Nanotubes Using Shell Models

N. Silvestre and D. Camotim

Abstract. This paper presents some remarks on the use of shell models to analyse the stability behaviour of single-walled NTs under compression. It is shown that there are three different categories of critical buckling modes of NTs under compression: while the axi-symmetric mode is critical for very short NTs, the flexural buckling mode is critical for long tubes. While the former exhibits cross-section contour deformation but no warping deformation, the later is characterised by the opposite situation (warping deformation but no contour deformation). Additionally, a third category exists (distortional buckling): it takes place for NTs with moderate length, it is related to the transitional buckling behaviour between the shell (axi-symmetric mode) and the rod (flexural mode) and it is characterised by both cross-section contour deformation and warping deformation. Concerning the distortional buckling behaviour of moderately long NTs, it is also shown that the well known Donnell-type theory of shells leads to erroneous results.

1 Introduction

After the seminal work of Iijima [5], much research has been done on carbon nanotubes (NTs). Since then, most of the studies performed were based on molecular dynamics approach to simulate the NT non-linear behaviour. Nevertheless, it is known that molecular dynamics analyses are computationally expensive and time consuming and are often limited to a maximum number of atoms. Yakobson et al. [11] analysed the buckling behaviour of single-walled NTs under compression, bending and torsion and compared the results (critical measures and buckling modes) obtained by molecular dynamics simulations with those determined by continuum shell models. They showed that shell model results agreed fairly well with those obtained from molecular dynamics simulations. Since then, a large

N. Silvestre and D. Camotim
Department of Civil Engineering and Architecture, ICIST/IST, Technical University of Lisbon, Lisboa, Portugal
e-mail: nunos@civil.ist.utl.pt

amount of investigations has been carried out, most of them confirming the accuracy of continuum shell analyses. However, the use of shell models to analyse the NT buckling behaviour should be carefully addressed, as different shell theories lead to very dissimilar results (Silvestre [9]).

2 Buckling Modes, Critical Strains and Shell Theories

The local buckling behaviour of cylindrical shells and tubes, which involve deformation of the circular section contour, can be classified in two categories:

- *Local buckling (axi-symmetric) modes*: they occur for very short lengths and are characterised by null warping deformation of the circular section (Fig. 1(a)).
- *Distortional buckling (diamond-shape) modes*: they occur for short lengths and are characterised by significant warping deformation of the circular section. Figure 1(a) depicts the in-plane deformed configuration of a distortional buckling mode with two circumferential half-waves (m=2) and its warping displacement profile (in perspective).

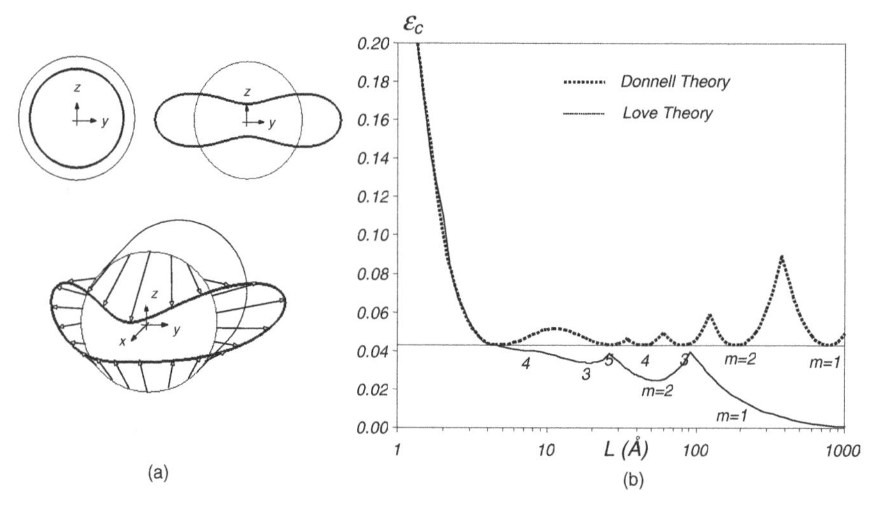

Fig. 1 (a) Local mode (m=0) and two-wave distortional mode (m=2) configuration and warping displacement profile, (b) Variation of the critical strain ε_c with the length L of NT(23,0)

In the theoretical investigations of NTs under compression, it is always assumed that the critical strain ε_c and the buckling half-wavelength values of *both local and distortional* buckling modes are based on the *Donnell theory* of shells and are given by

$$\varepsilon_c = \frac{h/r}{\sqrt{3(1-\nu^2)}} \qquad L_c = \frac{n\pi\sqrt{h\,r}}{\sqrt[4]{12(1-\nu^2)}} \tag{1}$$

where h and r are the shell thickness and radius, respectively, and n is the number of longitudinal half-waves. Less well known than *Donnell theory* is the *Love theory* of shells. For the local buckling modes, both Donnell and Love theories lead to the same expressions (Eq. (1)). However, for the distortional modes, the Love theory of shells gives the expressions for the critical strain ε_c and the buckling half-wavelength L_c,

$$\varepsilon_c = \frac{h/r}{\sqrt{3(1-\nu^2)}}\frac{m^2-1}{m^2+1} \qquad L_c = n\pi\,r\sqrt{\frac{r}{h}}\frac{\sqrt[4]{12(1-\nu^2)}}{m\sqrt{m^2-1}} \tag{2}$$

For illustration purposes, figure 1(b) depicts the variation of the critical strain ε_c with L and m, for the NT(23,0) with a single longitudinal half-wave (n=1). In this analysis, one adopted ν=0.19, r=9Å and h=0.66Å. It is seen that the two curves exhibit several local minima. The critical strain obtained from the Donnell theory (ε_c=0.043) does not depend on the NT length L. Unlike the Donnell-type theory, the critical strain ε_c obtained from Love theory for the distortional buckling modes (m>1) decreases with increasing lengths. The number of circumferential half-waves of the critical buckling mode (m) also decreases. From this example, it is concluded that the use of Donnell theory is exact for local buckling of very short NTs (axi-symmetric buckling mode with m=0) but the results for long and moderately long tubes (distortional buckling modes) become unsafe, with differences in ε_c values reaching up to 40%. The Donnell-type theory is unable to predict accurately both the warping and tangential displacement profiles due to the omission of some important terms in the kinematic relations. The non accurate estimation of warping and tangential displacements (u and v) has far reaching implications (errors) in the results obtained from the Donnell-type theory for the distortional buckling modes of moderately long NTs. For more detailed information, see Ref. [9].

3 Shell Models and Results for NTs under Compression

The critical value of the compressive force (P_c) applied to the NT corresponds to critical strain by means of $\varepsilon_c = P_c / EA$, where EA is the NT axial stiffness. With the objective of investigating the buckling behaviour of NTs under compression, let us first consider the NT(7,7) with armchair helicity under uniform compression (r=4.75Å). This NT was investigated by Yakobson et al. [11] using both molecular dynamics simulations and analytical formulae derived from shell models. The NT has fix-ended rigid supports (cross-section deformation and global rotation is restrained at both supports and the axial translation of one end section was left free, in order to allow the axial shortening of the NT under large compression) and, like Yakobson et al. [11], the following properties were adopted: ν=0.19, E=5.5TPa, h=0.66Å. Using a shell model based on finite element simulation, the results shown in figure 2 were obtained, where it is depicted the variation of the critical strain ε_c and the buckling mode configuration with L. From the observation of figure 2, the following remarks can be drawn:

- For very short NTs (L<3Å), buckling takes place in *local modes* (m=0) – note that the length is only 30% of the NT diameter and, thus, its buckling behaviour is more similar to that of a short (shallow) cylindrical shell. The critical strain ε_c=0.082 at the local minimum and the corresponding half-wavelength (L_c=3Å) are in perfect agreement with the values obtained from Eq. (1).
- For small to moderate lengths (3Å<L<100Å), the NT buckles in *distortional modes* with two circumferential waves (m=2). The ε_c vs. L variation exhibits several local minima, corresponding to an increasing number n of longitudinal half-waves (from 1 to 5). The critical strain at the local minima (ε_c=0.049) and the corresponding half-wavelength (L_c=21.0Å) are in perfect agreement with the values obtained from Eq. (2). Notice that $\varepsilon_c^D=0.049=0.6\times0.082=0.6\varepsilon_c^L$ and $L_c^D=21Å=3Å\times4.75/0.66=L_c^L\times r/h$ (superscripts L and D denote local and distortional modes).

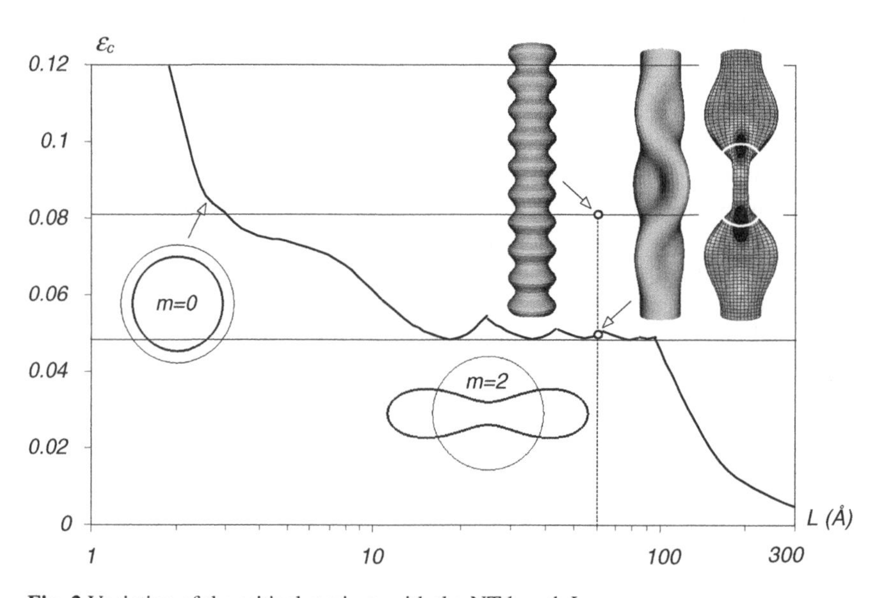

Fig. 2 Variation of the critical strain ε_c with the NT length L

- Using molecular dynamics simulations, Yakobson et al. [11] studied a NT(7,7) with L=60Å and obtained a critical strain ε_c=0.050 (see the top white dot in figure 2), which matches virtually with the numerical value obtained here (ε_c=0.049). Moreover, the critical mode shape determined by these authors also agrees very well with the two-wave distortional mode obtained herein. Using Eq. (1), Yakobson et al. [11] obtained ε_c=0.077 (the small difference with respect to ε_c=0.082 is due to the fact that they used r=5Å, instead of r=4.75Å) and

argued that this value is "*close to the value obtained in MD simulations*", i.e., ε_c=0.050. However, this is not the case, since there is a 50% difference between those values (ε_c=0.077 and ε_c=0.049). This discrepancy is due to the fact that the value ε_c=0.077 is valid only for local (axi-symmetric) mode and not valid for the mode they found (pure distortional), exhibiting identical "...*flattenings perpendicular to each other...*" – see figure 1 in Yakobson et al. [11]. The main difference between these two buckling modes is that the local (axi-symmetric) mode does not exhibit warping (axial) displacements while the distortional one displays warped cross-sections at x=L/3 and x=2L/3 (see white lines in figure 2). These warped cross-sections are qualitatively similar to those represented in figure 1(a).
- For long NTs (L>100Å), buckling takes place in a *bending mode* (m=1) with one longitudinal wave (n=1), and the strain always decreases with L.

Finally, the figure 3 shows the variation of the critical strain with the NT radius, for the local (m=0 – top curve) and distortional (m=2, 3, 4 – bottom curves) buckling modes. Also depicted are the results obtained by several researchers, who carried out atomistic simulations of NTs with different radius and lengths. Very rencently, Cao and Chen [2] performed an extensive study on the buckling behaviour of NTs using molecular mechanics simulations. They investigated NTs with rather small (but similar) radii (NT(5,5), NT(6,4), NT(7,2), NT(8,2) and NT(9,0)) but with different lengths (L=50, 100, 190, 390Å). In order to study the diameter effects, these authors also performed analyses for NT(10,10), NT(14,6), NT(17,0) with equal length (L=50Å) and moderate radii. Using the molecular dynamics method, Liew et al. [6] also carried out extensive numerical invetigations on the buckling behaviour of compressed NTs. They determined the buckling loads of 32 zig-zag NTs and 32 armchair NTs with small, moderate and large radii. Conversely, Cornwell and Wille [3] investigated the buckling behaviour of NTs with larger radii. Additionally, also shown are the results by Yakobson et al. [11], Buehler et al. [4], Arroyo and Belytschko [1], Srivastava et al. [10] and Ni et al. [7].

From the observation of figure 3, it is possible to mention that the great majority of the molecular dynamics estimates and atomistic simulation results are very close to the curve corresponding to the local (axi-symmetric) buckling mode (top curve). This is the case of the NT(8,0) studied by Srivastava et al. [10], which undoubtedly buckled in a local (axi-symmetric) mode. However, there is a non negligible number of results that fall well below this local buckling curve. These results correspond to several points located close to the curves with m=2, 3, 4, which suggests that the corresponding NTs buckle preferably in distortional buckling modes. This clearly occured in the case of the NTs studied by Yakobson et al. [11], Buehler et al. [4], Arroyo and Belytschko [1] and Ni et al. [7]. Pantano et al. [8] also simulated numerically (using continuum shell finite elements) the behaviour of the NT(10,10) investigated by Ni et al. [7] and they found that it buckled in

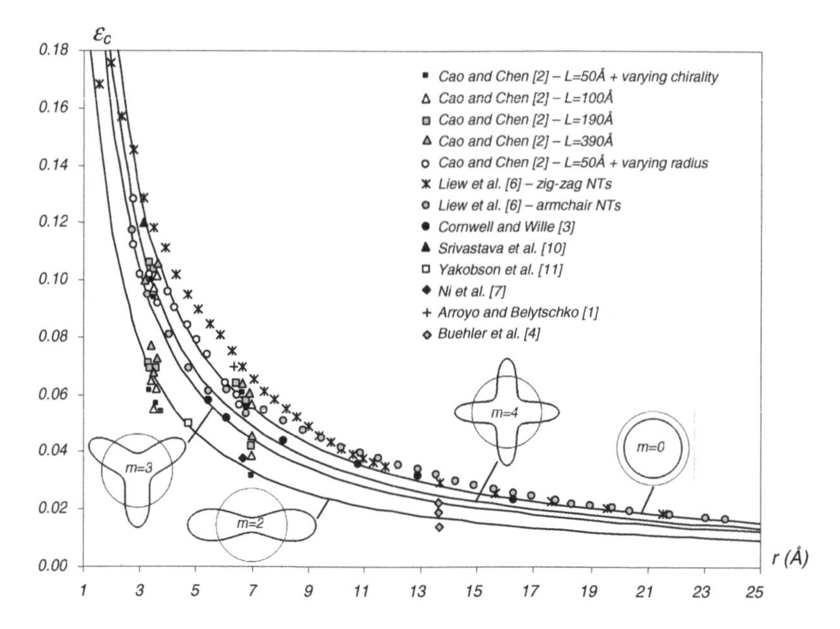

Fig. 3 NT data and variation of the critical strain with the radius

a clear distortional mode (m=2) configuration similar to that depicted in figure 1(a). Moreover, figure 3 shows that several points are very close to the curve corresponding to the distortional buckling curve (m=2, bottom curve). Finally, it becomes visible that the absolute difference between the local (top curve) and distortional (m=2, bottom curve) critical strains increases substantially for diminishing radii. However, remember that the relative difference is always 40%.

References

1. Arroyo, M., Belytschko, T.: Finite element methods for the non-linear mechanics of crystaline sheets and nanotubes. Int. J. Numer. Meth. Eng. 59, 419–456 (2004)
2. Cao, G., Chen, X.: The effect of chirality and boundary conditions on the mechanical properties of single-walled carbon nanotubes. Int. J. Solids Struct. 44, 5447–5465 (2007)
3. Cornwell, C.F., Wille, L.T.: Elastic properties of single-walled carbon nanotubes in compression. Solid State Commun. 101, 555–558 (1997)
4. Buehler, M.J., Kong, Y., Gao, H.: Deformation mechanisms of very long single-wall carbon nanotubes subject to compressive loading. J. Eng. Mat-T ASME 126, 245–249 (2004)
5. Iijima, S.: Helical microtubules of graphitic carbon. Nature 354, 56–58 (1991)
6. Liew, K.M., Wong, C.H., He, X.Q., Tan, M.J., Meguid, S.A.: Nanomechanics of single and multiwalled carbon nanotubes. Phys. Rev. B 69, 115429(1-8) (2004)

7. Ni, B., Sinnott, S.B., Mikulski, P., Harrison, J.: Compression of carbon nanotubes filled with C_{60}, CH_4 or Ne: predictions from molecular dynamics simulations. Phys. Rev. Lett. 88, 205505(1-4) (2002)
8. Pantano, A., Parks, D.M., Boyce, M.C.: Mechanics of deformation of single and multi-wall carbon nanotubes. J. Mech. Phys. Solids 52, 789–821 (2004)
9. Silvestre, N.: Length dependence of critical measures in single-walled carbon nanotubes. Int. J. Solids Struct. 45, 4902–4920 (2007)
10. Srivastava, D., Menon, M., Cho, K.: Nanoplasticity of single-walled carbon nanotubes under uniaxial compression. Phys. Rev. Lett. 83, 2973–2976 (2006)
11. Yakobson, B.I., Brabec, C.J., Bernholc, J.: Nanomechanics of carbon tubes: instabilities beyond linear response. Phys. Rev. Lett. 76(14), 2511–2514 (1996)

Bending Instabilities of Carbon Nanotubes

N. Silvestre and D. Camotim

Abstract. This paper presents an investigation on the buckling behaviour of single-walled carbon nanotubes (NTs) under bending and unveils several aspects concerning the dependence of critical bending curvature on the NT length. The buckling results are obtained by means of non-linear shell finite element analyses using ABAQUS code. It is shown that eigenvalue analyses do not give a correct prediction of the critical curvature of NTs under bending. Conversely, incremental-iterative non-linear analyses provide a better approximation to the molecular dynamics results due to the progressive ovalization of the NT cross-section under bending. For short NTs, the limit curvature drops with the increasing length mostly due to the decreasing influence of end effects. For moderate to long tubes, the limit curvature remains practically constant and independent on the tube length. An approximate formula based on the Brazier expression is proposed to predict the limit curvature.

1 Bifurcation (Eigenvalue) Analysis of Carbon Nanotubes

Let us start by consider the NT(13,0) under uniform bending, previously investigated by Yakobson et al. [8]. The NT is modelled with rigid end sections (the flexural rotations at both supports are free and the axial translation of one end section was left free, in order to enable the axial shortening of the tube under large bending displacements) and the following properties were adopted: ν=0.19, E=5.5TPa, h=0.66Å, r=5.09Å. Performing bifurcation (eigenvalue) analyses, it is found that the NT buckling is triggered by a *local mode*, characterized by the deformation of the top compressed zone. This local mode is very similar to the *axi-symmetric* mode of NTs under compression, both displaying a large number of half-waves. The critical curvature at the local minimum is κ_c=0.0160Å^{-1} and the corresponding half-wavelength is L_c=3.4Å. These values are close to the ones (κ_c=0.0150Å^{-1} and L_c=3.1Å) obtained from

N. Silvestre and D. Camotim
Department of Civil Engineering and Architecture, ICIST/IST, Technical University of Lisbon, Lisboa, Portugal
e-mail: nunos@civil.ist.utl.pt

$$\kappa_c = \frac{h/r^2}{\sqrt{3(1-\nu^2)}} \tag{1}$$

where h and r are the shell thickness and radius, respectively. The minor discrepancy is due to the fact that the top compressed zone of the NT under bending is partially restrained by its bottom tensioned zone. Using molecular dynamics simulations and adopting Brenner's empirical potential for the atomic interactions, Yakobson et al. [8] studied the NT(13,0) with L=80Å and obtained a critical curvature κ_c=0.0155Å^{-1}, which is close to the numerical value obtained (κ_c=0.0160Å^{-1}). Nevertheless, it seems fair to say that the critical mode obtained by the current analysis looks very different to that identified by Yakobson et al. [8]. The later displays only one kink at the middle of the tube and the deformed configuration of the mid-span section is very similar to the two half-wave (m=2) *distortional* mode.

In order to shed light on this subject, several data on NTs under bending obtained by other authors were collected. Figure 1 shows the variation of the critical curvature with the NT radius and includes several dots corresponding to the mentioned data. The solid curve is obtained from Eq. (1), which is related to the critical *local* mode of the top compressed zone. The first remark is that the critical curvature decreases with the NT radius. However, the decreasing rate of the solid curve is much more pronounced than that of the data. The carefull observation of this figure shows that, apart from the Yakobson's result (black dot), all data (white dots) are located below and far from the solid curve. This evidence leads to us to question the accuracy of Eq. (1), which is extensively used in NT analysis (Iijima *et al.* [4], Buehler *et al.* [1]), and to have some reservations on the performance of pure bifurcation (eigenvalue) analysis of NTs under bending.

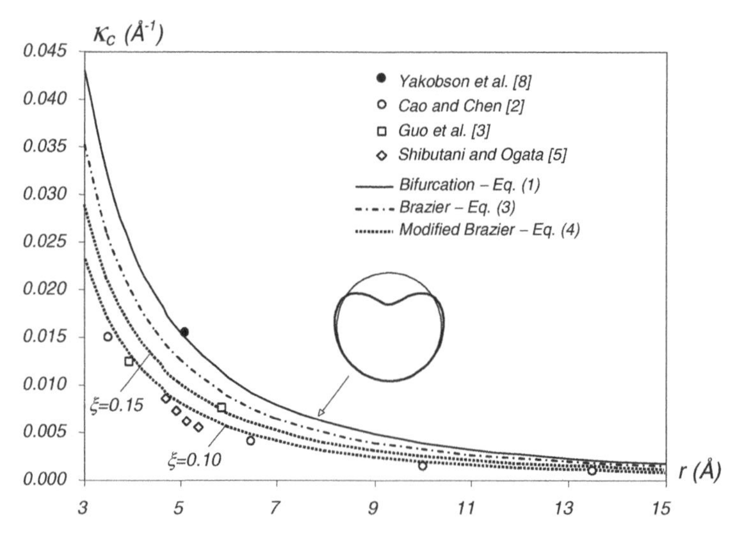

Fig. 1 Variation of the critical curvature with the NT radius

With the aim of investigating the above mentioned difference between the critical curvatures, one is aimed to perform fully *non-linear analysis*. Details of the numerical model can be found in the recent work by the author [6]. First, the non-linear analysis of the NT(13,0) with L=80Å was performed. Figure 2 shows (i) the non-linear equilibrium path M(θ) (or M(κ)) obtained from the incremental-iterative analysis (solid curve) and (ii) the linear equilibrium path (dashed line). Moreover, while several points (**A** to **M**) are located along the non-linear path, point **P** is located on the top of the linear equilibrium path and corresponds to the NT bifurcation in the local mode (the values κ_c=0.0150Å^{-1} and θ_c=0.0150Å^{-1}×80Å=1.20rad are obtained by means of Eq. (1), with linear pre-buckling path). The observation of figure 2 deserves the following comments:

- Until point D is achieved, the equilibrium path is almost linear. After point D, the non-linear path starts to deviate from the linear one and reaches a limit point at G. After that, the bending moment always decreases and the bending angle (or curvature) increases after a small snapback. In the descending branch of the M(θ) curve, the increase in the bending angle leads to a final deformed configuration (point M) with three kinks in the mid-span zone. From this non-linear analysis, it is seen that the NT under bending never reaches the critical bending moment value M_c=22.5TPaÅ3, which is associated with the local mode (point P) determined from the bifurcation analysis. It should be stressed that the same non-linear trend of the ascending branch of the M(θ) curve was also unveiled by Yakobson et al. [8] and Vodenitcharova and Zhang [7].

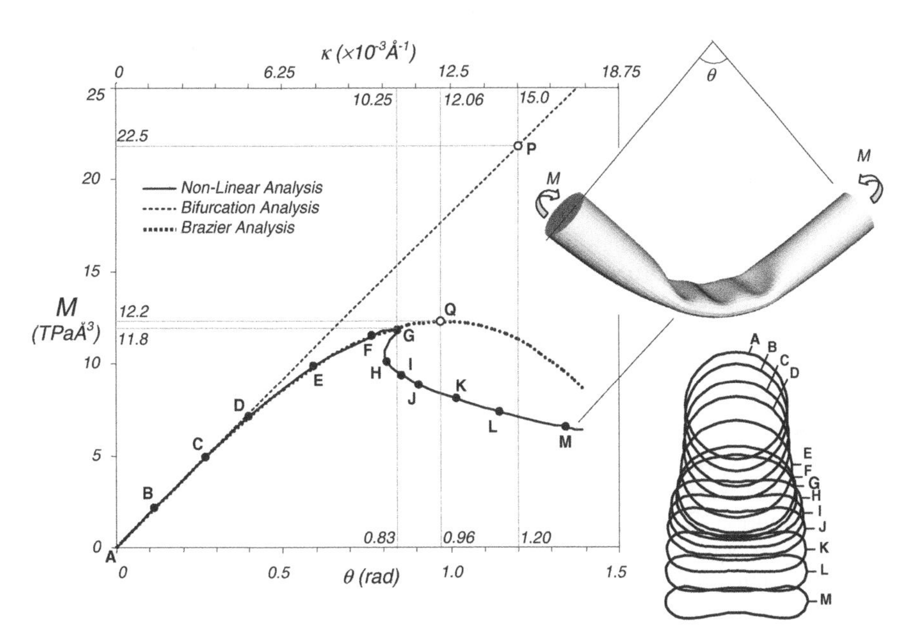

Fig. 2 Non-linear behaviour and progressive ovalization and collapse of NT(13,0) with L=80Å under uniform bending

- The roundness of the non-linear path is related to the well known Brazier effect, which is due to the action of normal (longitudinal) stresses on the curvature of the bended NT, thus resulting in transverse (vertical) pressure directed towards the NT neutral axis. This pressure leads the top (compressive) and bottom (tensioned) zones to move towards the neutral axis, thus resulting in the NT ovalization. It is obvious that the ovalized shape of the cross-section coincides perfectly with the two-wave distortional mode represented in figure 1(b). The NT flattening (ovalization) is responsible to a decrease in the cross-section second moment of area ($I(\kappa)$), which depends on the curvature. Consequently, it also leads to a drop in the actual bending moment ($M=EI(\kappa)\kappa$). The Brazier equilibrium path is shown in figure 2 (dotted curve) and is given by

$$M = EI\kappa(1-\tfrac{3}{2}\xi) \qquad \xi = \kappa^2 \frac{r^4}{h^2}(1-\nu^2) \tag{2}$$

where $I=\pi r^3 h$ is the second moment of area of the circular section and ξ is the ovalisation parameter (oval minor axis width / 2r). The Brazier curve exhibits a local maximum (point **Q**) given by

$$M_{BR} = \frac{2\pi\sqrt{2}}{9}\frac{Erh^2}{\sqrt{1-\nu^2}} \qquad \kappa_{BR} = \frac{\sqrt{2}}{3}\frac{h/r^2}{\sqrt{1-\nu^2}} \tag{3}$$

- The Brazier curve (dotted line) approximates very well the non-linear path until the limit point **G** is reached. While the limit point **G** is associated with M_{lim}=11.8TPaÅ^3 and κ_{lim}=0.01025Å^{-1}, the Brazier curve local maximum point **Q** is characterised by M_{BR}=12.2TPaÅ^3 and κ_{BR}=0.01206Å^{-1}. It is fair to say that both points lead to similar values of the bending moment M and to a much lower curvature κ than the critical local mode one (κ_c=0.01500Å^{-1}).
- From the Brazier analysis, the limit value of the bending moment (M_{lim}=11.8TPaÅ^3) is reached for an ovalization parameter ξ=0.157, which means that the circular section vertical axis width decreased about 16%. Vodenitcharova and Zhang [7] found a value of the ovalization parameter ξ=0.14, very close to the one obtained here. Moreover, the picture in the right of figure 10 shows the variation of the ovalization parameter ξ with the bending angle θ. For the maximum Brazier bending moment (M_{BR}=12.2TPaÅ^3 – point **Q**), the circular cross-section exhibits an ovalization parameter equal to 21%.
- For the several coloured points (**A** to **M**), figure 2 also shows the corresponding deformed configurations of the NT mid-span section. While **A** is the underformed circular section configuration, the configurations **B**, **C** and **D** remain almost circular after bending. Nevertheless, the configurations **E**, **F** and **G** exhibit clearly visible ovalized configurations, where the two-wave distortional mode is prevalent. In particular, it should be noticed that the (black) deformed configuration **G**, corresponding to the limit situation, exhibits an ovalization parameter equal to *19%*. This value is located between those mentioned before in the context of the Brazier analysis (*16%* and *21%*), thus reflecting the relative accuracy and usefulness of the later.

Figure 3(a) shows the non-linear $M(\theta)$ equilibrium paths obtained for the NT(15,0), for several length values (10Å<L<120Å). Obviously, the inclination values of the initial branches (M/θ) are proportional to the NT bending stiffness values (EI/L). It is also visible that shorter NTs possess almost linear (straight) equilibrium paths until the limit moment M_{lim} is reached. Conversely, the equilibrium path of the longer NTs is more rounded near the point of limit bending moment M_{lim}. This fact proves that the ovalization phenomenon (Brazier effect) is more evident in the longer NTs. Moreover, it is also seen that the limit bending moment M_{lim} decreases abruptly for the shorter NTs but remains nearly constant for longer lengths. A possible explanation for this M_{lim} decrease in the shorter NT behaviour resides in the influence of the boundary conditions, which is absent in the longer NT behaviour: the mid-span section of longer NTs, where collapse takes place, is too distant from their end supports, a fact that does not occur in the shorter NTs. The transition between these two (shorter and longer NT) behaviours is different in the case of NT(15,0). It is clear that the NT(15,0) length-independence of M_{lim} (≈13.7 TPaÅ^3) occurs for L>40Å. As a first approach, one can state that this limit length depends on the NT radius (r) and is relatively well approximated by the NT perimeter (p): L>p=36.9Å for the NT(15,0).

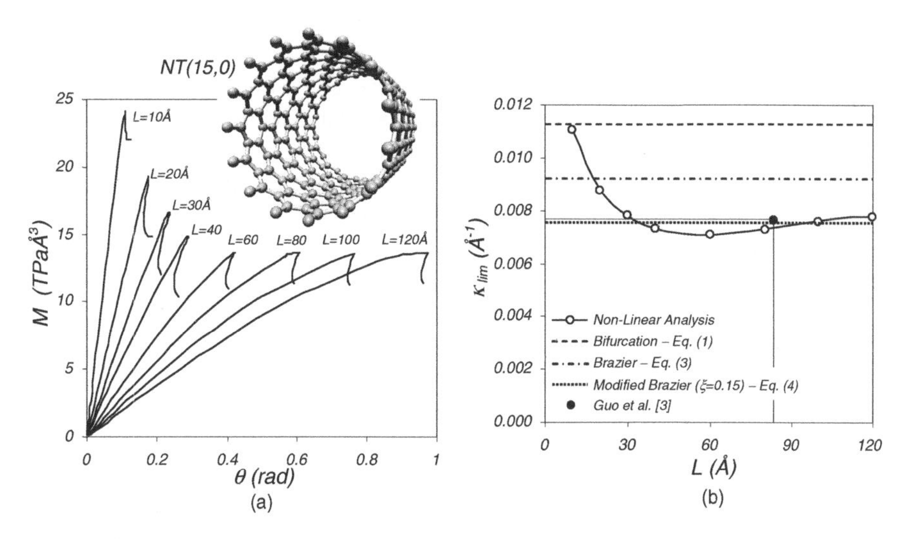

Fig. 3 Buckling behaviour of NT(15,0) under bending: (a) M-θ equilibrium paths and (b) variation of limit curvature with the length

The figure 3(b) shows the variation of the limit curvature (κ_{lim}) with the length of the NT(15,0). The limit bending curvature κ_{lim} also decreases for the shorter NTs, exhibits a local minimum and then augments sligthly for increasing lengths. It is interesting to mention that Cao and Chen [2], using molecular dynamics to simulate NTs under bending, also found that the curve κ(L/d) exhibited a "kind" of local minimum for very short NTs (i.e., with very low aspect ratio L/d). Moreover, the black dots in figure 3(b) represents the critical curvatures (κ_c) obtained by Guo et al. [3], using the atomist-scale finite element method, for the NT(15,0) with L=83.5 Å. The NT(15,0) critical curvature (κ_c=0.0077 Å^{-1}) is rather close to the limit curvature calculated from non-linear analysis (κ_{lim}=0.0070 Å^{-1}). Moreover, figure 3(b) also shows four horizontal lines corresponding to the curvature values obtained from (i) bifurcation analysis (dashed line – Eq. (1)), (ii) Brazier analysis (dashed-doted line – Eq. (3b)) and (iii) modified Brazier analysis (dotted line). This modified Brazier analysis is based on the following expression,

$$\kappa = \sqrt{\xi}\,\frac{h/r^2}{\sqrt{1-\nu^2}} \tag{4}$$

This expression can be used to evaluate the limit curvature as a function of the ovalization parameter ξ. While the dotted line in figure 3(b) correspond to the value ξ=0.15, the bifurcation and Brazier lines correspond to the adoption of ξ=3/9 and ξ=2/9, respectively. From the observation of figure 3(b), it is possible to conclude that both bifurcation (κ=0.0113 Å^{-1}) and Brazier (κ=0.0092 Å^{-1}) curvature estimates are too high, in comparison with the non-linear values (white dots). However, the modified Brazier curvature values with ξ=0.15 (κ=0.0075 Å^{-1}) lead to lower and more accurate estimates of the critical curvature. Finally, let us look at figure 1, where the bifurcation (ξ=3/9), Brazier (ξ=2/9) and modified Brazier (ξ=0.15) curves are represented and compared with available data. Despite the modified Brazier with ξ=0.15 curve is the one that gives more accurate estimates, it is also obvious that it does not fit well with available data. Therefore, one proposes the use of a modified Brazier analysis with ξ=0.10, which leads to very accurate results for all data (bottom dotted curve), with the exception of Yakobson's result. For a more detailed discussion of the results, the reader is referred to a recent work by the author [6].

References

1. Buehler, M.J., Kong, Y., Gao, H.: Deformation mechanisms of very long single-wall carbon nanotubes subject to compressive loading. J. Eng. Mat. Tech. ASME 126, 245–249 (2004)
2. Cao, G., Chen, X.: Buckling of single-walled carbon nanotubes upon bending: molecular dynamics simulations and finite element method. Phys. Rev. B 73, 155435(1-10) (2006)
3. Guo, X., Leung, A.Y.T., He, X.Q., Jiang, H., Huang, Y.: Bending buckling of single-walled carbon nanotubes by atomic-scale finite element. Composites Part B: Eng. 39, 202–208 (2008)
4. Iijima, S., Brabec, C.J., Maiti, A., Bernholc, J.: Structural flexibility of carbon nanotubes. J. Chemical Phys. 104, 2089–2092 (1996)
5. Shibutani, Y., Ogata, S.: Mechanical integrity of carbon nanotubes for bending and torsion. Modelling and Simulation in Materials Science and Engineering 12, 599–610 (2004)
6. Silvestre, N.: Length dependence of critical measures in single-walled carbon nanotubes. Int. J. Solids Struct. 45, 4902–4920 (2007)
7. Vodenitcharova, T., Zhang, L.C.: Mechanism of bending with kinking of a single-walled carbon nanotube. Phys. Rev. B 69, 115410(1-7) (2004)
8. Yakobson, B.I., Brabec, C.J., Bernholc, J.: Nanomechanics of carbon tubes: instabilities beyond linear response. Phys. Rev. Lett. 76, 2511–2514 (1996)

Effect of Surface Roughness on the Steel Fibre Bonding in Ultra High Performance Concrete (UHPC)

T. Stengel

Abstract. Micro steel fibres are currently added to UHPC to improve ductility. In UHPC under load, the high fibre strength is only utilised partly when the fibres are pulled out of the UHPC matrix. Using steel fibres with a nano- and micro-roughened surface is one measure to enhance the bond between the UHPC matrix and the fibre. In this contribution the effect of surface roughness on the bond behavior of steel fibres is shown. Single fibre pull-out tests were performed to investigate the bond behavior. Aim is to improve the bond behavior so that fibre pull-out takes place under a high load level with respect to the fibre strength.

1 Introduction

Ultra high performance concrete (UHPC) typically has a compressive strength of 200 to 250 MPa. The high compressive strength, the high durability as well as a reduced maximum grain size diameter of about 2 mm opens up a field of new construction methods and applications for concrete. However, the extreme brittleness of UHPC, characterized by explosive failure under critical load, restricts the exploitation of the otherwise good properties of this material. Currently, 1.5 to 3.0 % by volume of micro steel fibres 0.15 -0.20 mm in diameter with tensile strength of up to 3.500 MPa are used to improve ductility. Short fibre length is necessary for the typically slender and light UHPC structures. Therefore the micro steel fibres usually have an aspect ratio (L_f / $Ø_f$) of about 50 to 100. High costs as well as a considerable contribution to the environmental impact of UHPC are caused by the aforementioned amount of micro steel fibres [1, 2]. It is therefore aim of a research project to reduce the amount of micro steel fibres necessary or to use more economic and environmental friendly fibres. The amount of micro steel fibres

T. Stengel
Centre for Building Materials (cbm), Technische Universität München (TUM)
e-mail: stengel@cbm.bv.tum.de
www.cbm.bv.tum.de

could be reduced e.g. by increasing the efficiency of every single fibre. If the fibres could be utilized to a higher extent a smaller amount of fibres would lead to the same composite performance. This can be achieved for example with an optimized bond between the UHPC matrix and the fibres when the fibres are pulled out under a load near the fibre breaking force. Fibres with a larger diameter of e.g. 1.0 mm, which are much more economic, would lead to unfeasible fibre length of 50 to 100 mm when the same aspect ratio as mentioned before is considered. In that case an optimized bond behavior may lead to aspect ratios lower than 50 providing for a feasible application of these fibres. This contribution deals with the effect of surface roughness on the bond behavior of steel fibres in UHPC. Preliminary single fibre pull-out tests using steel fibres 0.15 and 0.30 mm in diameter with a roughened surface have shown that the bond between the UHPC matrix and the fibre could be enhanced significantly [3]. Following that in this study three different surface roughnesses and two different UHPC compositions were investigated together with steel fibres 0.98 mm in diameter.

2 Investigations

The fibre surfaces were mechanically treated by abrasive papers with three different roughnesses (grit 180, 240, 400). The surface roughening was done rubbing a 2 cm piece of abrasive paper for ten times along the fibre axis. Surface roughness parameters of the fibre surfaces were determined according to ISO 4287 with a confocal LED microscope using a fifty fold objective (vertical accuracy 2 nm). The compressive and splitting tensile strength of the plain UHPC was tested on cylinders with a diameter of 50 mm and an aspect ratio of 1 and 2 respectively. The load controlled tests were done at a rate of 0.5 MPa/s in the case of compressive strength and 0.5 kN/s in the case of splitting tensile strength. The pull-out tests were performed according to [3] but using a laser sensor (accuracy: 0.001 mm) for the displacement measurement. An embedded length of 7 mm was chosen for all pull-out specimens. Plain fibres as well as roughened fibres (grit 180 and 400) together with two different UHPC compositions were tested. From the pull-out curves the bond strength (assuming constant bond stress over the whole embedded length) as well as the bond energy (integral of load slip curve between a slip of 0 and 0.5 mm) was derived. The steel fibres used in this study had a diameter of 0.98 mm and a tensile strength of 2,600 MPa. All single fibres were alcohol cleaned when fixed in the moulds. Two different UHPC compositions – with and without PVA fibres – were used, Table 1. The PVA fibres (tensile strength appr. 780 MPa) had a diameter of about 15 μm and a length of 12 mm. For both mixtures a well-cement strength class 42.5 and a water cement ratio of 0.28 was chosen. The mixing was done according to [3] but applying a vacuum (50 mbar) during the last 60 s. Within 60 s after the end of mixing, the workability was tested with a mortar cone (H_{cone} = 50 mm, $Ø_{upside}$ = 70 mm, $Ø_{bottom}$ = 100 mm) on a glas plate. The workability was 32 cm and 26 cm for C4 and C5 respectively. Specimens were casted without any compaction.

Table 1 Composition of the two UHPC mixtures [kg/m³]

Concrete	Cement	Silica Fume	Quartz Powder	Quartz Sand	Water	Super-plast.	PVA-Fibres
C4	680	138	360	990	166	35	0
C5	674	137	352	975	158	54	3

3 Results

Figure 1a to Figure 1d shows parts of the confocal microscope pictures (each appr. 305 × 229 μm²) of the different fibre surfaces. The height scale lasts from 0 (white colour) to 10 μm (black colour). It can be seen that with increasing grit number the depth and the distance of the scratches is decreasing while the number of scratches is increasing.

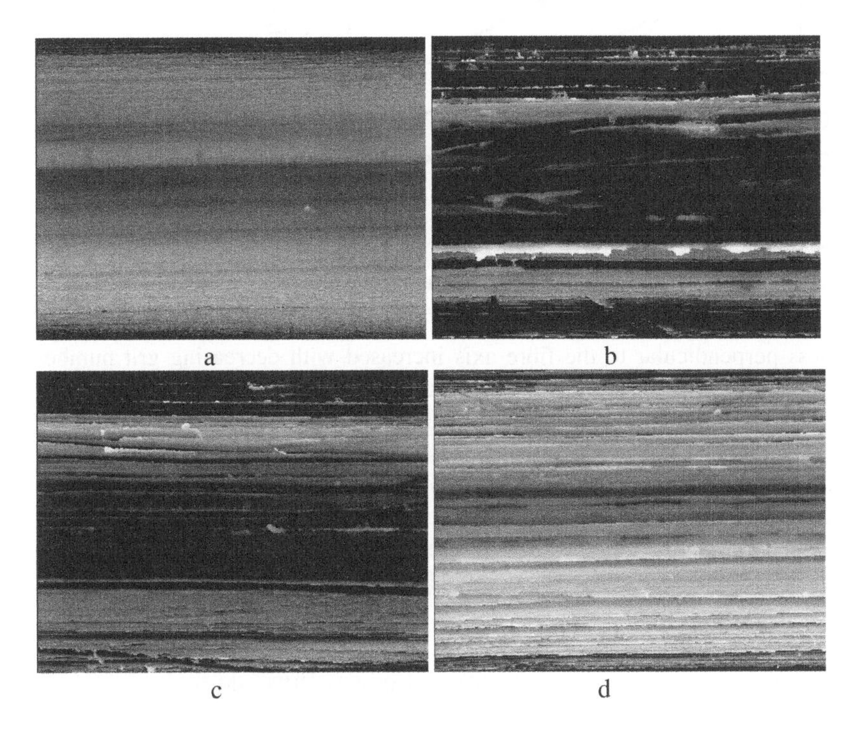

Fig. 1 Confocal microscope pictures of fibre surfaces: a: plain surface, b: grit 180 roughened, c: grit 280 roughened, d: grit 400 roughened

The roughness parameters parallel and perpendicular to the fibre axis (measured length: 250 μm) of the four fibre surfaces are given in Table 2. Only small differences in surface roughness were found parallel to the fibre axis while the

Table 2 Roughness parameters of the different fibre surfaces parallel (‖) and perpendicular (⊥) to the fibre axis

Roughness parameter	Plain Surface	Grit 180 Roughened	Grit 280 Roughened	Grit 400 Roughened
Rq ‖ [μm]	0.034	0.051	0.053	0.037
Rq ⊥ [μm]	0.293	0.879	0.703	0.516
Rt ‖ [μm]	0.224	0.379	0.351	0.326
Rt ⊥ [μm]	3.241	5.556	4.793	3.254

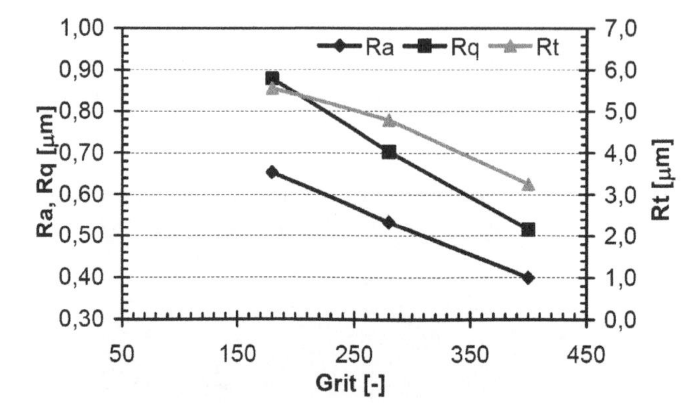

Fig. 2 Effect of abrasive paper grit number on roughness parameters of steel fibre surfaces

roughness perpendicular to the fibre axis increased with decreasing grit number linear dependent, Figure 2. The root mean square roughness was between 0.293 μm for plain and 0.879 μm for grit 180 roughened fibre surface.

The mean compressive and splitting tensile strengths of the two concretes are shown in Table 3. Concrete C4 had a 9% higher compressive strength than concrete C5; the splitting tensile strength was even 58% higher. This may be due to the diverse workability which causes different deariation behavior. Further investigations will be done on that. The coefficient of variation was in every case less than 7% (three specimen tested).

Table 3 Mean Compressive and Splitting Tensile Strength of the UHPCs [MPa]

Concrete	Compressive Strength	Standard Deviation	Splitting Tensile Strength	Standard Deviation
C4	235	3	15.8	1.1
C5	216	9	10.0	0.7

Table 4 shows the derived mean bond properties for the two concretes and the different fibres.

Table 4 Mean bond properties evaluated from single fibre pull-out tests

Concrete / Fibre	Bond Strength [MPa]	Bond Energy [10^{-3} J]
C4 / plain	5.83	31.9
C4 / 180	9.10	78.7
C4 / 400	3.69	31.1
C5 / plain	4.27	39.9
C5 / 180	11.5	93.3
C5 / 400	5.94	45.2

For both concretes the grit 180 roughened fibres showed the highest bond strength as well as the highest bond energy. This is obviously due to the surface roughness which was the highest for the grit 180 roughened fibre surfaces. Despite the lower compressive and splitting tensile strength the addition of PVA fibres seems to affect the bond behavior of steel fibres beneficially when the fibre surface is rough. Using PVA fibres the bond behavior was enhanced significantly for both surface roughnesses.

4 Discussion

The surfaces of steel fibres were roughened with abrasive paper of different grits (180, 280, 400) along the fibre axis. The fibres roughened with the lowest grit showed the best bond properties. This is due to the highest surface roughness which was obtained with the lowest grit number. The grit 400 roughened fibres had a similar bond behavior as the plain fibres when using UHPC without PVA

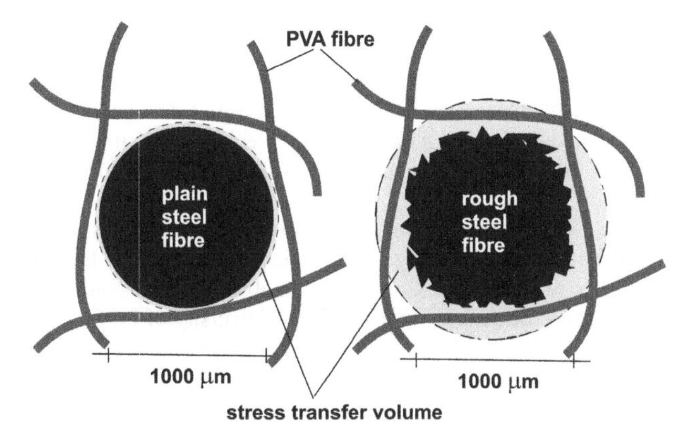

Fig. 3 Reinforcing effect of PVA fibres within the UHPC matrix surrounding a steel fibre

fibres. An addition of PVA fibres affected the bond behavior of all roughened fibres beneficially. The higher pull-out load in the case of roughened fibres may result in a higher volume of UHPC matrix involved in the stress transfer, Figure 3.

When using surface roughened fibres the stress transfer volume has an extension beyond the mean roughness and lasts into the bulk matrix where the micro reinforcing PVA fibres can be found, Figure 3. In that case the PVA fibres can bridge micro cracks and therefore strengthen the matrix volume involved in stress transfer. This affects the pull-out behavior in a positive way.

5 Conclusions

Surface roughening done with abrasive paper affects the bond behavior of steel fibres in UHPC significantly. The lower the grit number of the abrasive paper the higher is the roughness of the steel fibre surface. Surface roughness depends on the grit number in a linear way. Due to the higher surface roughness steel fibres treated with grit 180 paper showed the best bond behavior in this study. Bond properties were enhanced by roughly 200% compared to plain fibres. Even better bond behavior of the roughened steel fibres is achieved when PVA fibres are added to the UHPC matrix. Bond properties were enhanced by another 20% compared to the plain UHPC without PVA fibres. This may be because of a micro reinforcing effect of the PVA fibres to the UHPC matrix surrounding the steel fibre. Due to higher pull-out loads when using roughened steel fibres the stress transfer volume around the steel fibre is significantly larger than in the case of plain steel fibres. Therefore PVA fibres can bridge microcracks and strengthen the matrix within the stress transfer volume. To verify this, further investigations on the effect of PVA fibres like SEM analyses of the interfacial zone will be done.

Acknowledgments. The author would like to thank the German government for the financial support within the NanoTecture research programme and the FRT GmbH for the surface measurements.

References

1. Stengel, T., Schiessl, P.: Sustainable Construction with UHPC – from Life Cycle Inventory Data Collection to Environmental Impact Assessment. In: Fehling, E., Schmidt, M., Stuerwald, S. (eds.) Proc. of the 2nd Int. Symposium on UHPC. Kassel University Press, Kassel (2008)
2. Stengel, T., Schiessl, P.: Life Cycle Assessment of UHPC Bridge Constructions: Sherbrooke Footbridge, Kassel Gärtnerplatz Footbridge and Wapello Road Bridge. In: Kotynia, R., Gawin, D. (eds.) Proc. of the 6th Int. Conference Analytical Models and New Concepts in Concrete and Masonry Structures AMCM 2008. University of Lodz, Lodz (2008)
3. Stengel, T.: Optimisation of Steel Fibre Bonding in Ultra High Performance Concrete. In: Gettu, R. (ed.) Proc. of the 7th Int. RILEM Symposium on Fibre Reinforced Concrete: Design and Application BEFIB 2008. RILEM Publications S.A.R.L., Bagneux (2008)

Geotechnical Properties of Soil-Ball Milled Soil Mixtures

M.R. Taha

Abstract. Laboratory experiments were conducted to study the fundamental geotechnical properties of mixtures of natural soils and its product after ball milling operation. The product after ball milling process is termed nano-soil herein. SEM analysis showed that much more nano size particles were obtained after the milling process. Testing and comparison of the properties of original kaolinite, montmorillonite and UKM soil with regard to its liquid limit, plastic limit, plasticity index, and specific surface and after addition of its nano-soil were also conducted. Laboratory tests results showed that the values of liquid limit and plastic limits were higher after nano-soil addition. However, its plasticity index reduces which is advantageous in many geotechnical constructions. Compressive strength of original soil-cement-1% nano-soil mixture showed almost double its value without nano-soil. It demonstrated that a small amount of these crushed particles or nano-soil can provide significant improvement in the geotechnical properties of soil. Thus, nanoparticles are potentially suitable for improving the properties of soil/clay for various applications.

1 Introduction

As the demand for land for development purposes increases, soil improvement techniques are sought after to convert land in which construction are impossible, such as soft clays and peat into one where structures can be built. One of the most common techniques is soil stabilization with additives or admixtures. In this technique the geotechnical properties of soil is enhanced by mixing it with another material. The additives that have been used in the past include cement, lime, calcium chloride, fly ash, bitumen, etc. In addition, especially for soft clays and peaty deposits which are composed of fine particles with high moisture content, granular materials such as sand and industry by-products have also been used. Mixing these

---

M.R. Taha
Universiti Kebangsaan, Malaysia
e-mail: profraihan@gmail.com
http://pkukmweb.ukm.my/~jkas/webjkas/cv/ProfRaihan.html

additives in the ground served the following purposes [1]: strength increase, deformability/settlement reduction, volumetric stability (control of shrinking and swelling), reduction of permeability, reducing erodibility, increasing durability, and control of variability.

The use of additives is advantageous because one of the fundamental requirement for civil engineering projects is that it must be inexpensive. At times, land to be improved covers a huge area and also involving large distances (for example highways). Thus, the need to keep the project economical calls for low cost materials. In order to meet this requirement, industrial by-products or waste have also been extensively studied. However, these raised another important question, i.e. toxicity of the materials. Leaching of toxic chemicals to the environment posed health related issues and is therefore another basic requirement for candidate materials for soil improvement.

In this study, experiments were conducted to evaluate the behavior of soil upon mixing with nano-soil. The nano-soil is actually a product of milling of a natural soil in which a greater portion of its particles was pulverized into nano sizes (1-100 nm). The use of such materials almost entirely eliminates the toxicity questions, as usually raised for nanomaterials applications, since the material o be used is originally a natural soil. Much study has been conducted and documented on the use of nanoparticles or nanotechnology in conceivably all fields of knowledge except geotechnical engineering. Although it is possible to speculate the general properties of soil when nanoparticles are added but real data are lacking due to inexistence of such research. This study could also pave the way to more extensive research on nanotechnology and nanoparticles in geotechnical engineering.

2 Materials and Methods

The soils studied and involved in this study include the following:

i. Metasediment (a sedimentary residual soil) obtained from a site within the campus of Universiti Kebangsaan Malaysia (UKM) in which this type of soil dominates its surficial geology. This soil is termed as UKM soil.
ii. Commercial kaolinite and montmorillonite. These are clay minerals quite commonly used in geotechnical engineering to represent in-situ clays.

In order to obtain the nano-soil, the soils were pulverized using a ball mill. The soil specimens were loaded in small amounts in a ball mill of type "Planetary Mono Mill Pulverisette 6" produced by Fritsch, Germany. The 20 balls used for grinding have a diameter of 5 mm each and is made of sintered corundum. The time taken for milling each batch is between 10-13 hours. In between the cyclonic milling operation, de-aired water was added to the bowl to prevent excessive heating. The original soil and the soil material from the milling process were then analyzed with an electron microscope at the UKM Electron Microscope Unit.

Almost all basic geotechnical properties were mainly conducted in accordance to BS 1377:1990 [2] in which the liquid limit was tested using the cone penetration test. The soil-nano soil mixtures were set at 98 % original soil and 2 % nano-soil by weight due to the very limited amount of the nano-soil obtained. The specific surface determination was done using the ethylene glycol monoethyl ether (EGME) method [3].

The compressive strength test was conducted by compacting the soil mixtures in a mortar cube with dimensions of 50 x 50 mm. The control mixtures consist of 94 % UKM soil and 6 % Portland cement. In order to evaluate the effects of nano-soil on the strength of the soil-cement mixtures, 1 and 2 % nano-soil was added. The specimens were compacted at the same energy and upon releasing it from the mold, the specimens were cured for 7 day by covering it with a damp cloth and spraying it daily with water. Finally, it was tested in a universal testing machine.

3 Results and Discussion

The SEM images of the original sample of kaolinite and the corresponding samples which were milled for a period of 10 to 13 hrs is shown in Fig. 1 (magnified 20,000 times). It generally shows that the milled samples have greater portion

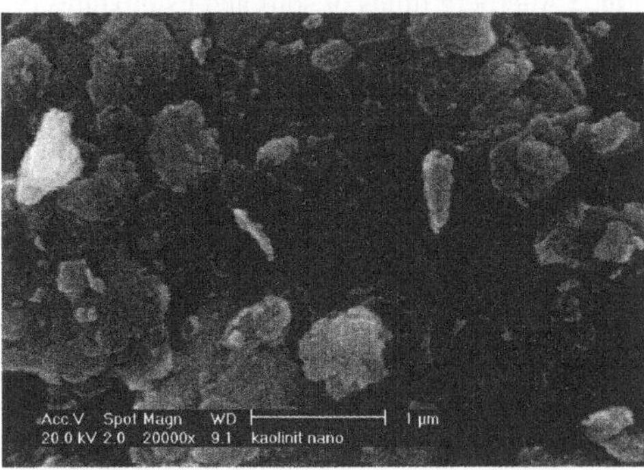

Fig. 1 Kaolinite before (top) and after (bottom) milling process

nano size particles in which the 40-80 nm particles are clearly evident. This is actually the target of these exploratory experiments.

The Atterberg limits (ie. plastic and liquid limits) result which shows the change in plasticity characteristics of the original soils and its mixtures mixtures (98 % original and 2 % milled soil) are shown in Fig. 2. It generally shows that the plastic limit (PL) and liquid limit (LL) increases upon mixing with the finer materials. However, since the increase in the LL is less that of PL, the plasticity index, PI (PI=LL-PL) reduces for all samples. This is important in geotechnical engineering as compaction of high plasticity soils will generally result in high shrinkage upon drying [4]. Eventually the soil will have high hydraulic conductivity which is a disadvantage for structures such as landfill liners and caps. Thus, the addition of finer particles such as nano-soil, even at low doses, could enhance the properties of soils.

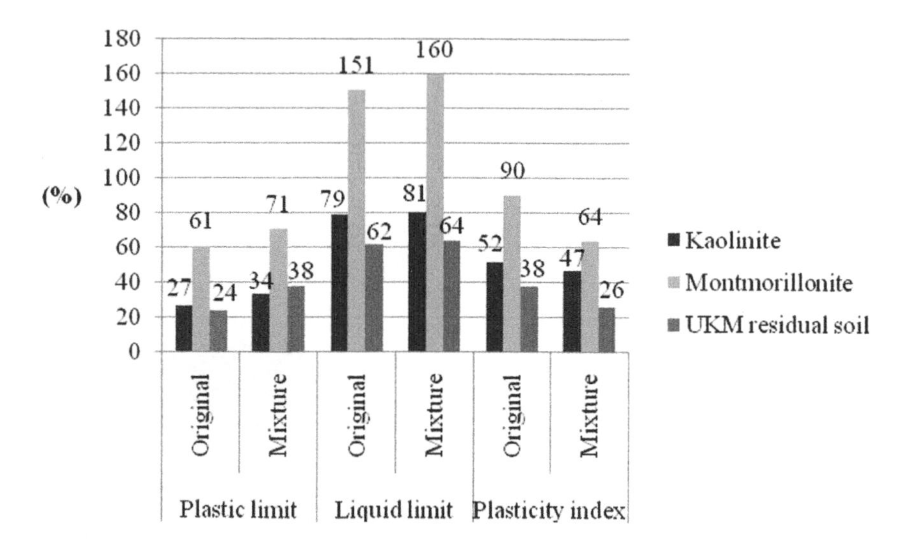

Fig. 2 Atterberg limits of soils and its mixtures (98% original soil dan 2 % nano-soil)

Specific surface of the soils evaluated using the EGME method is given in Table 1. It shows that the specific surface increases which explained the changes in the plasticity characteristics as shown by the Atterberg limits. With the increase of specific surface, the water required to cover the particles increases and hence increasing plastic and liquid limits.

The results from strength tests of UKM soil-cement-UKM nano-soil is shown in Fig. 3. The addition of cement to soil is a common soil improvement technique. As previously mentioned, the control mixtures consisted of 94 % UKM soil and 6 % Portland cement. In order to evaluate the effects of nano-soil on the strength of

Table 1 Specific surface area of the soils

Kaolinite (m^2/g)		Montmorillonite (m^2/g)		UKM soil (m^2/g)	
Original	After milling	Original	After milling	Original	After milling
25.3	39.8	730.1	792.7	2.4	3.9

the mixtures, 1 % and 2 % nano-soil was added. The results show significant improvement in the strength of the nano-soil mixtures over the control test albeit the amount of nano soil added was minimal or rather quite low. The strength of the soil-cement mixture almost doubled when 1 % UKM nano-soil was added. In addition, the mixture which is usually termed as cement-modified soil in geotechnical engineering will have lower tendency of volume change and PI in addition to increase in load bearing capacity [5].

This study has demonstrated that even a small addition of nanoparticles will show marked enhancement in soil behavior. It is also possible to engineer nanoparticles to improve behavior of soils to suit design and practice requirements. This research is the first effort to introduce nanoparticles a soil improvement material. More intensive research is needed before the materials can be utilized for its intended use in the field.

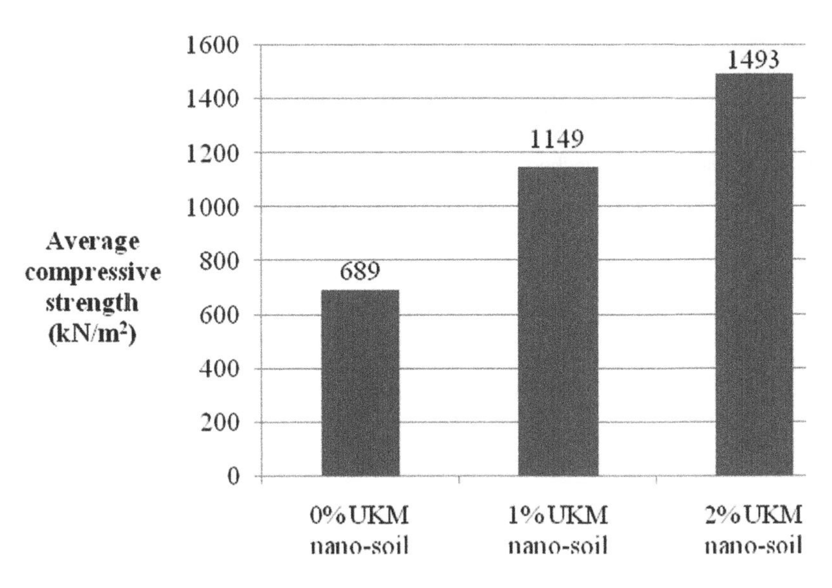

Fig. 3 Compressive strength of soil-cement-UKM nano-soil mixtures

4 Conclusions

Nano-soil was used as a possible soil improvement material. It was obtained from ball milling operations of a parent soil. From SEM analysis, it is evident that after between 10 to 13 hours of cyclonic milling operations, there were more nano particles compared to the original soils. The plastic and liquid limits of soil mixtures consisting of 98 % original soil and 2% nano-soil increases compared to the values of 100 % original soil. However, the plasticity index reduces which is advantageous in geotechnical construction such as landfill liners and caps. Mixing 96 % UKM soil, 4 % cement and 2 % UKM nano-soil results in almost doubling the compressive strength compared with samples without nano soil. Thus, nano-soil or more generally nanoparticles is an excellent candidate material for soil improvement. However, research need to be intensified beyond this initial study before real field applications can be realized.

References

1. Haussmann, M.R.: Engineering principles of ground modification. McGraw-Hill, New York (1990)
2. Taha, M.R., Lim, S.Y., Chik, Z.: Ciri-ciri asas beberapa tanah yang dikisar (Basic properties of milled soils). Jurnal Kejuruteraan (Engineering Journal) (2009) (accepted and in-print)
3. Cerato, A.B., Lutenegger, A.J.: Determination of surface area of fine-grained soils by the Ethylene Glycol Monoethyl Ether (EGME) method. Geotech. Test J. 25(3), 314–320 (2002)
4. Mitchell, J.K., Soga, K.: Fundamentals of soil behavior, 3rd edn. John Wiley & Sons, New York (2005)
5. Winterkorn, H.F., Pamukcu, S.: Soil stabilization and grouting. In: Fang, H.Y. (ed.) Foundation Engineering Handbook, 2nd edn. Van Nostrand Reinhold, New York (1991)

Mortar and Concrete Reinforced with Nanomaterials

J. Vera-Agullo, V. Chozas-Ligero, D. Portillo-Rico, M.J. García-Casas, A. Gutiérrez-Martínez, J.M. Mieres-Royo, and J. Grávalos-Moreno

Abstract. In this work, several nanomaterials have been used in cementitious matrices: carbon nanofilaments (either multiwall nanotubes or nanofibers), nanosilica and nanoclays. The physico-chemical behavior of these nanomaterials at three different levels has been analyzed: cement paste, mortar and concrete. It has been determined the setting times, the workability, the mineralogical structure and the dispersion of the nanomaterials in the cement matrix by ESEM/EDX , the percentage of hydration by TGA and the mechanical properties of mortar and concrete at 3, 7, 28 and 56 days. It has been found that almost all the nanomaterials used in this study accelerate the hydration process (with a proper dispersion), obtaining reinforcements in compression and flexural strength at 3 and 7 days (between 20 and 40 %). At 28 days, it has been observed that carbon nanotubes and nanofibers exhibit a reinforcement in the flexural strength (more than 25%), due to their fibrilar structure. Reinforcements, either in compression or flexural strength, have been reached with nanosilica (between 20 and 40 %); indeed, pozzolanic activity has been confirmed with nanosilica.

1 Introduction

One of the most desired properties of nanomaterials in the construction sector is their capability to confer a mechanical reinforcement to cement based structural materials. Three main objectives are considered: a very high-strength concrete for specific applications can be obtained with the use of nanomaterials. Reduce the amount of cement needed in concrete in order to obtain similar strengths, decreasing the cost and the environmental impact of construction materials. And moreover, by means of the addition of several types of nanomaterials a high-strength concrete can be obtained at shorter times, reducing the construction periods.

J. Vera-Agullo, V. Chozas-Ligero, D. Portillo-Rico, M.J. García-Casas,
A. Gutiérrez-Martínez, J.M. Mieres-Royo, and J. Grávalos-Moreno
ACCIONA Infraestructuras
e-mail: jgravalo@acciona.es
http://www.acciona.es/

Besides these objectives, there are other functionalities that can be obtained with the addition of nanomaterials to concrete. Self-cleaning facade with the addition of titania nanoparticles to the mortar [1]; and self sensing cementitious composites [2] or EMI shielding [3] by the addition of conductive nanomaterials (carbon nanotubes, carbon nanofibers or carbon black). Nanomaterials have a huge potential in the construction sector and new applications appear every year.

Mechanical improvements in cementitious matrices by the addition of nanomaterials have been obtained and have been reported in the literature. The compressive strength at early ages and the flexural modulus increased with the addition of nanoalumina to cementitious composites [4, 5]. Abrasion resistance [6] and flexural fatigue performance [7] can be improved with the use of titania nanoparticles and nanosilica.

Nanosilica is being used recently in the construction sector because of its best performance compared to the conventional silica fume [8]. It has been demonstrated by several authors that nanosilica particles increased the hydration process [9, 10], thus increasing the mechanical properties at 3 days [9]. Nanosilica works as a nucleation site for the early C-S-H precipitation and shortens the induction period [8-10]. Moreover, the pozzolanic effect of nanosilica takes place in a large extent in the paste-aggregate interface, thus improving the interface structure [8].

Unlike the nanosilica, carbon nanofilaments (either carbon nanotubes [11] or carbon nanofibers [12, 13]) are not being used yet in cementitious composites with structural purposes in the construction sector. Researchers are doing an increasing effort to elucidate the potential of carbon nanofilaments as reinforcement in cementitious matrices; however, there are still very few works related to cementitious composites reinforced with carbon nanofilaments compared to the huge amount of works regarding to polymer composites. Ying Li et al. [14] found that the compressive strength increases up to 19%, while the flexural strength increases up to 25% in mortars by the means of the addition of carbon nanotubes at 0.5% in weight with respect to the cement content; moreover, the failure strain was increased and the porosity reduced.

This work shows the main results of mechanical reinforcement obtained with the use of nanomaterials (nanosilica, carbon nanofibers and nanotubes). The analysis of the influence of the nanomaterials in the cementitious matrices has been carried out at three different levels: cement paste, mortar and concrete.

2 Experimental

2.1 Materials

The nanosilica used in this work is currently under development in a pilot plant; the silica nanoparticles have a diameter between 3 and 15 nm and a BET surface area between 20 and 1000 m^2/g. The two types of carbon nanofilaments used in this study are: GANF carbon nanofibers and Graphistrength C100 MWNTs.

GANF material is a commercial helical-ribbon "stacked-cup-like" CNFs, produced by Grupo Antolín Ingeniería (Burgos, Spain). This sample has been deeply characterized elsewhere [13, 15]. Graphistrength C100 MWNTs consisting of high purity (> 90%) multi-wall carbon nanotubes are produced in semi-industrial quantities by Arkema. The cementitious materials used in the tests were ordinary Portland cement (CEM I 42,5R and CEM I 42,5R/SR). The fine aggregate used in the mortar assays was standard silica sand (European standard) in the mortar assays. Quartize was used as the sand and the coarse rounded aggregates at concrete level.

2.2 *Characterization Techniques*

The percentage of hydration in the cement has been analyzed by thermogravimetric analysis (TGA). The method is explained elsewhere [16]. Mechanical properties have been measured according to the standards UNE-EN 196 in the case of mortars and UNE-EN 12390 in the case of concrete.

2.3 Challenges in the Preparation of Nanocomposites

One of the most important challenges in the nanocomposite research is to disperse properly the nanomaterials into the matrices. Nanomaterials like carbon nanofilaments are usually entangled and grouped in nest. Thus, a great effort has been done in this work in order to break these agglomerates and disperse the nanofilaments indiviadually. High energy mixing machines have been used to disperse the nanomaterials into the cementitious matrices.

Other problem related with the use of nanomaterials in cementitious matrices is the fact that they adsorb a high quantity of water due to their high surface area that can be hydrophilic in some cases. The challenge is to obtain a good workability without adding an extra quantity of water.

3 Results and Discussion

Table 1 shows the results of the degree of hydration in cement paste with nanomaterials. It can be seen that the percentage of hydration increases strongly with the addition of medium % of NS. Carbon nanofibers accelerate the hydration too, but in a minor extent.

Table 1 Percentage of hydration with and without NS

Sample	% of hydration at 7 days	% of hydration at 28 days
Cement Paste	69.9	75.6
Cement Paste medium % NS	82.2	90.7
Cement Paste low % CNF	71.4	80.2
Cement Paste medium % CNF	75.4	79.0
Cement Paste high % CNF	75.6	82.2

Moreover, the pozzolanic effect of NS has been confirmed by the thermogravimetric analysis; there is more C-S-H gel and less portlandite than the cement paste without NS. The portlandite content does not increase at 28 days with respect to the results at 3 days.

The pozzolanic effect and the acceleration of the hydration process improve the mechanical properties at concrete levels as can be seen in Fig. 1.

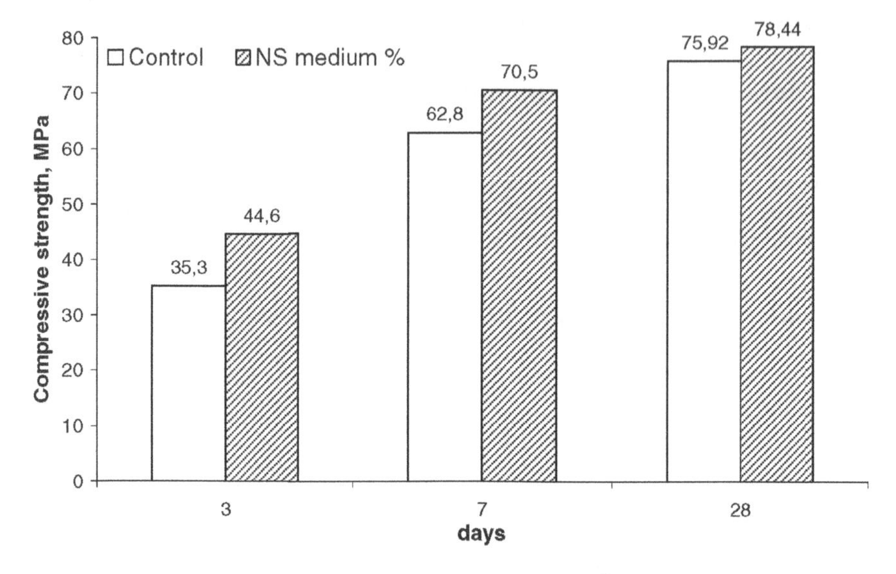

Fig. 1 Concrete with a medium % of NS. 350 kg of cement/m^3

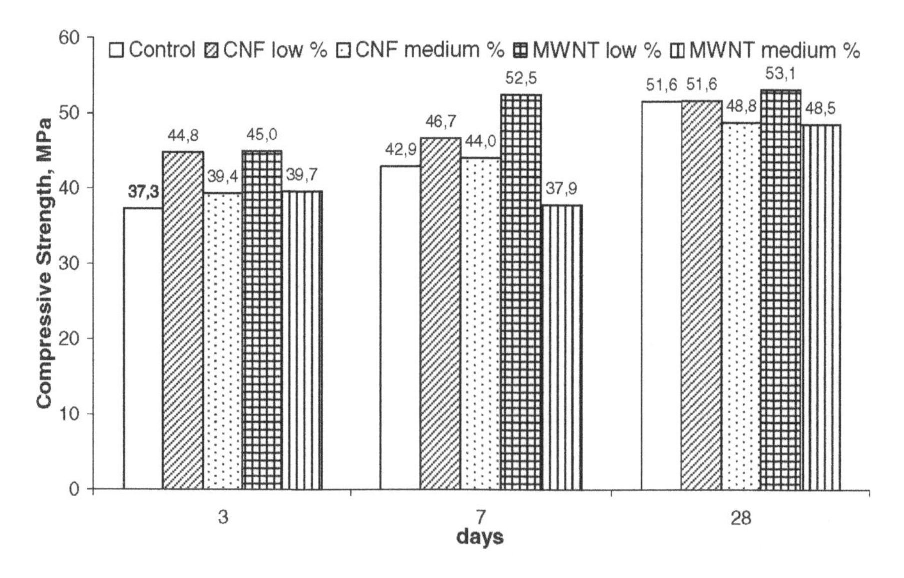

Fig. 2 Compressive strength. Mortars loaded with carbon nanofilaments

Regarding to carbon nanofilaments, it can be seen that both nanofibers and nanotubes increases the compressive strength at early ages in mortar, especially at low % in weight (21% at 3 days). However, this increment in the compressive strength is not kept at 28 days (Fig. 2).

In the case of flexural strength in mortar with carbon nanofilaments (Fig. 3), the reinforcement is permanent and not only at early ages. There is an outstanding result with a low % in weight of MWNTs at 28 days (27% at 28 days). The strong fibrilar structure of carbon nanofilaments improves the flexural behaviour of mortars.

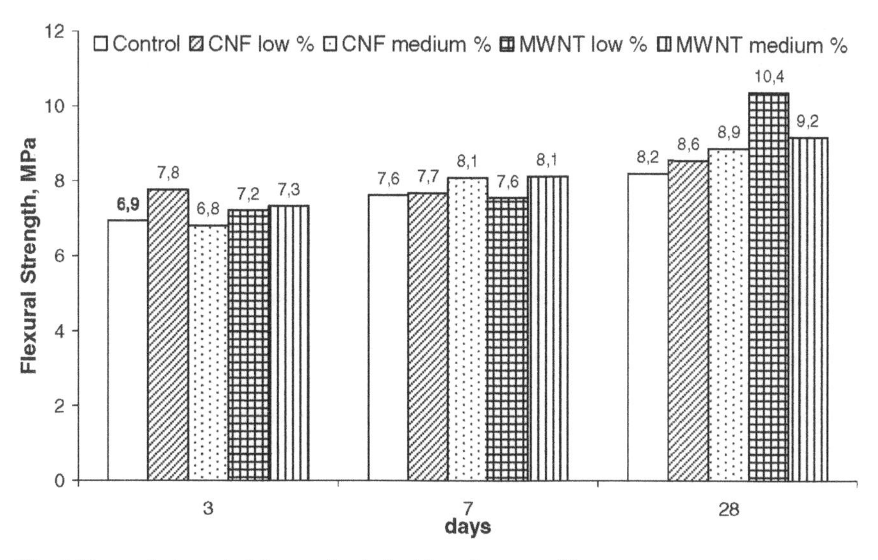

Fig. 3 Flexural strength. Mortars loaded with carbon nanofilaments

4 Conclusions

As conclusions we can say that:

- Nanosilica fills the voids and makes the cement structure denser, accelerates the hydration process, has a pozzolanic effect, improves reology and increases the compressive strength.
- Carbon nanofilaments (either carbon nanofibers or nanotubes) accelerate the hydration process and the compressive strength at early ages. And they improve the flexural strength at 28 days.

Acknowledgments. This work has been partially supported by the CENIT project DOMINO [http://www.cenitdomino.es], which is funded by Spanish Ministry of Science and Innovation. The views expressed in this paper are not necessarily those of the DOMINO consortia.

References

1. Vittoria Diamanti, M., Ormellese, M., Pedeferri, M.: Characterization of photocatalyticnext term and superhydrophilic properties of mortars containing titanium dioxide. Cem. Concr. Res. 38, 1349–1353 (2008)
2. Ying Li, G., Ming Wang, P., Zhao, X.: Pressure-sensitive properties and microstructure of carbon nanotube reinforced cement composites. Cem. Concr. Compos. 29, 377–382 (2007)
3. Chung, D.D.L.: Comparison of submicron-diameter carbon filaments and conventional carbon fibers as fillers in composite materials. Carbon 39, 1119–1125 (2001)
4. Campillo, I., Guerrero, A., Dolado, J.S., et al.: Improvement of initial mechanical strength by nanoalumina in belite cements. Mater. Lett. 61, 1889–1892 (2007)
5. Li, Z., Wang, H., He, S., et al.: Investigations on the preparation and mechanical properties of the nano-alumina reinforced cement composite. Mater. Lett. 60, 356–359 (2006)
6. Li, H., Zhang, M.-h., Ou, J.-p.: Abrasion resistance of concrete containing nanoparticles for pavement. Wear 260, 1262–1266 (2006)
7. Li, H., Zhang, M.-h., Ou, J.-p.: Flexural fatigue performance of concrete containing nano-particles for pavement. Int. J. Fatigue 29, 1292–1301 (2007)
8. Qing, Y., Zenan, Z., Deyu, K., et al.: Influence of nano-SiO2 addition on properties of hardened cement paste as compared with silica fume. Constr. Buildings Matt. 21, 539–545 (2007)
9. Tobón, J.I., Restrepo Baena, O.J., Payá Bernabeu, J.J.: Portland cement blended with nanoparticles. Dyna 152, 277–291 (2007)
10. Korpa, A., Kowald, T., Trettin, R.: Hydration behaviour, structure and morphology of hydration phases in advanced cement-based systems containing micro and nanoscale pozzolanic additives. Cem. Concr. Res. 38, 955–962 (2008)
11. Oberlin, A., Endo, M., Koyama, T.: Filamentous growth of carbon through benzene decomposition. J. Cryst. Growth. 32, 335–349 (1976)
12. Baker, R.T.K., Barber, M.A., Harris, P.S., et al.: Nucleation and growth of carbon deposits from the nickel catalyzed decomposition of acetylene. J. Catal. 26, 51–62 (1972)
13. Martin-Gullon, I., Vera, J., Conesa, J.A., et al.: Differences between carbon nanofibers produced using Fe and Ni catalysts in a floating catalyst reactor. Carbon 44, 1572–1580 (2006)
14. Ying Li, G., Ming Wang, P., Zhao, X.: Mechanical behaviour and microstructure of cement composites incorporating surface-treated multi-walled carbon nanotubes. Carbon 43, 1239–1245 (2005)
15. Vera-Agullo, J., Varela-Rizo, H., Conesa, J.A., et al.: Evidence for growth mechanism and helix-spiral cone structure of stacked-cup carbon nanofibers. Carbon 45, 2751–2758 (2007)
16. Rivera-Lozano, J.: Hidratación de pastas de cemento con adiciones activas, PhD, Universidad Autónoma de Madrid (2004)

Experimental Study and Modeling of the Photocatalytic Oxidation of No in Indoor Conditions

Q.L. Yu, H.J.H. Brouwers, and M.M. Ballari

Abstract. Heterogeneous photocatalytic oxidation (PCO) has shown to be a promising air purifying technology. Nitrogen monoxide (NO) is one common indoor air pollutant. The present paper addresses the PCO reaction in indoor conditions using NO as target pollutant with the gypsum plasterboard as a special substrate and carbon-doped TiO_2 as photocatalyst. A photocatalytic reaction setup is introduced for the assessment of the indoor air quality. The PCO effect of the carbon-doped TiO_2 is evaluated using different light wavelengths. Furthermore, the influence of the reactor volume on the PCO rate is studied. The Langmuir-Hinshelwood model is applied to describe the photocatalytic reaction mechanism. Experimental results show the validity of the L-H model in the present research. Using this model, a mathematical expression is proposed to describe the concentration change in the reactor.

1 Introduction

To indoor air quality (IAQ) has been paid much attention because of the important role it plays on human beings. NO_x (NO and NO_2) and Volatile Organic Compounds (VOCs) in indoor air can be emitted from building materials, furniture, heating sources, cooking, tobacco smoke, and the pollutant can even come from outside. Photocatalytic oxidation (PCO) has been studied since 1970s [2, 6, 7] and shown as an effective technology for outdoor air pollution control in building materials [8, 9].

The present research addresses indoor air quality using the photocatalytic technology. Gypsum plasterboard is used as substrate material for the application

---

Q.L. Yu, H.J.H. Brouwers, and M.M. Ballari
Department of Construction Management & Engineering, Faculty of Engineering Technology, University of Twente, The Netherlands
e-mail: q.l.yu@ctw.utwente.nl
www.utwente.nl

of photocatalyst in this research, because it is used widely as an indoor wall board owing to its good fire resistance and aesthetics properties. A PCO setup for indoor air purifying research was developed. NO is chosen as the target pollutant in the first research stage.

2 Experimental

TiO_2 is widely used as photocatalyst because of its excellent properties like safety, low price, stability, and high photocatalytic efficiency [11]. However, TiO_2 can only be activated by UV light which is only 0.1-5 μW/m^2 in indoor illumination [10]. A modified photocatalyst is used in this research to utilize the visible light, which is carbon-doped TiO_2 (Kronos, Germany) with a cut-off wavelength of 535 nm (band gap of 2.32eV) that corresponds to bluish green light [3]. The carbon-doped TiO_2 was deposited onto the glass fibres, which were then sprayed onto the gypsum plasterboard paper with a good bonding between them.

The schematic diagram of the PCO setup for indoor air assessment is shown in Fig. 1. The setup, made from non-adsorbing plastic materials with a size of 100 × 200 mm^2 (W × L), is developed according to standard ISO 22197-1 [1]. The experimental setup is composed of the reactor, visible light source, gas supply, analyzer, parameters controller. A detailed description can be found in [9]. The applied visible light source is composed of three cool day light lamps of each 25 W (Philips, The Netherlands), emitting a visible radiation in the range of 400 to 700 nm. The light intensity can be adjusted with a light intensity controller, is measured using a VIS-BG radiometer (Dr. Gröbel UV-Elektronik GmbH, Germany).

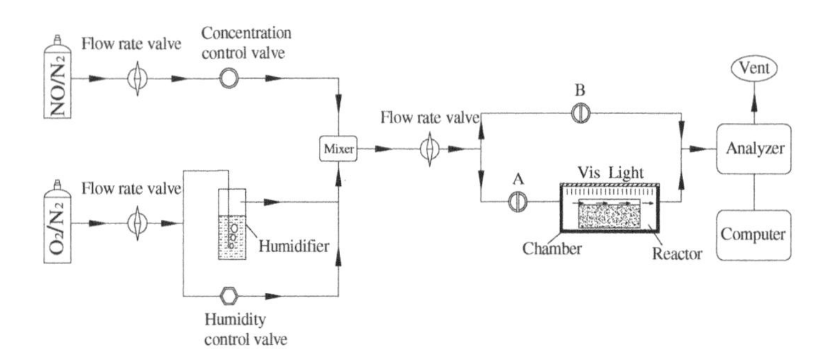

Fig. 1 Schematic diagram of photocatalytic oxidation setup

3 Results

PCO can be divided into three main steps: (1) mass transport and adsorption of pollutants from the bulk air to the surface of catalyst; (2) PCO on the catalyst; (3) desorption and mass transport of the reaction products from the surface of catalyst to air. The theory behind the PCO of NO can be found [4, 5, 12].

Recovery test of the samples was carried out by letting NO flows through the reactor in the dark. Results indicate that no photocatalytic reaction takes place in the dark and the average recovery efficiency is 97.6% shown as Table 1.

The photocatalytic reaction takes place immediately when the sample is exposed to the visible light. A total reaction time of 30 minutes was employed in the present research. The conversion of NO is calculated using equation (1).

$$Degradation(\%) = \frac{C_{in} - C_{out}}{C_{in}} \times 100\% \quad (1)$$

where C_{in} is the initial NO concentration and the C_{out} is the average pollutant concentration at exit during the last 5 minutes in the measurement time. According to PCO reactions, the NO_2 produced during the PCO could not turn into NO_3^- completely, so the exit pollutant concentration is calculated as the total of exit NO and NO_2 concentrations. Table 1 shows some PCO results.

Table 1 PCO of NO. Experimental conditions: flow rate: 3 L/min; light intensity: 10 W/m^2; relative humidity: 50 %; room temperature (21.1-22.9 °C). Concentration unit: ppm

C_{in}	C_{NO} at 10 min	Exit C_{NO}	Exit C_{NO2}	Recovery (%)	Degradation (%)
0.0992	0.0963	0.0280	0.0107	96.3	61.30
0.3003	0.2929	0.1519	0.0306	97.6	39.17
0.5002	0.4928	0.3158	0.0401	98.6	28.82
1.0010	0.9767	0.7785	0.0524	97.7	16.91

4 Kinetic Studies

The Langmuir-Hinshelwood (L-H) rate model has been employed to describe the reaction in gas-solid phase for heterogeneous photocatalysis [8]:

$$r = \frac{kKC}{1 + KC} \quad (2)$$

where r is the oxidation rate (mg/m^{2}s), k is the rate constant (mg/m^{2}s), K is the adsorption constant (m^3/mg), and C is the initial NO concentration (mg/m^3).

Along the longitudinal of the reactor, the NO mass balance reads:

$$r_{NO} = v_{air} \frac{dC}{dx} = -a_v r = -a_v \frac{kKC}{1 + KC} \quad (3)$$

where v_{air} (m/s) is the linear air flow rate along the reactor, and a_v (1/m), with a_v=A/V, as the active surface area per reactor volume because the PCO reaction only occurs in the sample surface, A (m^2) is the active surface area of the sample (in the present study, A=LW) and V (m^3) is the reactor volume (V=LWH), L, W, and H are the length, the width, and the height of the reactor, respectively.

Expression (4) is obtained by solving differential equation (3) considering the boundary condition C(x = 0) = C_{in}, yielding:

$$\frac{WL}{Q(C_{in}-C_{out})}=\frac{1}{k}+\frac{1}{kK}\frac{Ln(C_{in}/C_{out})}{(C_{in}-C_{out})} \tag{4}$$

where Q (m^3/s) is the volumetric flow rate of the pollutant (Q=v_{air}WH), and C_{out}=C(x = L).

The experimental data is in good agreement with the L-H model as shown in Fig. 2 which indicates that the L-H model is suitable to be used to in this research. The k and K obtained from the intercept and slope of the trend line in Fig. 2 are 5.75 $\times 10^{-4}$ (mg/m^2s) and 7.593 (m^3/mg), respectively. The accuracy of the trend line is also reflected by the accuracy of the initial NO concentration which is addressed in the first column of the Table 1.

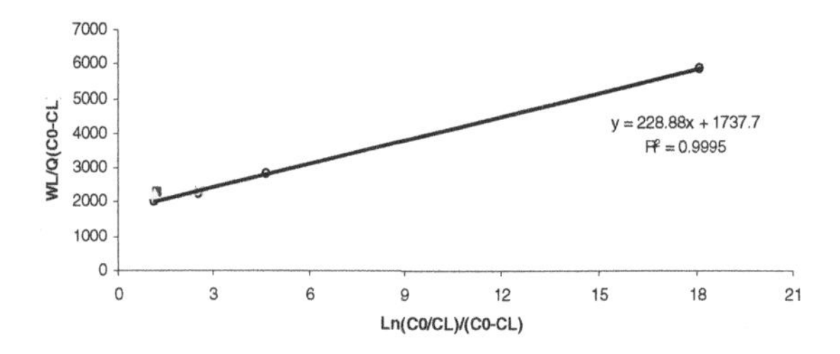

Fig. 2 Plot of NO photocatalytic degradation using L-H model

5 Discussion

To assess the effect of the light source, experiments with UVA have been executed as well. The obtained conversion rate is then 52.6%, with as experimental conditions: initial NO concentration: 1.0 ppm; flow rate: 3.0 L/min; UVA light intensity: 10.0 W/m^2; relative humidity: 50%; room temperature (21.5°C) which is in accordance with [3]. The results indicate that the carbon-doped TiO_2 has a similar effect as the undoped TiO_2 in UV light region, which is more than three times of that of the visible light with the same experimental conditions.

The PCO reactor size also plays an important role in PCO study. The present reactor has a fixed surface size with the height, the distance between the paralleled surface of the testing sample and the covering glass plate, can be adjusted by the screws in the bottom of the reactor. In the present study, 3 mm is used as a standard height. Equation (4) shows the final concentration only relates with the initial concentration with a fixed volumetric flow rate. To study the influence of the reactor height, tests were deployed shown in Fig. 3. The results indicate that the PCO has no relation with the reactor height. To improve the PCO rate, one method is to increase the contact area of the sample with the pollutant, e.g. to increase the surface area of the photocatalyst.

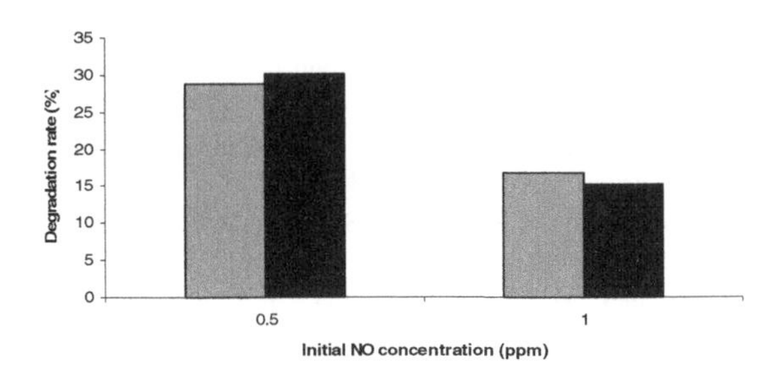

Fig. 3 PCO results. Reactor height: Initial NO concentration (0.5 ppm): 5 mm (right) and 3 mm (left); initial NO concentration (1.0 ppm): 5 mm (right) and 3 mm (left)

6 Conclusions

With the application of plasterboard as substrate and carbon-doped TiO_2 as photocatalyst and NO as pollutant, PCO has been proved to be a promising air purifying technology for indoor air quality improvement in the present research.

The kinetic L-H model is applied to describe the PCO reaction mechanism. A mathematical expression is obtained to express the concentration change in the reactor. The L-H model turns out to agree well with the experimental results.

The study of the reactor shows that only the surface area of the reactor influences the PCO rate. The effect of the carbon-doped TiO_2 is studied and UV light has a rather better PCO effect than that of visible light. So how to use the UV light in indoor conditions will be an important topic in the next research stage.

Acknowledgments. The authors gratefully express their appreciation to Dipl.-Ing. M. Hunger, Dipl.-Ing. G. Hüsken, and Ir. A.C.J.de Korte for their help with the measurement and analysis. Moreover, the authors wish to express their thanks to the European Commission (I-SSB Project, Proposal No.026661-2) and the following sponsors of the research group: Bouwdienst Rijkswaterstaat, Rokramix, Betoncentrale Twenthe, Graniet-Import Benelux, Kijlstra Beton, Struyk Verwo Groep, Hülskens, Insulinde, Dusseldorp Groep, Eerland Recycling, ENCI, Provincie Overijssel, Rijkswaterstaat Directie Zeeland, A&G maasvlakte, BTE, Alvon Bouwsystemen, and V. d. Bosch Beton (chronological order of joining).

References

1. ISO 22197-1. Fine ceramics (advanced ceramics, advanced technical ceramics) ÷ Test method for air purification performance of semiconducting photocatalytic materials ÷ Part 1: Removal of nitric oxide, 1st edn. (2007)

2. Ao, C.H., Lee, S.C.: Enhancement effect of TiO_2 immobilized on activated carbon filter for the photodegradation of pollutants at typical indoor air level. Appl. Catal. B-Environ. 44, 191–205 (2003)
3. Blöß, S.P., Elfenthal, L.: Doped titanium dioxide as a photocatalyst for UV and Visible Light. In: Proceedings International RILEM Symposium on Photocatalysis, Environment and Construction Materials TDP, Florence, Italy (2007)
4. Carneiro, J.O., Teixeira, V., Portinha, A., et al.: Study of the deposition parameters and Fe-dopant effect in the photocatalytic activity of TiO_2 films prepared by dc reactive magnetron sputtering. Vacuum. 78, 37–46 (2005)
5. Choi, W., Termin, A., Hoffmann, M.R.: The role of metal ion dopants in quantum-sized TiO_2: correlation between photoreactivity and charge carrier recombination dynamics. J. Phys. Chem. 98, 13669–13679 (1994)
6. Fujishima, A., Honda, K.: Electrochemical photolysis of water at a semiconductor electrode. Nature 238, 37–38 (1972)
7. Hoffmann, M.R., Martin, S.T., Choi, W., et al.: Environmental applications of semiconductor photocatalysis. Chem. Rev. 95, 69–96 (1995)
8. Hunger, M., Brouwers, H.J.H.: Self-cleaning surfaces as an innovative potential for sustainable concrete. In: Proceedings international conference excellence in concrete construction-through innovation, London, United Kingdom, September 9-10 (2008)
9. Hüsken, G., Hunger, M., Brouwers, H.J.H.: Comparative study on cementitious products containing titanium dioxide as photo-catalyst. In: Baglioni, P., Cassar, L. (eds.) Proceedings International RILEM Symposium on Photocatalysis, Environment and Construction Materials-TDP 2007, pp. 147–154. RILEM Publications, Bagneux (2007)
10. Kuo, C.S., Tseng, Y.H., Huang, C., et al.: Carbon-containing nano-titania prepared by chemical vapor deposition and its visible-light-responsive photocatalytic activity. J. Mol. Catal. A-Chem. 270, 93–100 (2007)
11. Obee, T.N., Brown, R.T.: TiO_2 photocatalysis for indoor air applications: effects of humidity and trace contaminant levels on the oxidation rates of formaldehyde, toluene, and 1, 3-butadiene. Environ. Sci. Technol. 29, 1223–1231 (1995)
12. Turchi, C.S., Ollis, D.F.: Photocatalytic degradation of organic water contaminants: Mechanisms involving hydroxyl radical attack. J. Catal. 122, 178–192 (1990)

Spray Deposition of Au/TiO_2 Composite Thin Films Using Preformed Nanoparticles

W. Wang, K. Cassar, S.J. Sheard, P.J. Dobson, P. Bishop, I.P. Parkin, and S. Hurst

Abstract. A single-step process to deposit a composite film of Au nanoparticles in a TiO_2 matrix has been investigated by a spray deposition technique. A preformed gold colloid was used as a precursor along with the titanium precursor to deposit gold/titanium composite films onto glass. The composite films were deposited onto Pilkington float glass using spray coating. The deposition temperature was varied from 200 °C up to 550 °C. UV-vis spectra of the films showed that the surface plasmon absorption maximum was red-shifted from 544 nm to 600 nm with increasing substrate temperatures, corresponding to a colour transition from red to blue in transmission. This process, based on a spray technique, offers a simple, rapid and low-cost approach to large-area deposition. The single-step route leads to a homogeneous composite film and controllable properties of the spray solution.

1 Introduction

There has been intense research interest in nanocomposites of metal nanoparticles in a dielectric matrix [1-3]. Noble metal nanoparticles, e.g. Au, Ag, generally exhibit a strong absorption peak in the visible range of the spectrum, due to the surface plasmon resonance effects, which has led to application in optical devices [4, 5]. The tuneability of the plasmon resonances can be achieved by careful

---

W. Wang and S.J. Sheard
Department of Engineering Science, Oxford University, Oxford, United Kingdom

K. Cassar and P. Bishop
Johnson Matthey Technology Centre, Reading, United Kingdom

P.J. Dobson
Oxford University Begbroke Science Park, Oxford, United Kingdom

I.P. Parkin
Department of Chemistry, University College of London, United Kingdom

S. Hurst
Pilkington Technology Centre, Lathom, United Kingdom

design of the nanostructure such as its size, shape and composition. Dielectric materials, especially transition metal oxides, have been extensively investigated due to their functional properties, such as photocatalysis, electrochromism and solar control [6-9]. Nanocomposites comprising the matrix and embedded metal particles therefore bear the functionalities of both phases and could be carefully designed for giving the desired material properties. Such kind of nanocomposites has shown potential application to glass surfaces to modify the optical properties for window glass products used for solar control in buildings [10]. Here we describe an approach for the deposition of a Au-nanoparticle/TiO_2 composite thin film using preformed gold nanoparticles. The method is based on the spray deposition technique which is simple, flexible and capable of scaling up for large area deposition with potentially a low processing cost. Au nanoparticles can be deposited alone, or together with a matrix precursor such as TiO_2 for a composite thin film.

2 Experimentals

Apparatus set-up. The set-up for the laboratory-scale spray deposition comprises three major parts: the syringe pump as the injecting system, the spray head and the hotplate. Basically, the precursor liquid is first injected by the syringe pump and projected from the needle which is underneath the nozzle. The atomisation of the liquid droplets happens at the meeting point of the needle tip and the nozzle which ends up with an aerosol that is sprayed and impacts onto the substrate heated by a hotplate. The hotplate is integrated with an x-y table for fast, large-area uniform deposition. By varying and optimising the reaction conditions such as the injected liquid flow rate, compressed air pressure, hotplate temperature, and raster scanning speed, it is possible to fabricate a uniform thin film on a large area substrate.

Materials and experimental procedure. Gold nanoparticles were synthesised and dispersed in water. A one-pot dispersion for depositing the composites of Au nanoparticles in TiO_2 was prepared by mixing 0.1 M titanium precursor with gold colloids in water at a molar ratio of 10:1 to 20:1.

Typically, the precursor solutions were injected by the syringe pump under a constant flow rate (1-3 ml/min) and spread out by the compressed air at a pressure of 20-30 psi (138-206 kPa). The atomised droplets were sprayed onto the Pilkington float glass substrates (4mm thick, with SiO_2 barrier layer on top) which were heated at temperatures in the range of 150-550 °C. The coated glass samples received further heat treatment in most cases by placing them in a furnace at temperatures in the range of 200-600 °C for 30-60 minutes to remove any remaining organic ligands and further enhance the decomposition of the precursor.

Characterisation. UV-vis spectra were obtained using a Varian Cary 5000 UV-visible-NIR spectrometer. Thickness measurements were carried out using a Veeco Dektak 6M stylus profilometer. In order to obtain information about the morphology of the particles and their size and distribution in the nanocomposite films, transmission electron microscopy (JEOL 2000FX, JEOL 4000HR) and scanning electron microscopy (JSM 6300, JSM 840F) micrographs were taken.

3 Results and Discussion

The initial gold colloids were transparent and a deep red colour. The UV-vis spectra of the gold colloid showed a surface plasmon resonance peak at 533 nm. TEM imaging of the colloid on a holey carbon film showed the gold nanoparticles

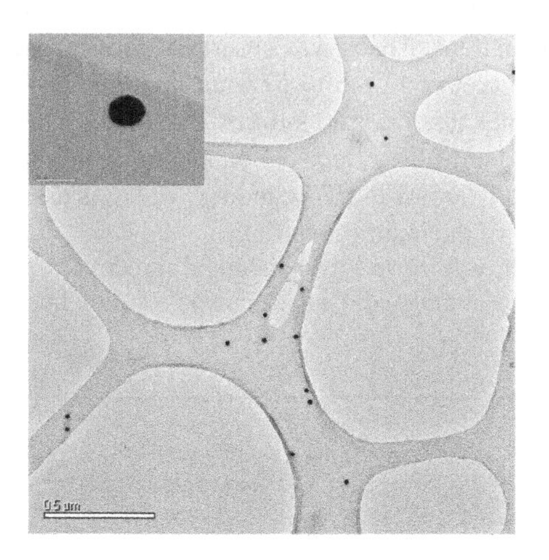

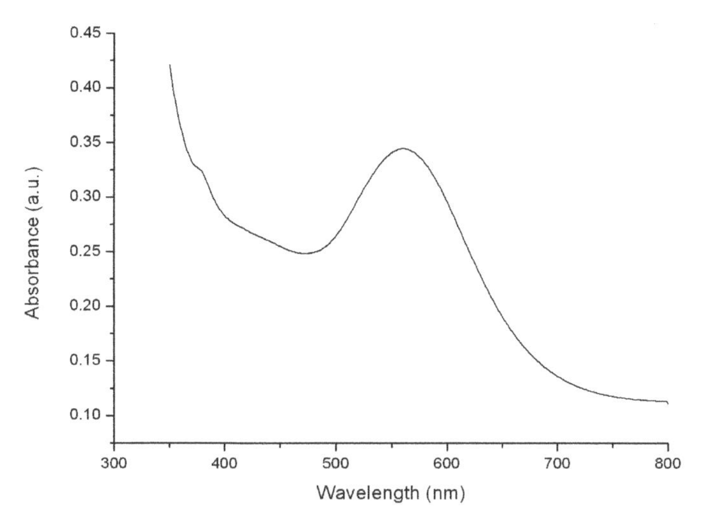

Fig. 1 TEM images of gold nanoparticles of a mean size of 20 nm and UV-vis spectrum of spray deposited thin film of gold nanoparticle on Pilkington float glass

have a mean diameter of 20 nm with a narrow size distribution. Gold nanoparticle thin films were spray deposited onto glass using this solution. The thin films were deposited at 180 °C and without further annealing. The films were red in appearance when viewed in transmission and an intense surface plasmon absorption peak at 542 nm was displayed in the UV-visible spectrum (Fig. 1).

The one-pot solution containing gold colloids and the titanium precursor in water was used for spray deposition of Au/TiO_2 nanocomposite thin films. The solution shows a clear deep red colour and the stability was maintained for over two weeks. Coating deposition was first carried out at 200 °C on Pilkington float glass, followed by annealing at 400 °C or 600 °C respectively. Colour changes from red to purple and to blue in transmission was observed. Optical spectroscopy shows an absorption peak shift from 543 nm to 572 nm and then to 599 nm, respectively (Fig. 2). The film thickness decreases from 728 nm to 320 nm, 240 nm respectively. This is probably due to the effects of annealing which removes organic ligands as well as consolidation of the TiO_2 matrix, which is assumed to result in a higher refractive index of the matrix material.

In order to compare the "spray followed by annealing" process, a direct spray deposition without annealing was conducted using the same solution but at different

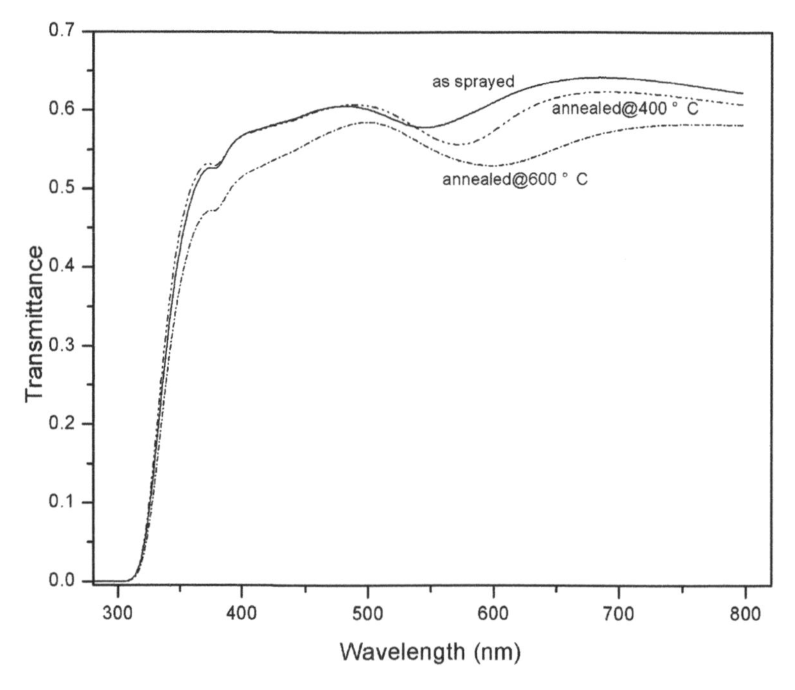

Fig. 2 Optical absorbance spectra of gold nanoparticle/TiO2 matrix thin film on Pilkington float glass before and after annealing. The small feature at 380-400nm is believed to be associated with the tin on the float glass

deposition temperatures ranging from 200 °C to 550 °C. The composite films were successfully deposited at all these temperatures with strong adherence to the glass substrates. Surface plasmon absorption peaks obtained from the UV-vis absorbance spectra show a red-shift from 544 nm to 600 nm with increasing substrate temperature (Fig. 3). We attribute this to the formation and densification of TiO_2 anatase in the matrix material with the increasing substrate temperature which gives an increase of the refractive index of the matrix and hence a red shift for the surface plasmon condition [9].

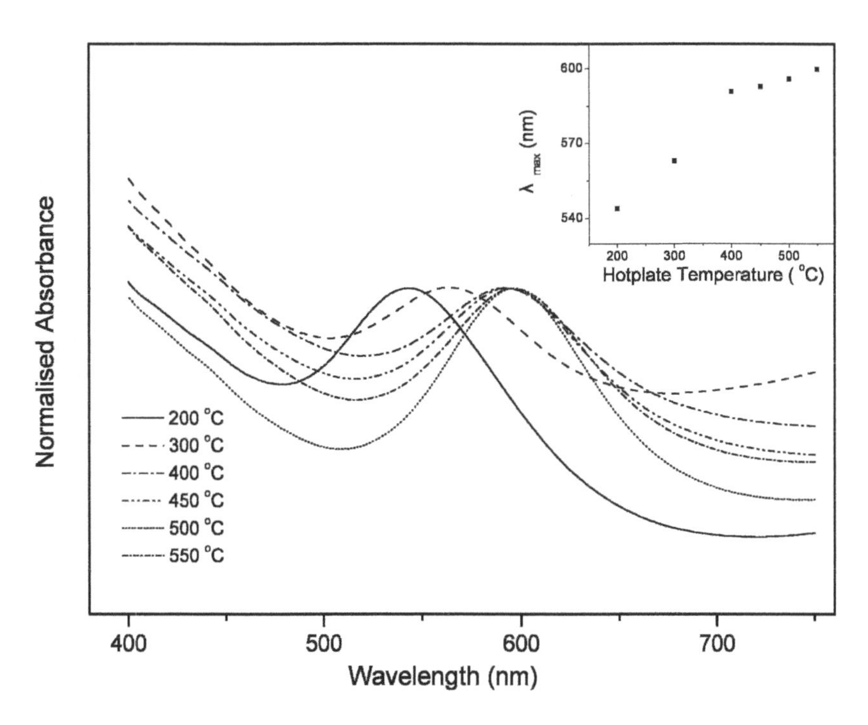

Fig. 3 Measured normalised absorbance spectra of gold nanoparticles/TiO2 thin films deposited with substrate temperatures from 200 °C to 550 °C which shows a shift of the surface plasmon peak resonance with increasing substrate temperature (see inset). The peaks show a red-shift from 545 nm to 600 nm

The Au/TiO_2 composite film deposited at 400 °C was characterised and analysed for comparison. The UV-vis spectrum shows a surface plasmon absorption peak at 589 nm, compared to the one that was sprayed at 180 °C and annealed at 400 °C showing a peak at 572 nm. This is probably because better consolidation of TiO_2 can be achieved when sprayed onto a hotter substrate. The film is transparent and blue in transmitted light. It is also robust under washing and has good adherence to the glass substrate. The gold nanoparticles could not be removed from the TiO_2 matrix either, indicating the gold nanoparticles are strongly embedded in the film. SEM micrographs of the Au/TiO_2 thin film suggest that the gold nanoparticles are evenly distributed in the TiO_2 matrix. However, there is also evidence of the film cracking probably due the thermal expansion coefficient mismatch between the film and the substrate. Nevertheless, this experiment has demonstrated the possibility of a single process of using a stable dispersion containing both gold nanoparticles and matrix precursors and spraying at a high temperature which is compatible and favourable in an industrial on-line process.

4 Conclusions

A single step process for spray deposition has been demonstrated using a stable dispersion containing both gold nanoparticles and matrix precursors which can be deposited at a high temperature. The optical properties of the composite films can be tailored by varying the properties of metal nanoparticles and matrix precursors as well as the deposition temperature. This process, based on a spray technique, offers a simple, rapid and low cost approach to large area deposition. Deposition using preformed nanoparticles allows the design of nanocomposite by tailoring the properties of nanoparticles such as size and structure. The single-step route leads to more homogeneous composite film and controllable properties of precursors. This film deposition method can easily be scaled up and is compatible with current industrial on-line processes.

Acknowledgments. The work is funded by Technology Strategy Board of UK. We thank Guillermo Benito, Troy Manning and Paolo Melgari for helpful discussion as well as Professor Patrick Grant for courtesy of providing the spray equipment.

References

1. Palgrave, R., Parkin, I.P.: Aerosol assisted chemical vapor deposition using nanoparticle precursors: A route to nanocomposite thin films. J. Am. Chem. Soc. 128, 1587–1597 (2006)
2. Xu, X., Stevens, M., Cortie, M.B.: In situ precipitation of gold nanoparticles onto glass for potential architectural applications. Chem. Mater. 16, 2259 (2004)
3. Buso, D., Pacifico, J., Martucci, A., Mulvaney, P.: Gold-nanoparticle-doped TiO2 semiconductor thin films: Optical characterization. Adv. Funct. Mater. 17, 347 (2007)
4. Armelao, L., Barreca, D., Bottaro, G., Gasparotto, A., Gross, S., Maragno, C., Tondello, E.: Recent trends on nanocomposites based on Cu, Ag and Au clusters: A closer look. Coord. Chem. Rev. 250, 1294 (2006)
5. Ung, T., Liz-Marzan, L.M., Mulvaney, P.: Gold nanoparticle thin films. Colloid Surface A 202, 119 (2002)
6. Lahiri, D., Subramanian, V., Shibata, T., Wolf, E.E., Bunker, B.A., Kamat, P.V.: Photoinduced transformations at semiconductor/metal interfaces: X-ray absorption studies of titania/gold films. J. Appl. Phys. 93, 2575 (2003)

7. Ashraf, S., Blackman, C.S., Hyett, G., Parkin, I.P.: Aerosol assisted chemical vapour deposition of MoO_3 and MoO_2 thin films on glass from molybdenum polyoxometallate precursors; thermophoresis and gas phase nanoparticle formation. J. Mater. Chem. 16, 3575–3582 (2006)
8. He, T., Ma, Y., Cao, Y., Yang, W., Yao, J.: Improved photochromism of WO3 thin films by addition of Au nanoparticles. Phys. Chem. Chem. Phys. 4, 1637–1639 (2002)
9. Medda, S.K., De, S., De, G.: Synthesis of Au nanoparticle doped SiO_2-TiO_2 films: Tuning of Au surface plasmon band position through controlling the refractive index. J. Mater. Chem. 15, 3278–3284 (2005)
10. Walters, G., Parkin, I.P.: The incorporation of noble metal nanoparticles into host matrix thin films: synthesis, characterisation and applications . J. Mater. Chem. 19, 574–590 (2009)

Nanoindentation Study of Resin Impregnated Sandstone and Early-Age Cement Paste Specimens

W. Zhu, M.T.J. Fonteyn, J. Hughes, and C. Pearce

Abstract. Nanoindentation testing requires well prepared samples with a good surface finish. Achieving a good surface finish is difficult for heterogeneous materials, particularly those with weak and fragile structures/phases, which are easily damaged or lost during preparation. The loss of weak structures can be drastically reduced by impregnating the sample with a resin before cutting and polishing. This technique is commonly used in SEM microscopy but has not been used for nanoindentation-testing before. This paper reports an investigation to extract micro-mechanical properties of different phases in resin impregnated sandstone and 1-day hydrated cement samples. The results appeared to show that it is feasible to use resin impregnated specimens for nanoindentation study of both materials.

1 Introduction

Cement based materials and sandstone are among the most utilized materials, essential to the construction industry and built environment. Studies by microscopic techniques have revealed that both materials are complex heterogeneous composite materials, with a random microstructure at different length scales, from the nano to the macro scale. Such complex composites are made even more complicated by the time dependent nature of the cement hydration processes which begin at the mixing of cement clinker minerals with water and continue for months and even years. The engineering properties and durability performance of these materials at the macro scale are all significantly affected, if not dominated, by their structural features and properties at the micro/nano scale where the deterioration

W. Zhu and J. Hughes
School of Engineering and Science, University of the West of Scotland, UK
e-mail: wenzhong.zhu@uws.ac.uk, john.hughes@uws.ac.uk

M.T.J. Fonteyn
Department of Mechanical Engineering, Eindhoven University of Technology, Netherlands
e-mail: M.T.J.Fonteyn@student.tue.nl

C. Pearce
Department of Civil Engineering, University of Glasgow, UK
e-mail: c.pearce@civil.gla.ac.uk

and failure process starts. Over the past 10-15 years depth sensing nanoindentation technique has shown to provide a very useful tool to study properties of nano/micro-scale features in many different materials, including composites or multiphase materials [1-5]. Particularly, a statistical or grid/mapping nanoindentation method has recently been applied to determine mechanical properties (e.g. elastic modulus, hardness) of individual phases in complex materials, including cement paste, bones and natural rocks, thus providing vital information for understanding of materials behaviour and development of computational models [6-8].

Generally, nanoindentation testing requires well prepared samples with a good surface finish. However, achieving a good surface finish is difficult for heterogeneous materials, particularly those with weak, porous and fragile structures/phases, which are easily damaged or lost during preparation. Indeed, it has been reported that there is a limit in obtainable surface roughness for even aged cement paste due to the porous structure [9]. Roughness of the specimen surface can also have a significant effect on the nanoindentation test results [10]. Furthermore, it has been impossible to prepare sandstone and 1-day hydrated cement paste samples using the common preparation steps since almost all the weak structures and hydrated phases disappear from the prepared surface. Therefore, the application of nanoindentation to study such materials is hindered by not only the high roughness of the prepared sample surface but also an unrepresentative specimen. The loss of weak structures or phases can be drastically reduced by impregnating the sample with a resin before cutting and polishing the surface. This technique is commonly used in SEM microscopy but has not been used for nanoindentation testing before. The main objective of this study is to investigate the feasibility of using resin impregnated samples for nanoindentation, and particularly to extract micro-mechanical properties of different phases in resin impregnated sandstone and 1-day hydrated cement samples.

2 Experimental

The sandstone sample is Giffnock Sandstone, commonly used for buildings in the Glasgow area. The stone is a well sorted, medium grained, quartz-arenite of Carboniferous age, sub-angular quartz dominating, with minor amounts of mica and degraded k-feldspars. It contains significant amounts of secondary minerals; 5-10% of interstitial kaolinite and 10-20% carbonate cement, composed of Fe-Mg bearing carbonate mineral Ankerite ($Ca(Fe^{2+},Mg,Mn)(CO_3)_2$). The cement paste sample was prepared with a 42.5N Portland cement and a water-to-cement ratio of 0.40. The fresh cement paste was left in a small mould (10x10x40 mm) covered with a plastic bag at ~20°C for 24 hours before demoulding. After demoulding, the 1-day hydrated sample was placed in bottle of methanol to stop the hydration. Then the usual sample preparation steps [4] were followed, including resin embedding (but not impregnating), precision sectioning, grinding, polishing (down to 6 μm) to obtain a disc specimens (ϕ30x15 mm) for both the sandstone and the cement samples. These specimens were then resin impregnated using the EPOFIX epoxy resin of Struers in a vacuum chamber at room temperature. Further grinding and polishing (down to 1 μm) and ultrasonic cleaning were applied to the impregnated specimens to produce the final specimens for nanoindentation. Methanol-based liquid was used as a lubricant and for cleaning in all the procedures so as to avoid further cement hydration or possible dissolution of minerals/hydrates.

The methodology and operating principle for the nanoindentation technique have been reviewed and presented in detail elsewhere [1-2]. Briefly, the test consists of making contact between a sample surface and a diamond indenter of known geometry, followed by a loading-unloading cycle while continuously recording the load, P, and indentation depth, h. The P-h curve obtained is a fingerprint of the mechanical properties of the test area. Most commonly, the elastic modulus (E) and hardness (H) of the test area are determined by analysing the initial part of the unloading data according to a model for the elastic contact problem. For studying multiphase composite materials, a refined statistical indentation method has been used, which involves testing and statistically analyzing a large number of indentation points within a representative sample area [7-8].

The nanoindentation apparatus used in this study was Nanoindenter XP fitted with a Berkovich indenter. In this study, all testing was programmed in such a way that the loading started when the indenter came into contact with the test surface and the load maintained for 30 seconds at the pre-specified maximum value before unloading. In order to provide statistical analysis of the micro-mechanical properties of different phases in the specimen, two randomly selected areas each with a grid of 12x16 indentation points with indent spacing of 10 μm were used for the cement paste. For the sandstone sample, a grid of 12x20 indentation points with indent spacing of 30 μm were used due to its relatively large grain size. The E and H values were extracted from each test point with the indentation depth range of 100 – 300 nm for the cement specimen by varying the maximum load applied while for the sandstone specimen a fixed maximum load of 5 mN was used.

3 Results and Discussions

Fig.1 shows typical optical images of the polished surface of the resin impregnated specimens. As shown, good surface finish was successfully achieved with both specimens. Large resin-filled areas could clearly be seen on the sandstone surface, while no such resin-filled areas could easily be found on the cement paste specimen. Nanoindentation trials were carried out on the 1-day hydrated cement paste specimen with and without the resin impregnation. It was interesting to find that for the resin impregnated specimen over 50% of the test points had an elastic modulus value lower than 45 GPa while for the specimen without the impregnation only less than 20% of the test points showed elastic modulus value lower than 45 GPa. As the elastic modulus of cement hydration products is generally lower than 45 GPa the above results confirmed previous observations that a significant loss of the weak, hydrates phases occurred during the preparation processes if the sample was not resin impregnated.

The large number of test results obtained from the grid indentation were statistically analysed to extract mechanical properties of individual phases in the tested area using a method presented previously [7-8]. Basically, the experimental E and H values at all test point were statistically analysed to produce a frequency plot. Then, the best model fit to the experimental results with multi-modal normal distribution curves (or Gaussian distribution) was produced using non-linear least

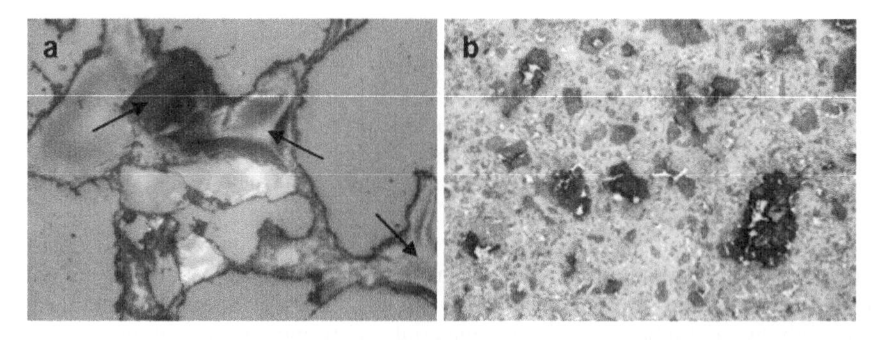

Fig. 1 Optical images of typical polished surfaces (magnification ~100) of the sandstone (a) and 1-day hydrated cement paste (b) specimens: the arrows showing large resin-filled areas

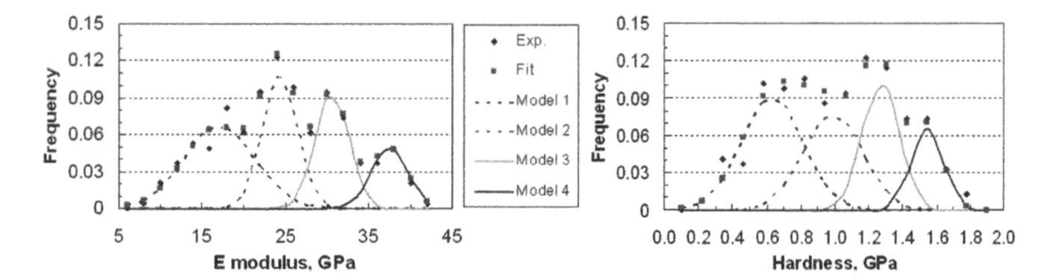

Fig. 2 Mechanical property frequency plots for test results of the resin impregnated 1-day hydrated cement specimen, together with model fits for individual hydrate phases

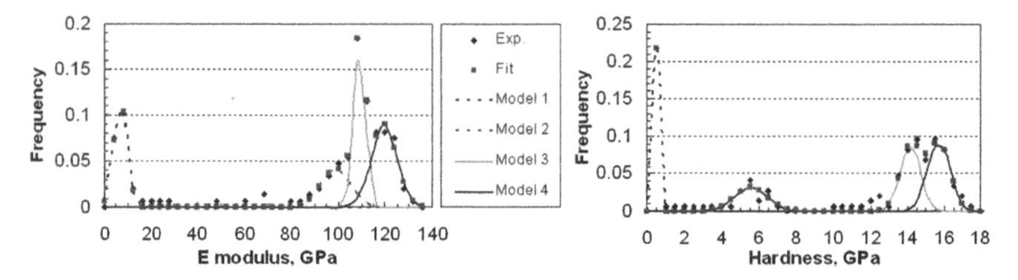

Fig. 3 Mechanical property frequency plots for test results of the resin impregnated sandstone specimen, together with model fits for individual phases

squares method. These curves are shown in Fig.2 and Fig.3 for the resin impregnated 1-day cement and the sandstone specimens respectively. From each model fit, the mean values of E and H were extracted, whose association with a specific hydrate or mineral phase could be ascertained by SEM-BSE and SEM-EDS analysis [8]. The area under each curve also provides an estimate of the volume fraction for the hydrate/mineral phase it associated with. Table 1 presents a summary of results of the mechanical properties of individual phases and their volumefractions for the 1-day cement and the sandstone specimens. For comparison, previous results obtained for a 28-day hydrated cement specimen without resin impregnation are also given. A SEM-BSE image of the indented area for the sandstone is shown in Fig.4. The mineral phases in the sandstone were also identified.

Table 1 Properties of different phases in the cement and sandstone specimens

	Sample	Resin impregnated 1-day hydrated cement paste			28-day hydrated cement without resin impregnation		
	Properties of phases	E, GPa	H, GPa	V%	E, GPa	H, GPa	V%
1	Loose-packed CSH	17.1	0.63	34.8	18.1	0.42	5.5
2	LD CSH	24.4	0.99	28.2	24.4	0.75	60.2
3	HD CSH	30.6	1.27	23.7	31.4	1.05	28.7
4	$Ca(OH)_2$ plus	37.4	1.54	13.3	37.5	1.27	5.6
	Sample	Resin impregnated sandstone					
	Properties of phases	E, GPa	H, GPa	V%	-	-	-
1	Epoxy resin plus	6.7	0.40	21.5	-	-	-
2	Ankerite	98.9	5.57	17.0	-	-	-
3	Quartz A	108.9	14.2	28.3	-	-	-
4	Quartz B	119.5	15.7	33.2	-	-	-

Fig. 4 SEM-BSE image of the tested area on sandstone (within the white box): Area includes quartz (Q), ankerite (A), kaolinite clay (K) and dark resin-filled porosity (P)

Results shown in Fig.2-3 and Table 1 appear to suggest that the epoxy resin, which has an E value of 3 - 4 GPa, could not be detected by the nanoindentation test in the 1-day cement specimen. This might be due to the small volume of resin present and pore sizes in the specimen. Generally the phases and properties obtained for the 1-day cement specimen are in good agreement with other published results for 28-day and 1-year cement samples [6-8]. Particularly, the E and H values obtained for the low and high density calcium silicate hydrates (i.e. LD-CSH and HD-CSH) were found to be almost the same for the 1-day and the 28-day cement specimens. These appeared to suggest that the mechanical properties of the CSH phases do not change with the hydration age. For the sandstone specimen, the epoxy resin phase was detected by nanoindentation. The E and H values of the quartz phases are in good agreement with the limited published data [3, 8] though the crystal orientation may have an effect. The results for Ankerite are reasonable. The kaolinite clay phase seen on SEM was not detected by the nanoindentation, which might be due to its soft/loose nature and thus merged with the resin peak.

4 Conclusions

The study seemed to indicate that it is feasible to use resin impregnated specimens for nanoindentation study of both sandstone and early age hydrated cement paste samples. The E and H values of different mineral phases in the sandstone sample determined by the statistical nanoindentation method appeared to be in good agreement with the limited published data. The results of the 1-day hydrated cement sample suggested that the mechanical properties of the individual hydrate phases (e.g. LD-CSH, HD-CSH, etc) did not change with cement hydration.

References

1. Oliver, W.C., Pharr, G.M.: An improved technique for determining hardness and elastic modulus using load and displacement sensing indentation experiments. J. Mater. Res. 7, 1564–1579 (1992)
2. Ficher-Cripps, A.C.: Nanoindentation. Springer, London (2002)
3. Broz, M.E., Cook, R.F., Whitney, D.L.: Microhardness, toughness, and modulus of Mohs scale minerals. American Mineralogist 91, 135–142 (2006)
4. Zhu, W., Bartos, P.J.M.: Application of depth-sensing microindentation testing to study of interfacial transition zone in reinforced concrete. Cem. Concr. Res. 30, 1299–1304 (2000)
5. Velez, K., Maximilien, S., et al.: Determination by nanoindentation of elastic modulus and hardness of pure constituents of Portland cement clinker. Cem. Concr. Res. 31, 555–561 (2001)
6. Constantinides, G., Ulm, F.J., Van Vliet, K.: On the use of nanoindentation for cementitious materials. Mater. Struct. 36, 191–196 (2003)
7. Ulm, F., Vandamme, M., Bobko, C., Ortega, J.A.: Statistical indentation techniques for hydrated nanocomposites: concrete, bones and shale. J. Am. Ceram. Soc. 90(9), 2677–2692 (2007)
8. Zhu, W., Hughes, J., et al.: Micro/Nano-scale mapping of mechanical properties of cement paste and natural rocks by nanoindentation. Mater. Charact. 11, 1189–1198 (2007)
9. Trtik, P., Dual, J., Muench, B., Holzer, L.: Limitation in obtainable surface roughness of hardened cement paste: 'virtual' topographic experiment based on focussed ion beam nanotomography datasets. J. Microscopy 232(2), 200–206 (2008)
10. Miller, M., Bobko, C., et al.: Surface roughness criteria for cement paste nanoindentation. Cem. Concr. Res. 38, 467–476 (2008)

Heterogeneous Photocatalysis Applied to Concrete Pavement for Air Remediation

M.M. Ballari, M. Hunger, G. Hüsken, and H.J.H. Brouwers

Abstract. In the present work the degradation of nitrogen oxides (NO_x) by concrete paving stones containing TiO_2 to be applied in road construction is studied. A kinetic model is proposed to describe the photocatalytic reaction of nitric oxide (NO) in a standard flow laminar photoreactor irradiated with UV lamps. In addition the influence of several parameters that can affect the performance of these stones under outdoor conditions are investigated, such as irradiance, relative humidity and wind speed. The kinetic parameters present in the NO reaction rate are estimated employing experimental data obtained in the photoreactor. The obtained model predictions employing the determined kinetic constants are in good agreement with the experimental results of NO concentration at the reactor outlet.

1 Introduction

Heterogeneous photocatalysis represents an emerging environmental control option for the efficient removal of chemical pollutants and it can be applied to water and air purification. This process involves a nano-solid semiconductor catalyst, regularly titanium dioxide (TiO_2), which is activated with ultraviolet light of the appropriate wavelength.

Nitrogen oxides (NO_x) is the generic term for a group of highly reactive gases, most of them emitted in air in the form of nitric oxide (NO) and nitrogen dioxide (NO_2). Nitrogen oxides form when fuel is burned at high temperatures, as is the case in combustion processes in automobiles. NO_x causes a wide variety of health and environmental impacts, like the formation of tropospheric ozone and urban smog through photochemical reactions with hydrocarbons. Furthermore, NO_x together with SO_x (sulfur dioxide and sulfur trioxide) is the major contributor to the "acid rain", one of the most serious environmental problems across the world.

The European Union (EU) has taken important steps over the past decade leading to a decrease in the emissions to air and water of a number of pollutants. One of its directives (1999/30/EC) establishes limit values for concentrations of

M.M. Ballari, M. Hunger, G. Hüsken, and H.J.H. Brouwers
Department of Construction Management & Engineering, Faculty of Engineering Technology, University of Twente, Enschede, The Netherlands
e-mail: M.Ballari@ctw.utwente.nl
www.utwente.nl

sulphur dioxide, nitrogen dioxide and oxides of nitrogen, particulate matter and lead in ambient air. Some of the pollutant emissions have since become more or less manageable; however particulates, NO_x and smog are still problematic.

The development of innovative materials that can be easily applied on structures, with both de-soiling and de-polluting properties, is a significant step towards improvements of air quality. The use of TiO_2 photocatalyst in combination with cementitius and other construction materials has shown a favorable effect in the removal of nitrogen oxides.

In the present work the degradation of NO_x compounds employing concrete paving stones with TiO_2 to be applied in road construction is studied. The experiments were carried out in a photoreactor designed according to the standard ISO 22197-1 (2007) to assess these kind of photocatalytic materials employing NO as the pollutant source. A kinetic model is proposed to describe the photocatalytic oxidation of NO_x and the influence of several parameters that can affect the performance of these stones under outdoor conditions, such as irradiance, relative humidity and wind speed. A reaction rate expression for the NO oxidation is postulated and the kinetic parameters are determined employing the experimental data. Finally the model predictions with the estimated kinetic constants are compared with the experimental results obtaining a good agreement between them.

2 Experimental Setup

The standard ISO 22197-1 (2007) serves as a sound basis for measurements, its recommendations were largely followed for the practical conduction of the present study. The applied apparatus is composed of a planar reactor cell housing the concrete stone sample, a suitable UV-A light source, a chemiluminescent NO_x analyzer, and an appropriate gas supply (Figure 1).

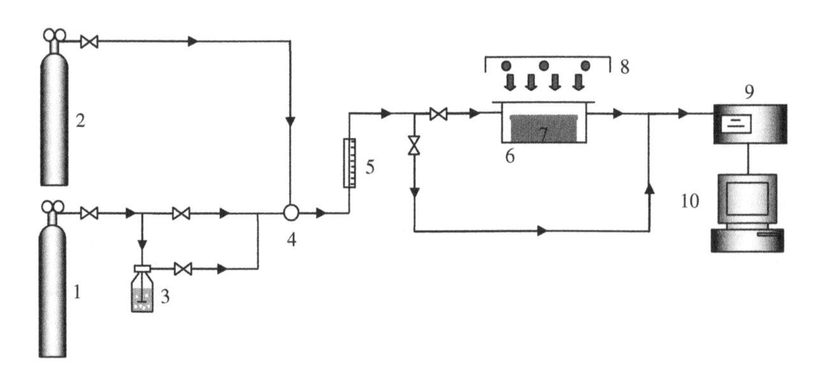

Fig. 1 Schematic representation of the experimental setup. 1. Synthetic air. 2. NO source. 3. Gas washing bottle. 4. Temperature and relative humidity sensor. 5. Flow controller. 6. Gas photoreactor. 7. Paving stone sample. 8 Light source. 9. NO_x analyzer. 10. Computer

Table 1 shows the main characteristics, dimensions and operating conditions of the experimental setup that were employed to carry out the photocatalytic NO degradation experiments.

Table 1 Experimental setup main characteristics and operating conditions

Description		Value
Reactor	Length (L)	2 dm
	Width (B)	1 dm
	Height (H)	0.02-0.04 dm
	Volume ($V_{reactor}$)	0.04-0.08 dm^3
Photocatalytic stone	Length (L)	2 dm
	Width (B)	1 dm
Lamps: Philips Compact S × 3	Input power	25 W
	Emission wavelength	300-400 nm
Flow rate (Q)		3-5 $dm^3\ min^{-1}$
Relative Humidity (RH)		10-80 %
Inlet NO concentration ($C_{NO,in}$)		0.1-1 ppm
Irradiance flux (E)		0.3-13 $W\ m^{-2}$

3 Theoretical Model

The kinetic expression proposed for the NO degradation reaction rate is the corresponding to the Langmuir-Hinshelwood model [1, 7], which is widely employed for the photocatalytic degradation of other contaminants [2, 6, 8]. However the reaction rate should be expressed as a superficial rate for a gas-solid heterogeneous system [3, 4]. In addition, water competes with NO for free active sites at the catalyst surface and therefore it can be considered as an additional reactant [8]. Following this model applied to a heterogeneous reaction, the Langmuir-Hinshelwood kinetic model for NO disappearance rate per unit area of active surface reads:

$$r_{NO} = -\frac{kK_{NO}C_{NO}}{1+K_{NO}C_{NO}+K_wC_w} \tag{1}$$

where r_{NO} is the superficial reaction rate (mole dm^{-2} min^{-1}) of NO. C_{NO} and C_w are the corresponding molar concentration (mole dm^{-3}) of NO and water. k is the reaction rate constant (mole dm^{-2} min^{-1}). K_{NO} and K_w are the adsorption equilibrium constant (dm^3 $mole^{-1}$) for NO and for water respectively.

Regarding the UV light effect, it is supposed that the irradiance only has an influence on the reaction rate constant (k). Therefore a mathematical expression of the reaction constant k in function of the radiative flux E is proposed:

$$k = k'\left(-1+\sqrt{1+\alpha E}\right) \tag{2}$$

with k' (mole dm^{-2} min^{-1}) and α (dm^2 W^{-1}) being factors to be fitted from the experiments. This expression takes account of the linear and the square root dependency of the reaction rate with the light intensity, that have announced in several publications [3] for high and low irradiance respectively. When UV-radiation is absent, i.e. E = 0, the reaction rate becomes zero. For small E, Eq. (2) tends $k'\alpha E/2$, and for large E it tends to $k'\sqrt{\alpha E}$.

The NO balance equations for a plug flow reactor reads:

$$v_{air} \frac{dC_{NO}}{dx} = a_v r_{NO} \tag{3}$$

where v_{air} is the air velocity (dm min^{-1}) in the reactor and a_v is the active surface area per unit reactor volume (dm^{-1}). The reactor inlet condition is:

$$C_{NO}(x=0) = C_{NO,in} \tag{4}$$

4 Kinetic Parameters Estimation

To solve the NO mass balance with the complete kinetic expression, a discretization of the differential equation (Eq. (3)) can be applied (Euler method). Then the optimization of all kinetic parameters present in the reaction rate can be achieved employing the "solver" tool of Excel. The result of this estimation is shown in Table 2.

5 Experimental Results Versus Model Prediction

It is possible to analyze the effect of different operating variables on the system by resorting the estimated kinetic parameters from the complete model. This analysis can include a comparison between simulated values obtained with the model and experimental measurements.

Table 2 Non linear parameters optimization (based on 36 experimental results) employing the Excel Solver tool and the numerical solution of the NO differential mass balance performing a forward discretization

Parameter	Value
k' (mole dm^{-2} min^{-1})	7.24×10^{-10}
α (dm^2 W^{-1})	1.40×10^{6}
K_{NO} (dm^3 mole^{-1})	2.50×10^{10}
K_w (dm^3 mole^{-1})	7.67×10^{5}

Figure 2 shows the model predictions and the experimental data corresponding to the NO outlet concentration in function of the NO inlet concentration to the reactor for two different flow rates. When the inlet concentration of NO increases, NO outlet concentration rise as well. However, as expected, decreasing the initial concentration of the pollutant the final conversion of the reacting system increases.

The effect of the flow rate is possible to analyze comparing Figures. 2(a) and (b). When the flow rate is increased the resident time in the reactor decreases. Therefore, for low flow rates a larger conversion of the pollutant is observed.

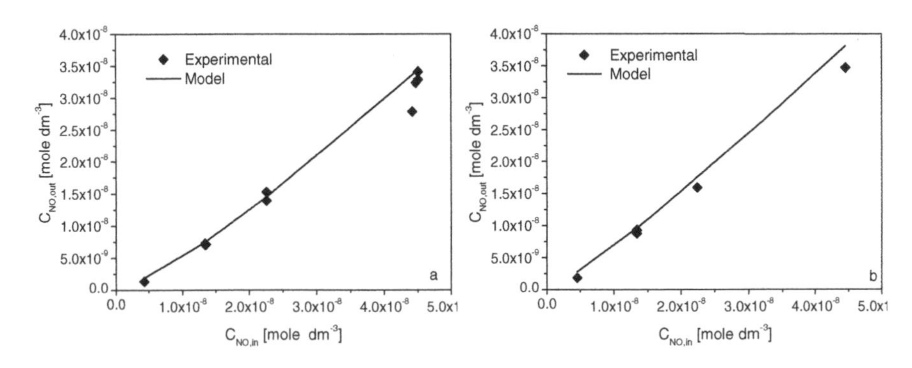

Fig. 2 Model prediction vs. experimental data. NO outlet concentration in function of the NO inlet concentration. RH = 50 %. E = 10 W m^{-2}. a) Q = 3 dm^3 min^{-1}. b) Q = 5 dm^3 min^{-1}

Regarding the irradiance and relative humidity effect, Figures 3(a) and (b) show the obtained results varying these two parameters, respectively. When the irradiance is increased a higher conversion of the systems is achieved. However when the relative humidity is enlarged water competes with NO for the same active site and the NO consumption declines.

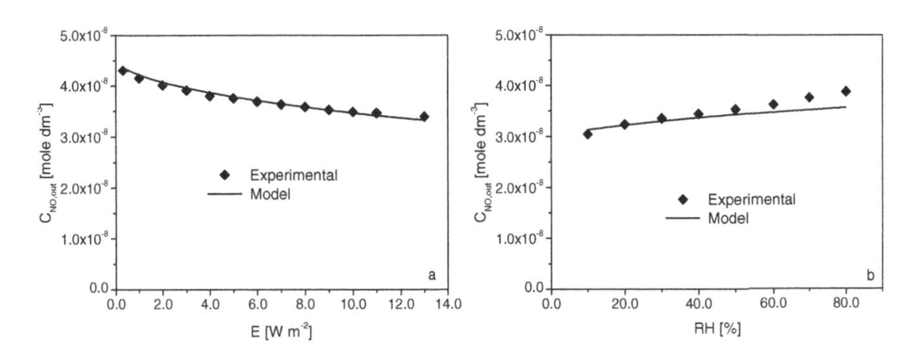

Fig. 3 Model prediction vs. experimental data. a) NO outlet concentration in function of the radiative flux. b) NO outlet concentration in function of the relative humidity. $C_{NO,in}$ = 1 ppm. Q = 3 dm^3 min^{-1}. H = 3 mm

6 Conclusions

In the present work, a kinetic study of the photocatalytic degradation of nitrogen oxides was conducted. A heterogeneous kinetic expression for the NO degradation was proposed. Several experiments were carried out according to a suitable ISO standard for photocatalytic materials assessment employing only NO as a contaminat. Different operating conditions were selected to perform the experiments (NO inlet concentration, reactor height, flow rate, irradiance flux and relative humidity). Employing these experimental data and the reaction rate expression, the kinetic parameters were estimated for a numerical solution of the governing equations in the reactor, based on Langmuir-Hinshelwood kinetics. In all cases, a very good correlation between the experimental data and the computer simulation with the estimated kinetic parameters was obtained.

Acknowledgments. The authors wish to express their thanks to the following sponsors of the research group: Bouwdienst Rijkswaterstaat, Rokramix, Betoncentrale Twenthe, Graniet-Import Benelux, Kijlstra Beton, Struyk Verwo Groep, Hülskens, Insulinde, Dusseldorp Groep, Eerland Recycling, ENCI, Provincie Overijssel, Rijkswaterstaat Directie Zeeland, A&G maasvlakte, BTE, Alvon Bouwsystemen, and V. d. Bosch Beton (chronological order of joining).

References

1. Devahasdin, S., Fan, C., Li, J.K., Chen, D.H.: TiO_2 Photocatalytic Oxidation of Nitric Oxide: Transient Behavior and Reaction Kinetics. J. Photochem. and Photobiol. A Chem. 156, 161–170 (2003)
2. Dong, Y., Bai, Z., Liu, R., Zhu, T.: Decomposition of Indoor Ammonia with TiO_2-Loaded Cotton Woven Fabrics Prepared by Different Textile Finishing Methods. Atm. Env. 41, 3182–3192 (2007)
3. Imoberdorf, G.E., Irazoqui, H.A., Cassano, A.E., Alfano, O.M.: Photocatalytic Degradation of Tetrachloroethylene in Gas Phase on TiO_2 Films: A Kinetic Study. Ind. Eng. Chem. Res. 44, 6075–6085 (2005)
4. Levenspiel, O.: Chemical Reaction Engineering. Willey, New York (1999)
5. Lin, Y.M., Tseng, Y.H., Huang, J.H., Chao, C.C., Chen, C.C., Wang, I.: Photocatalytic Activity for Degradation of Nitrogen Oxides over Visible Light Responsive Titania-Based Photocatalysts. Environ. Sci. Technol. 40, 1616–1621 (2006)
6. Ollis, D.F.: Photoreactors for Purification and Decontamination of Air. In: Ollis, D.F., Al-Ekabi, H. (eds.) Photocatalytic Purification and Treatment of Water and Air, pp. 481–494. Elsevier, Amsterdam (1993)
7. Wang, H., Wu, Z., Zhao, W., Guan, B.: Photocatalytic Oxidation of Nitrogen Oxides Using TiO_2 Loading on Woven Glass Fabric. Chemosphere 66, 185–190 (2007)
8. Zhang, J., Hu, Y., Matsuoka, M., Yamashita, H., Minagawa, M., Hidaka, H., Anpo, M.: Relationship between the Local Structures of Titanium Oxide Photocatalysts and their Reactivities in the Decomposition of NO. J. Phys. Chem. B 105, 8395–8398 (2001)

Synthesis of α-Al_2O_3 Nanopowder by Microwave Heating of Boehmite Gel

T. Ebadzadeh and L. Sharifi

Abstract. Alumina nanopowders were prepared by a sol-gel process from the mixture of boehmite gel and carbon black followed by microwave heating (2.45 GHz and 900 W) at different times. The average size of boehmite particles in sol was 25 nm. After heating, the products were characterized by powder x-ray diffraction. The results show that the main phase is γ-Al_2O_3 at low heating times (4 and 6 min, 610° and 790°C, respectively), while after 8 min heat-treatment (960°C), besides the strong peaks of γ-Al_2O_3, the weak peaks of α- Al_2O_3 were also appeared. α- Al_2O_3 was the only crystalline phase after 10 min heating (1050°C). Synthesized particles were observed by TEM and the average size of particles was measured to be 35 nm. The surface area of powder heated for 10 min was 51 m^2/g.

1 Introduction

It is generally desirable that the transformation of boehmite (AlOOH) to α-alumina occurs as fast as possible and at as a low temperature as possible to minimize the energy that is used in the process [1]. Microwave at the 2.45 GHZ frequently are used almost universally for industrial and scientific applications. Some reasons for the growing interest in using microwave energy in ceramic materials processing are the rapid kinetics (saving of time and energy) and synthesis of new materials. Compared to conventional heating, microwave leads to the more uniform and rapid heating followed by lowering process temperatures which are suitable to give nano-sized powders [2,3].

In this study, high specific surface area carbon black was used to form an obstacle during the thermal processing to produce alumina nanopowders from microwave heating of boehmite precursor.

2 Experimental Procedure

Aluminum nitrate ($Al(NO_3)_3.9H_2O$, MERCK) was used as a starting material for preparation of boehmite gel. In this manner aluminum salt was dissolved in

T. Ebadzadeh and L. Sharifi
Ceramic Division, Materials & Energy Research Centre, Tehran, Iran

deionized water in the ratios of 1 to 20 at 80 °C. The pH of the solution was reached to 8 by adding of NH_4OH. The evaporation of water from the solution was performed on a hot plate at 80 °C while stirred severely and dried at 100 °C. The precursor powder was heated in a microwave oven (900 W and 2.45 GHz) for 10, 15 and 20 minutes. The equivalent temperature of each heating time is shown in Table 1. The progress of formation of α-alumina was examined by x-ray analysis with a Philips' Diffractometer Simens D-500 system, using CuKα radiation. The average crystallite size (t) was calculated using Scherrer's equation:

$$t = -0.9\frac{\lambda}{\beta}\cos\theta \tag{1}$$

where λ is the wavelength of X-rays, β is the half-width corrected using the (111) line of the silicon as the standard. The morphology of synthesized powders was observed by TEM (CM 200 FEG-Philips). Surface area analysis of the alumina powders was performed by a BET technique with N_2 adsorption (Micrometrics, Gemini 2375).

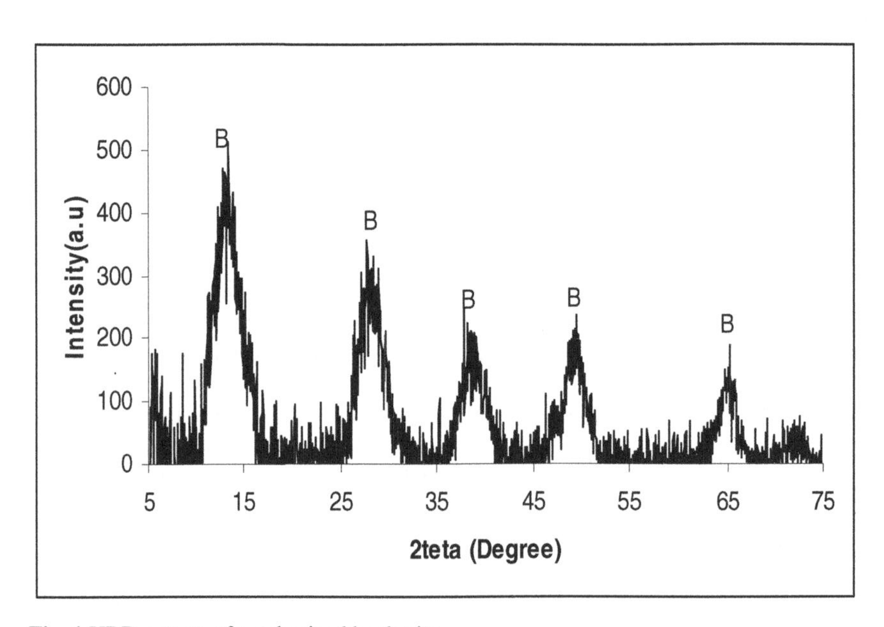

Fig. 1 XRD pattern of synthesized boehmite

3 Results and Discussion

The x-ray pattern of synthesized boehmite is shown in Figure 1. The broad peaks of synthesized boehmite reveal the small crystallite size. Figure 2 shows phase development of the boehmite precursor during microwave heating for different times. As observed, γ-alumina is the only crystalline phase after heating of precursor powder for 10 min. By increasing the heating time to 15 min the sharp peaks of α-alumina are appeared while the weak peaks of γ-alumina are still distinguishable. α-alumina is the only crystalline phase after 20 min heating. The crystallite size measured on x-ray peaks was about 35 nm which has in high correspondence with size of particles shown by TEM (Fig. 3).

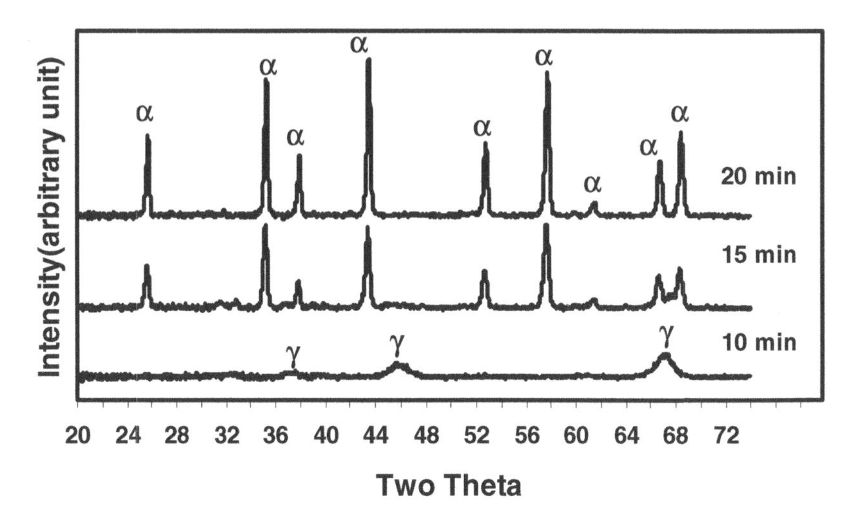

Fig. 2 XRD patterns of the alumina precursor heated for different times (α: α-alumina and γ: γ-alumina)

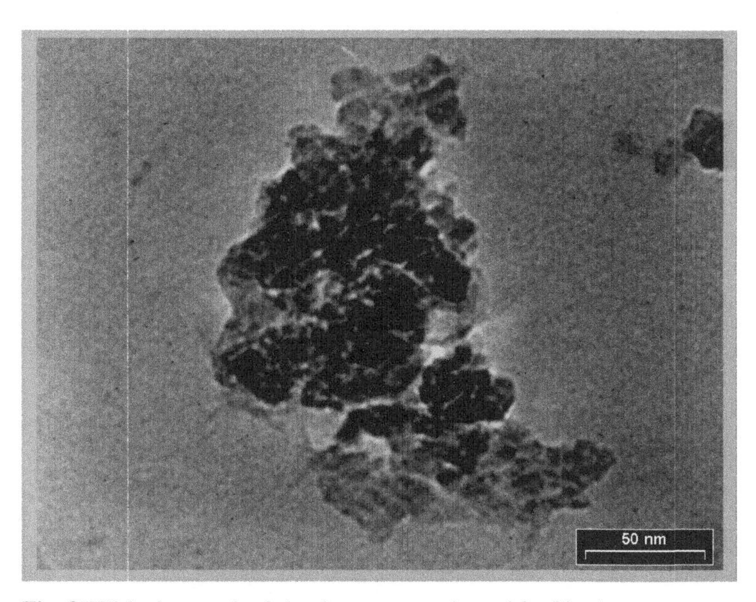

Fig. 3 TEM micrograph of alumina precursor heated for 20 min

Table 1 Temperature-heating time values

Time (min)	Temperature (°C)
10	784
15	1030
20	1205

References

1. Kumagai, M., Messing, G.L.: Controlled transformation and sintering of a boehmite sol-gel by α-alumina seeding. J. Am. Ceram. Soc. 68, 500–505 (1985)
2. Binner, J.G.P., Hassine, N.A., Cross, T.E.: Microwave synthesis of nanoxide ceramic powders. Ceramic Transactions 59, 565–572 (1995)
3. Gedye, R., Westaway, K., Smith, F.: The application of microwave to rapid chemical synthesis. Ceramic Transactions 59, 525–531 (1995)

Effects of Sabalan Tuff as a Natural Pozzolan on Properties of Plastic Concrete

R. Sadeghi Doodaran and M. Pasbani Khiavi

Abstract. One of the most common methods for watertighting in dams is to use cutoff or watertight walls. These types of walls should have high plasticity and low permeability [1]. Developing these walls with such qualities requires using plastic concrete with paneling method. In most cases, adding these materials to concrete improves some of its qualities such as consistency, viscosity of fresh concrete, decrease in permeability, increase in plasticity, long term increase in compressive strength, durability against sulphats and decrease in price. The main purpose of this project is using Sabalan natural tuff in plastic concrete for an access to a desirable modulus of elasticity without a decrease in compressive strength. In this paper, in addition to the impact of Sabalan natural pozzolan tuff on plastic concrete qualities, the impacts of pozzolan cement made of this kind of tuff (Ardebil pozzolan cement) on it has been considered. The results showed improvement on the qualities of plastic concrete such as decrease in permeability, decrease in modulus of elasticity and increase in durability against destructive factors with retaining compressive strength.

1 Introduction

Pozzolan is a substance that shows cement properties in the vicinity of lime. A standard definition for pozzolan is a siliceous or aluminates siliceous material that bears no cementation value in itself, however, in the form of tiny particles and in the vicinity of moisture with ordinary temperature it shows chemical redaction with calcium hydroxide and creates combinations that bear cementation properties [1]. Plastic concrete is a type of concrete which comes from the combination of water, cement, bentonite, gravel, sand and in some cases clay. For this purpose one can use admixture in cases where required. Since Sabalan tuff is well available in Ardabil region, and taking account of the economic considerations of pozzolan applications in concrete industry and the need to understand its effect on plastic concrete properties, pozzolanic cement of Ardabil cement plant and also

R. Sadeghi Doodaran
Islamic Azad University, Astara Branch, Islamic Republic of Iran

M. Pasbani Khiavi
University of Mohaghegh Ardabili, Ardabil, Islamic Republic of Iran

different percents for pozzolan (10, 20, 30) have been selected and studied as weight substitute for Soofian cement (control cement).

2 Material Characteristics

Water used in this project was fresh water of Sahand University of Technology in Iran. Cement used in this project can be classified into two kinds: Portland cement of Soofian cement plant with the relative density of 3.15 and Portland-pozzolan cement of Ardabil cement plant with the relative density of 3.1. The relevant abbreviations for five types of cement materials are as follows:

1. Scc for Sufian Portland cement
2. Apsc for Ardabil Portland-pozzolan cement
3. Scc+ 10% P for Soofian cement + 10 percent of pozzolan of Sabalan tuff
4. Scc + 20% P for Soofian cement + 20 percent of pozzolan of Sabalan tuff
5. Scc + 30% P for Soofian cement + 30 percent of pozzolan of Sabalan tuff

Added pozzolans are as weight substations for the consumed cement.

Washed river pea gravel, production of sand screwing plant of zafar Imamiyeh in Tabriz was selected and used. In this project washed river sand of Seram mine in Tabriz was used. Since the grading curve of the used sand was not within the recommended range by ASTM, the used sand was screened by NO.4 sieve. Bentonite used in this project was that of Afra Company (35km to Zanjan, Shahid Bahonar Industrial Town). Sabalan tuff used in Ardabil cement plant, whose mine is located at 5 kilometers to Namin, has been used in this project, then this pozzolan was ground and screened using NO.100 sieve in Azar Dash industry plant. According to the recommendation made by ASTM–C618 standard the sum of the three main oxides (SiO_2, Al_2O_3, and Fe_2O_3) must exceed 70% and the maximum shortage of ignition loss is limited to 10 percent. The fall of the used tuff ignition loss was 3.85 percent and the sum of the three main oxides (61, 17, 3.5) was 81.5 percent and this ranges within the ASTM–C618 standard limitations.

3 Mix Design

In this research 45 mix design with cement materials (C= 200kg/m^3), water to cement ratio (W/C= 1.5, 2, 2.5), Bentonite (B= 30 , 40 , 50 kg/m^3) and coarse aggregate to fine aggregate ratio equal to 1, were provided and curing in ordinary as well as sulfated environment (containing 10% sodium sulfate).

4 Results of the Experiments

Substituting the pozzolan for cement with high percentages (30%) leads to considerable compressive strength reduction (7 day, 32.22% and 150 day, 27.21%). This reduction in compressive strength, particularly in law water cement is more observable. This is so, perhaps because additional pozzolan does not take part in pozzolanic reactions and remains as filler in the mixture of plastic concrete. In other words, some part of substituted pozzolan cannot take part in pozzolanic reaction with cement hydration products and remains inactivated in the mixture and reduces its resistance. So, from the point of view of compressive strength and regarding the Fig1, it can be concluded that the 10% substitution is an optimized one and any additional one to it will cause considerable reduction in compressive strength.

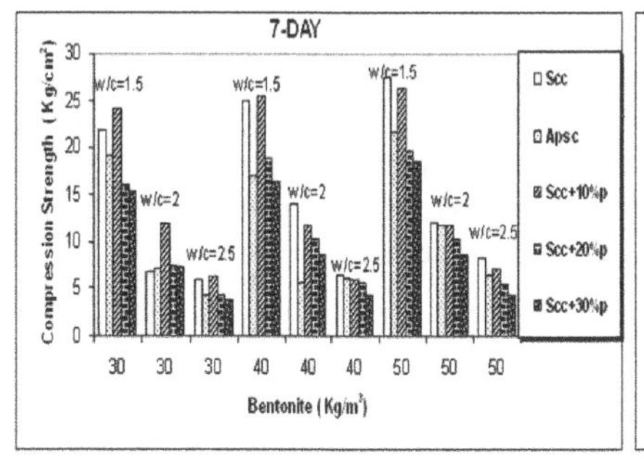

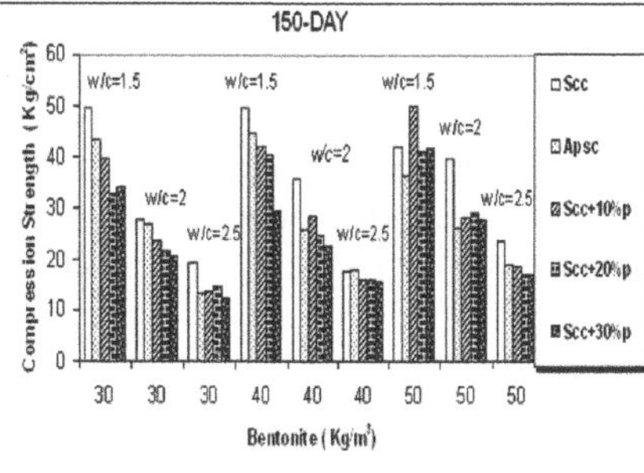

Fig. 1 Comparison of compressive strength for cements and variable percents of pozzolan with increasing measures of bentonite

Some part of the combination of Ardabil cement has been formed of natural pozzolan. As it is evident in the section about the compressive strength of specimen, this pozzolan in its early ages can not completely take part in giving resistant to the cement paste and it operates as filler, and accordingly bears weaker function than that of Soofian cement. As diagrams show, addition of pozzolan to Soofian cement reduces the elasticity module of the specimen. It should be noted that this reduction brings reduction in compressive strength and the more the percent of subsistent pozzolan, the more the reduction in specimens with low measures of bentonite and high water to cement ratio (Fig 2).

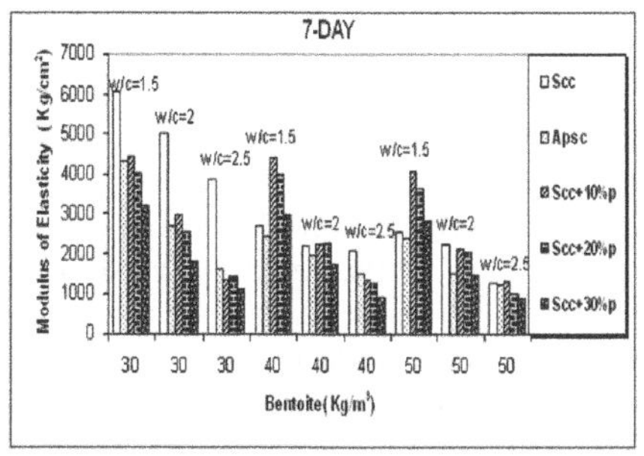

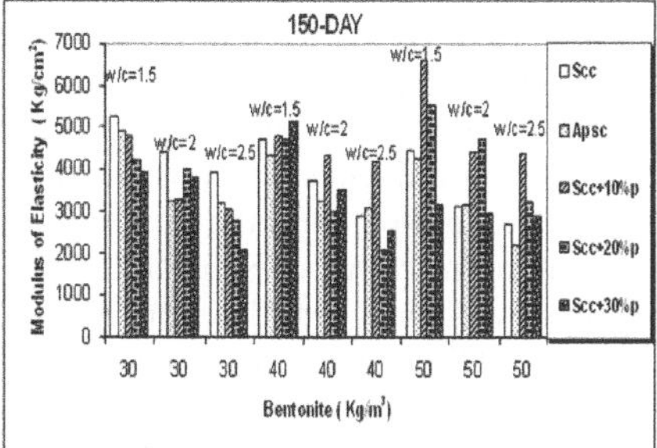

Fig. 2 Comparison of modulus of elasticity for cements and variable percents of pozzolan with increasing measures of bentonite

One of the objectives that are followed in plastic concrete production is increasing or retaining the compressive strength while reducing the elasticity module. In other words, ductile and flexible plastic concrete with good relative compressive strength will have better function. As Fig3 show, the lines representing the communication of compressive strength with elasticity module of specimen prepared by Ardabil pozzolanic concrete are lower than those representing the Soofian cement ones. In other words, for a case of fixed compressive strength, the elasticity module of plastic concretes prepared with pozzolanic cement of Ardabil is lower than those specimens relating to Soofian cement, and this is especially true in case of mixtures with less bentonite. By studying the lines representing the communication of compressive strength and elasticity module of plastic concretes prepared with different percentages of substitution of Soofian cement with natural pozzolan of Sabalan tuff, it can be seen that the more the rate of pozzolan increases, the better the functions of plastic concretes prepared for retaining the compressive strength -while decreasing the elasticity module- will be.

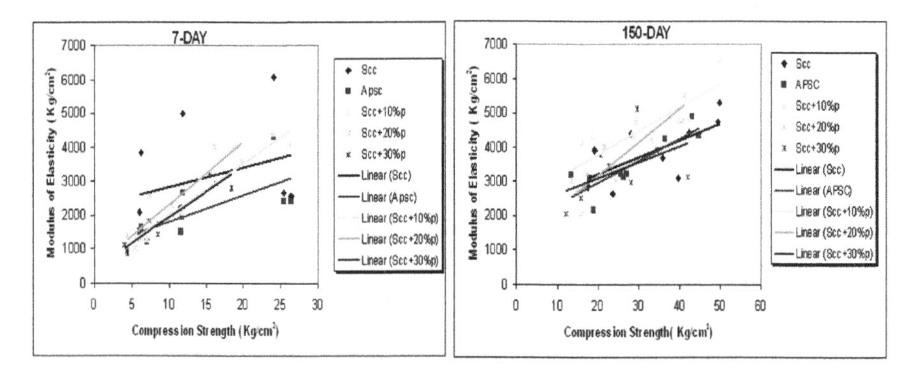

Fig. 3 Comparison of modulus of elasticity with compressive strength

Considering Fig. 4 and comparing the amounts of permeability for the three selective plans (Soofian cement (Scc1), Ardabil cement (Apsc 10), and substituting 10 percent of Soofian cement with Sabalan tuff (scc+p1 19)), we can say that ready – made plastic using the pozzolanic cement of Ardabil shows lower permeability coefficient in both ordinary and sulfated curing environments. Also, the concrete made of Soofian cement 10 percent of natural pozzolan of Sabalan tuff in sulfated environment showed better function. Studying Fig. 5, we can also see that in sulfated environment, the more the percent of substitution of Soofian cement with natural pozzolan of Sabalan tuff increases, the more the compressive strength of curing specimen in sulfate, increases. This is so, perhaps because, the molecular volume of CH (calcium hydroxide) is about 33.2 ccm, while in case of hydrated sodium sulfate it is 74.3 ccm [2], and with the presence of pozzolan and its reaction with CH and forming CHS the pores in concrete will be filled and the result will be a relative solid material that prevents the penetration of materials that increase sulfates, and it also adds to the compressive strength of the plastic cement.

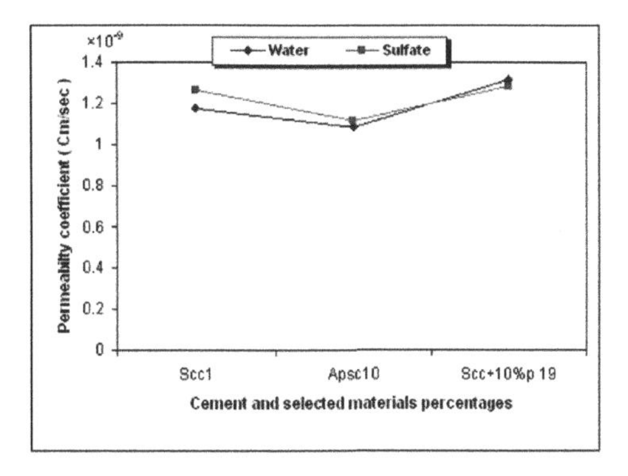

Fig. 4 Permeability of plastic concrete and comparison of them in sulphated environment

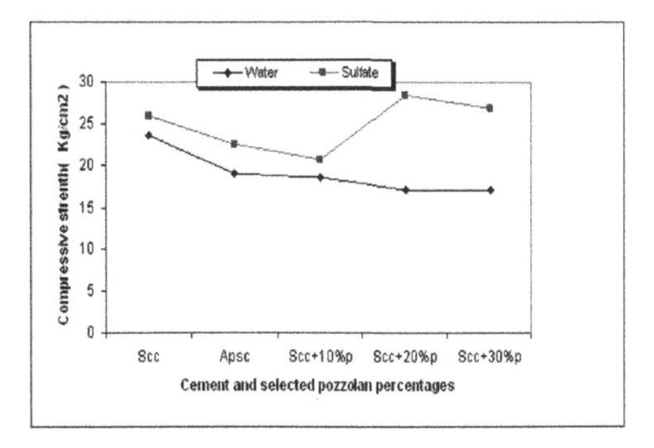

Fig. 5 Effect of sulphated curing on compressive strength of plastic concrete

5 Discussions

- Replacing 10% pozzolan of Sabalan tuff in the form of weight replacement is optimal and the replacement exceeding the afore-mentioned amount will cause more reduction in compressive strength.
- In case of adding pozzolan to Soofian cement, the elasticity module of mixtures will be decreased.
- For fixed compressive strength, elasticity module of plastic concretes prepared from pozzolanic Ardabil cement is lower than that of specimens pertaining Soofian cement.
- The more the rate of pozzolan, the better function of prepared plastic concretes, regarding the retention of compressive strength as well as the reduction in elasticity module will be.
- Plastic concrete made from Ardabil pozzolanic cement shows lower permeability coefficient in both normal and sulfated environments.
- In sulfated environment, the more the percent of replacement of soofian cement with natural pozollan of Sabalan tuff, the more the compressive strength of fostered specimens in sulfate will be.

References

1. ICOLD BULLETIN, Filling materials for waterting cut off walls. 51 (1985)
2. Janotka, I., Stevula, L.: Effect of bentonite and zeolite on durability of cement suspension under sulfate attack. ACI Mater. J. 95, 710–715 (1998)

Synergistic Action of a Ternary System of Portland Cement – Limestone – Silica Fume in Concrete

J. Zelić, D. Jozić, and D. Krpan-Lisica

Abstract. Some experimental investigations on a synergistic action when a ternary system of Portland cement – silica fume – limestone is used in mortar or concrete are present in this paper. Standard laboratory tests with respect to the pore size distribution, micromorphology, compressive strength and sulphate resistance in both sodium and magnesium sulphate solutions were performed on mortars made with 70% (by mass) of Portland cement (PC), type CEM II/B-S and 30% of cement replacement materials consisted of various combination of fine ground limestone filler (LF) and silica fume (SF). In addition to these ternary systems, binary blends, such as: PC-LF, as well as PC-SF, along with 100 % PC mortars, were investigated for comparison. It is found that SF-blends reach higher compressive strengths than LF-blends for the same replacement of cement. When SF was added together with LF, the mortars show considerable increase in the compressive strength and show a lower expansion than a control, sulphate-resisting mortar, independent of the type of sulphate solutions, due to pore size refinement microstructure of mortars.

1 Introduction

Concrete is a porous material with many different kinds of pores, ranging from the air voids to the nanometre-scale pores produced by the cement-water chemical reaction. Since, these nanoscale pores control the properties of the calcium-silicate-hydrates hydration products (C-S-H), which is the main "glue" that holds concrete together, concrete is in some ways a nanoscale materials [1]. According to literature data [2], the addition to concrete mixes of silica fume (also known as condensed silica fume or microsilica), a by-product of the production of metallic silicon or ferrosilicon alloys, with mean particles size over 100 nm, results in high-performance concrete, in terms of better chemical resistance, higher strength or better durability. The beneficial effects of adding silica fume (SF) are due to the filling action of its fine particles in the pores and the formation of an additional quantity of C-S-H by pozzolanic reaction between the SF and calcium hydroxide

J. Zelić, D. Jozić, and D. Krpan-Lisica
Faculty of Chemical Technology, University of Split, Split, Croatia
e-mail: zelic@ktf-split.hr

formed during cement hydration. Recently, a new pathway based on the addition of the colloidal silica nanoparticles (nano-SiO_2) to Portland cement has been explored. The research confirmed that the silica nanoparticles addition to Portland cement has provided notably increase both the mechanical properties of cement-based materials [3,4], and the durability against the calcium leaching [5], respectively. Limestone is used in cement and concrete for various purposes, namely as a raw material for clinker production and as coarse or fine aggregate. For long-time, ground limestone has been considered as inert filler. During the last decades, however, studies have pointed that limestone filler (LF) addition to Portland cement (PC) produces several effect on mechanism and kinetics of cement hydration; thus, LF addition increases the hydration rate of PC, especially of C_3S, at the early ages, improves the particle packing of the cementitious system, provides new nucleation sites for calcium hydroxide, and produces the formation of calcium carboaluminates as results of the reaction between $CaCO_3$ from limestone and C_3A from Portland clinker [6]. Today, modern cements often incorporate several mineral admixtures one of which is LF. European standard EN 197 identifies two types of Portland limestone cements (PLC): Type II/A-L and Type II/B-L, containing 6-20% and 21-35% LF, respectively.

The aim of this paper was to investigate a synergistic action when ternary system PC-SF-LF used in concrete. More specifically, an answer was sought for the question of whether a suitable combination of LF and SF would improve the properties of hardened concrete more than these materials would separately. It is anticipated that the results and conclusions obtained here on mortars will be transferable to concretes [7].

2 Experimental

2.1 Materials

Commercial blended PC, Type CEM II/B-S 42.5N (supplied by Dalmacijacement, Croatia), high purity LF (supplied by Konstruktor, Croatia) and SF (supplied by Elkem Co., Norway) were used. The sulphate-resisting PC (supplied by Dalmacijacement, Croatia), Type CEM I-HS, was used as the control cement for the sulphate-resistance tests. Chemical composition and physical properties of the used materials are given in Table1. Particle size distributions of the blended PC, LF and SF were determined using the Coulter Counter method (methanol-LiCl solution), as shown in Fig. 1. Salts $MgSO_4 \times 7H_2O$, p.a., and $Na_2SO_4 \times 10H_2O$, p.a., were used in preparation of sulphate solutions. The sulphate ion concentration in both solutions was 2.5% and was kept constant during sulphate-resistance test.

Mortar mixes. Two kinds of mortar samples were prepared: (I) for mechanical strength tests –prismatic specimens dimensions 40 x 40 x 160 mm, according to HRN EN 196-1:2005; and (II) for the sulphate-resistance tests – prismatic specimens dimensions 25 x 25 x 350 mm, with two stainless-steel inserts cast into the ends to facilitate accurate monitoring of the changes in length, according to the ASTM C 452. Mix proportion and designation of mortars prepared are given in Table 2. A different water-to-cementitious material ratio (*w/cm*) was used in each

Table 1 Chemical and physical properties of materials used

	Cement I[a]	Silica fume	Limestone	Cement II[b]
Chemical composition, mass %:				
SiO_2	21.03	89.68	0.66	21.61
Al_2O_3	6.56	1.15	0.32	3.71
Fe_2O_3	2.68	1.58	0.20	4.49
CaO	58.10	1.40	53.45	62.94
MgO	2.57	1.45	0.64	1.88
SO_3	2.54	0.17	-	1.73
Total Alkalis	1.02	-	-	0.76
Loss on Ignition	1.04	2.99	43.23	0.80
Carbon	-	1.83	-	-
Composition (Bogue), mass%:				
C_3S	62.10	-	-	55.65
C_2S	17.60	-	-	19.98
C_3A	11.20	-	-	2.23(max. 3.5)
C_4AF	8.20	-	-	13.64
CaO (free)	0.90	-	-	-
Finenes:				
Density, g/cm^3	3.18	2.26	2.70	3.26
Specific surface area (SSA), cm^2/g				
Blaine	3,530	-	4,200	3,021
BET	11,000	180,000	13,000	-

[a]Commercial Portland cement Type CEM II/B-S 42.5N
[b]The sulphate-resisting Portland cement Type CEM I-HS

case in order to keep the mixture's fluidity constant. No superplasticizer was added. After mixing, the cement mortars were cast into prismatic molds and compacted by vibration. The specimens were demolded after of 24 h at 90% relative humidity and were then cured in water at 20 °C until testing.

Table 2 Mix proportion and designation of mortars prepared for the study

	Composition (mass%)				
Designation	Cement I	Silica fume	Limestone	Cement II	*w/cm*
PO	100	0	0	0	0.57
PL15	85	0	15	0	0.57
PL15S8	77	8	15	0	0.61
PL15S15	70	15	15	0	0.65
PS8	92	8	0	0	0.60
PS15	85	15	0	0	0.65
SPC[a]	0	0	0	100	0.53

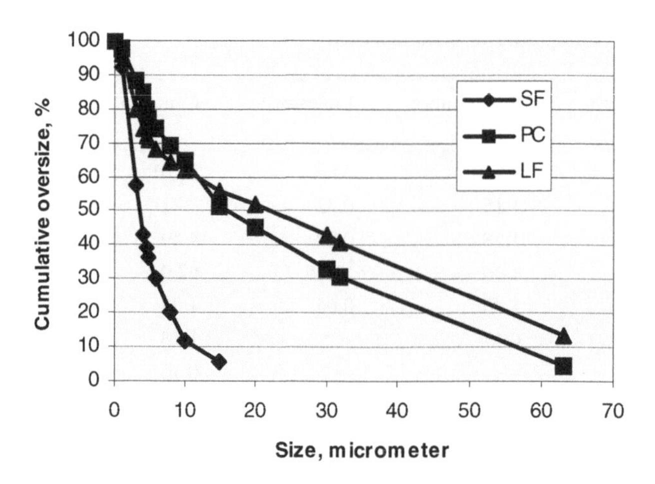

Fig. 1 Particle size distribution of PC, SF and LF used

2.2 *Test Methods*

Mechanical properties. The mechanical properties of produced blend mortars (I series) were determined by compressive strength measurements at ages of 3, 7, 14, 28, 90 and 120 days. The strength value was the average of three specimens.

Sulphate resistance. After 28-days curing in tap water, each blended mortars and the control mortars based on the sulphate-resisting cement (II series) were individually immersed in a plastic tank containing both a sodium and magnesium sulphate solution, respectively. At 30, 60, 90, 120, 150, and 180 days of sulphate immersion, the specimens were tested to determine the expansion of mortars.

The structure and particle shape of silica fume were identified using X-ray diffraction (RXD Philips PW 1010, with graphite monochromator) and transmission electron microscopy (TEM Philips EM 301), respectively. The morphologies of the hydration products were studied by using a scanning electron microscopy (SEM Leitz AMR 1600T). At certain curing ages, mortars were crushed and treated with acetone to stop the hydration. The fresh broken surface of mortars was gold coated. The pore structure, both the pore volume and pore size distribution, in the 28-day-old hydrated mortars were determined by a mercury intrusion porosimetry (MIP Carlo Erba series 200). The volume of mercury intrusion at the maximum pressure was considered to be the total porosity. It is important to note that only connected capillary pores can be reached by MIP.

3 Results and Discussion

The particle size distributions, physical and chemical properties of material used are present in Fig. 1 and Table 1, respectively. The particles sharp and the amorphous nature of the SF used are shown in Fig. 2 (a) - (b), respectively. The results reveal that SF, in form supplied, contains 89.68 mass% of silica, and consists of amorphous and extremely fine spherical particles. Fig. 2 (a) also shows that SF exists almost completely in the form of fine spheres linked together into chains of clusters (the particle agglomerate), rather than isolated spheres. The particles range from fraction of 2 to 15 μm (Fig. 1), although silica fume has high nitrogen BET SSA (Table 1). It is obviously that the reason for this phenomenon is that nitrogen can penetrate into space of original SF particle inside the agglomeration; thus, the Coulter Counter method measures the agglomeration size while nitrogen measures the original size [8].

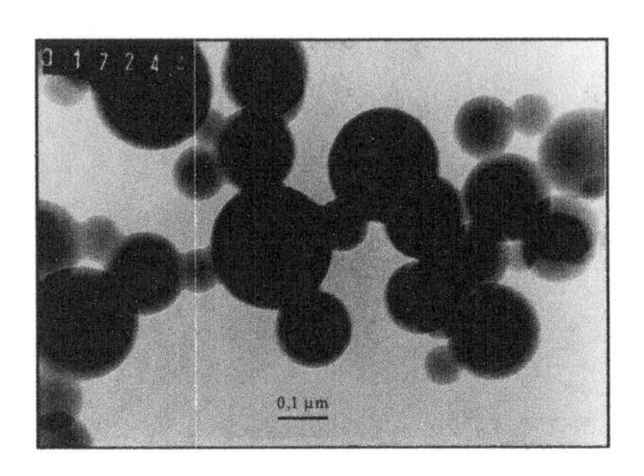

a)

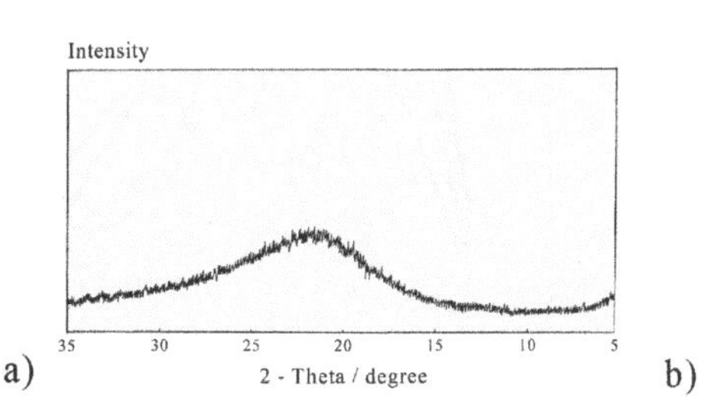

b)

Fig. 2 (a) TEM micrograph (magnification 130,000 x) and (b) X-ray diffraction curve of the SF used

The derivate plots of cumulative pore volume of $dV / d\ln r$ versus pore radius, r for 28-day-old mortars are present in Fig. 3. At a 15 mass% replacement of PC by the equal mass of LF, the derivate plot shows discontinuous pore structure with the first maximum of occurrence of pores radii at 0.02 μm, and second one at 0.03 μm, respectively. As both PC and LF show similar characteristics (the BET SAA, Table 1), their different behavior perhaps indicates the LF activity during cement hydration, which is responsible for the development of pore structure with a pore size distribution covering many orders of magnitude. For mortars containing SF together with LF, a larger number of small pores is seen than in analogous mortar samples containing no LF. A discontinuous pore structure with two derivate maximums; the first maximum at a pore radii of about 0.015 μm and the second one, between 0.025-0.030 μm, are also perceptible for the binary PC-SF (with 8 mass% of SF) and ternary the PC – SF – LF blends, respectively. According to Feldman [9], a discontinuous pore structure will result in a lower permeability, and better resistance to sulphate attack.

Figure 4 shows the total porosity of the 28-day-old mortars measured by MIP. Comparing the porosity of the control PC mortar, all mortar samples show an increase of the total porosity with higher w/cm ratios. When SF in the amount of 8 and 15 mass% is added to the mortars containing LF the total porosity of mortars first decreases (the PL15S8), and then the porosity increases (the PL15S15), respectively.

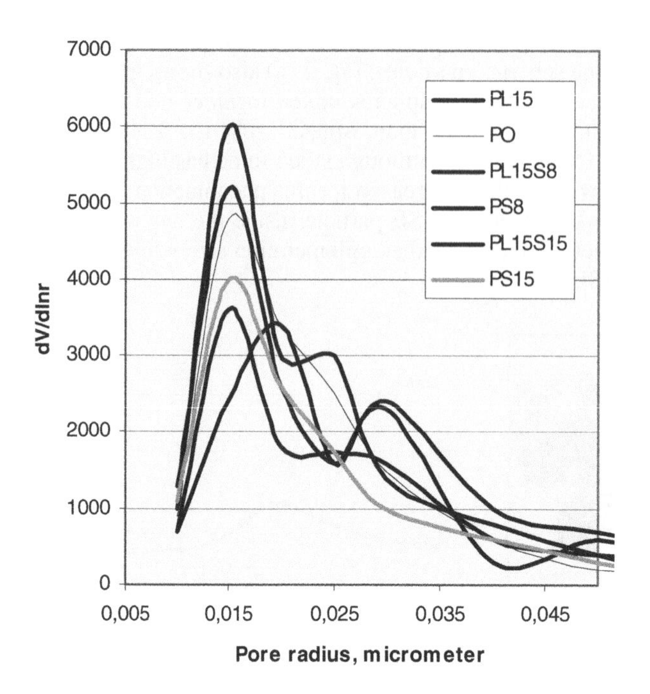

Fig. 3 The derivate plots pore volume *vs.* pore radius

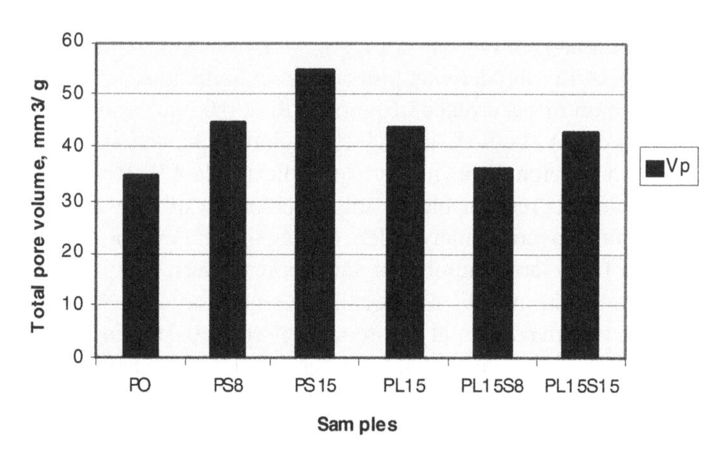

Fig. 4 Total pore volume in the 28-day-old mortars

Figure 5 presents the compressive strength of mortars for ages of 3, 7, 28, 90 and 120 days, respectively. The highest compressive strength shows PC mortar containing 8 mass% of SF. The SF addition of 15 mass% increases the w/cm ratio (Table 2) and, therefore, decreases the mortar strength due to the increase porosity (Fig. 4). For the same replacement of PC, SF gives higher strength than LF. Added

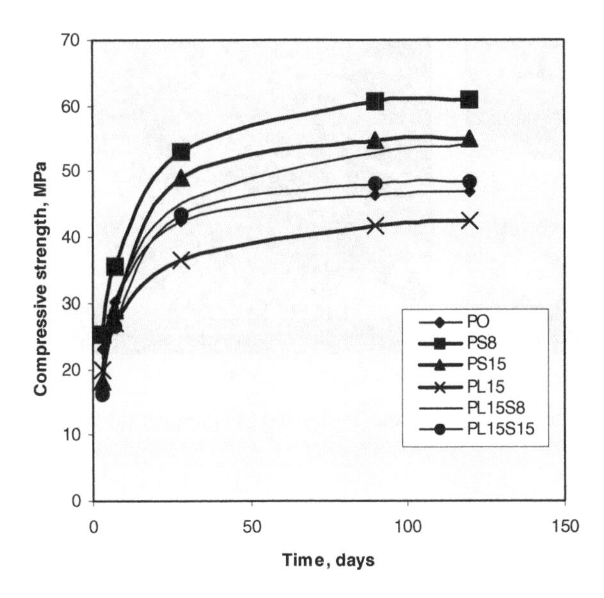

Fig. 5 Compressive strength developments of mortars studied

together with LF, SF increases the compressive strength, although the strength is lower than that of mortars containing no LF. It appears that certain LF-SF combination can improve the strength of mortars more than LF alone.

Figure 6 (a)-(b) shows the SEM micrographs of the fracture surface of the 28-day-old mortars containing SF alone (Fig. 6a), and SF together with LF (Fig. 6b). The recticular network formation of C-S-H layer precipitated on the surface of silica grains and the calcium hydroxide hexagonal plates is detected for the PS8 sample (Fig. 6a). The mortar containing together 15 mass% of LF and 8 mass% of SF (the PL15S8) does not exhibit recticular network features and shows occasional evidence of hexagonal plates (Fig. 6b). A distinct different between this sample and the sample not containing LF (the PS8) is that its surface seems to be consolidated and uniform, formed as an interconnected network.

Figures 7 and 8 show the length changes of mortars immersion in both sodium (N) and magnesium (M) sulphate solutions for 30, 60, 120, 150, and 180 days, respectively. By comparing the expansion values of mortars exposed to the attack of both sulphate solution, it can be observed that specimens not containing SF have the higher initial expansion after 30 days of sulphate immersion, and the intensive trend of breakdown; the P0 samples were disintegrated after only 90 days, while the PL broke down after 150 days.

Mortars containing SF show better sulphate resistance, although they have larger initial total porosity associated with the higher w/cm ratios. These results indicate that pore size refinement caused by pozzolanic reaction has a beneficial effect on the sulphate resistance. Other authors [10,11] came to similar conclusions. The presence of brucite, $Mg(OH)_2$ (which suggests a rapid attack by magnesium sulfate solution), detected in the SF-mortars after 120 days of immersion, had no

Fig. 6 SEM micrographs of the fracture surface of the 28-day-old PC mortars: (a) the PS8 sample, and (b) the PL15S8 sample

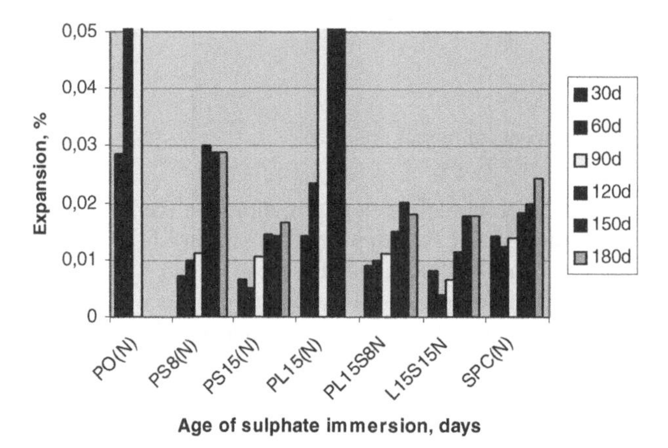

Fig. 7 Length changes of mortars immersion in sodium sulphate solution

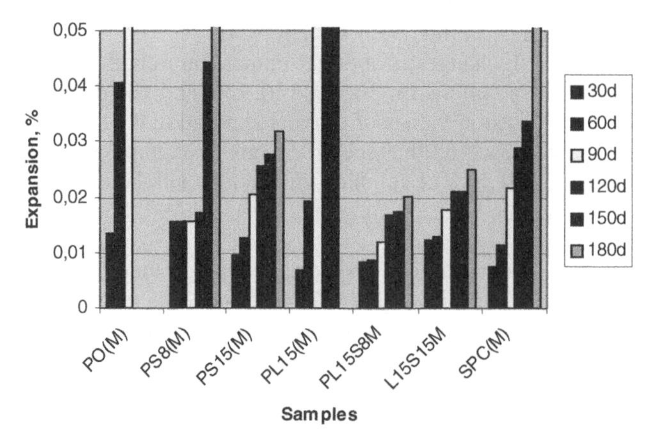

Fig. 8 Length of mortar bars on immersion in magnesium sulphate solution

great effect on the reduction of compressive strengths [11]. It is obvious that the effect of the pozzolanic reaction (reduction in permeability and refinement of the pore structure) overcame this negative effect of the sulfate attack. The smallest expansion, even smaller then controls mortars on the basis of sulphate-resisting cement (SPC), independent of type of sulfate solution, is shown by the samples containing 15 mass% of SF, and the SF-LF blends. Results from this work show that the LF activity during cement hydration [6] in combination with the pore refinement caused by the pozzolanic activity of the SF addition may increase the sulphate resistance of blended PC mortars or concrete.

4 Conclusions

The highest compressive strength is shown by PC mortar containing 8 mass% of SF. The replacement of PC by 15 mass% of LF causes a significant reduction in the compressive strength. Blending SF and LF simultaneously with PC is most effective on strength development, especially in the later ages, compared to a PC alone. Both SF and LF additions used alone or in combination with PC, modify the PC mortar microstructure. The 28-day-old PC-SF-LF mortars show a discontinuous structure with two derivate maximums: the first maximum at pore radii of about 0.015 μm and the second one, between 0.025-0.030 μm, respectively. These mortars are also characterised by a good sulphate resistance and show lower expansion than a control, sulphate-resistant mortar, regardless of the type of sulphate solutions, due to refinement pore structure.

Acknowledgments. The authors would like to acknowledge the Commissioners of the European Union for funding under the REINTRO Project (the 5FP), No. ICA2-CT-2002-10003.

References

1. Chong, K.P., Garboczi, E.J.: Smart and designer structural material systems. Prog. Struct. Engng. Mater. 4, 417–430 (2002)
2. Malhotra, P.K., Monteiro, P.J.: Concrete, microstructure, properties and materials. McGraw-Hill, New York (2006)
3. Li, G.: Properties of high-volume fly ash concrete incorporating nano-SiO_2. Cem. Concr. Res. 34, 1043–1049 (2004)
4. Dolado, J.S., Campillo, I., Erkizia, E., Ibáñez, J.A., Porro, A., Guerrero, A., Goñi, S.: Effect of nanosilica additions on belite cement pastes held in sulfate solutions. J. Am. Ceram. Soc. 90, 3972–3976 (2007)
5. Gaitero, J.J., Campillo, I., Guerrero, A.: Reduction of the calcium-leaching rate of cement paste by addition of silica nanoparticles. Cem. Concr. Res. 38, 1112–1118 (2008)
6. Irassar, E.F., González, M., Rahhal, V.: Sulphate resistance of type V cements with limestone filler and natural pozzolana. Cem. Concr. Compos. 22, 361–368 (2000)
7. Popovics, S.: Portland cement - fly ash - silica fume systems in concrete. Adv. Cem. Bas. Mat. 1, 83–91 (1993)
8. Cook, R.A., Hover, K.C.: Mercury porosimetry of hardened cement pastes. Cem. Concr. Res. 29, 933–943 (1999)

9. Feldman, R.F.: Pore structure damage in blended cements caused by mercury intrusion. J. Am. Ceram. Soc. 67, 30–33 (1984)
10. Sideris, K.K., Savva, A.E., Papayianni, J.: Sulfate resistance and carbonation of plain and blended cements. Cem. Concr. Compos. 28, 47–56 (2006)
11. Zelić, J., Radovanović, I., Jozić, D.: The effect of silica fume additions on the durability of Portland cement mortars exposed to magnesium sulfate attack. Mater. Tehnol. 41, 91–94 (2007)

Author Index

UNIVERSITIES AT MEDWAY LIBRARY